W0107674

Springer-Verlag Berlin Heidelberg GmbH

Progress in Colloid and Polymer Science

Editors: F. Kremer, Leipzig and G. Lagaly, Kiel

Volume 117 · 2001

Adsorption and Nanostructures

Volume Editor:
Imre Dékány

Springer

The series Progress in Colloid and Polymer Science is also available electronically (ISSN 1437-8027)

- Access to tables of contents and abstracts is *free* for everybody.
- Scientists affiliated with departments/institutes subscribing to Progress in Colloid and Polymer Science as a whole also have full access to all papers in PDF form. Point your librarian to the LINK access registration form at http://link.springer.de/series/pcps/reg-form.htm

ISBN 978-3-662-14669-9 ISBN 978-3-540-45405-2 (eBook)
DOI 10.1007/978-3-540-45405-2

ISSN 0340-255X

http://www.springer.de

Originally published by Berlin, Heidelberg, New York in 2002
Softcover reprint of the hardcover 1st edition 2002

Typesetting: SPS, Madras, India
Cover Design: Estudio Calamar, F. Steinen-Broo, Pau/Girona, Spain
Cover Production: design & production, D-69121 Heidelberg

SPIN: 10719392

Printed on acid-free paper

Progr Colloid Polym Sci (2001) 117: V

PREFACE

The Third International Conference of the Kolloid-Gesellschaft (Germany) was held in Budapest at September 25–28, 2000 and was jointly organized together with the Colloid Committee of the Hungarian Academy of Sciences.

"Adsorption and Nanostructures – from Theory to Application"

The Hungarian Academy of Sciences was founded 175 years ago by Count István Széchenyi. Since this time, the Academy was the center of sciences in Hungary. The scientific cooperation between Hungary and Germany goes back to 1925 when Aladár von Buzágh joined the group of Wolfgang Ostwald in Leipzig and the group of Herbert Freundlich in Berlin. In 1935 he established the first laboratory of colloid chemistry in Hungary at the University of Budapest. Buzágh became a member of the Kolloid-Gesellschaft and enjoyed many friendly relationships with German colloid scientists. Since this time a strong cooperation between Hungarian and German scientists survived all political troubles and hard times.

The subject of this cooperation is related to two fundamental topics of colloid science: adsorption and nanostructured materials. The lectures and posters in this conference were, therefore, related to dispersions, nanoparticles, nanocomposites, adsorption processes, microemulsions, and environmental aspects.

In discussing adsorption phenomena in Budapest, two scientists have to be remembered: Géza Schay and Lajos György Nagy. Professor Schay was born in 1900 and died at the age of 91. Professor Nagy passed away in 1999. The important contribution of both scientists to colloid science was acknowledged during the conference.

G. Lagaly

Progr Colloid Polym Sci (2001) 117: VI–VIII

CONTENTS

Dispersions

Progr Colloid Polym Sci (2001) 117: 1–4

Peter Ulbig
Johannes Seippel

Development of a group contribution method for liquid-phase adsorption onto activated carbons

P. Ulbig · J. Seippel
University of Dortmund,
Department of Chemical Engineering,
Institute for Thermodynamics,
Emil-Figge-Strasse 70,
44221 Dortmund, Germany

P. Ulbig (✉)
Physikalisch-Technische Bundesanstalt,
Section 3.14, Bundesallee 100,
38116 Braunschweig, Germany
e-mail: peter.ulbig@ptb.de
Tel.: +49-531-5923142
Fax: +49-531-5923205

Abstract The description of liquid-phase adsorption equilibria onto activated carbons was investigated in terms of group contributions. The adsorbate–solid solution theory was used in order to introduce the active sites of the adsorbent as a component in a multicomponent mixture with the liquids. The active sites and the liquids are subdivided into functional groups and the corresponding group contribution parameters were fitted to experimental data. By using these parameters predictions of liquid-phase adsorption equilibria of different systems, especially with other activated carbons, are possible.

Key words Liquid-phase adsorption · Group contribution · Activated carbon · Surface-excess isotherms

Introduction

Nowadays, the use of computational methods in order to calculate whole separation processes is common practice in chemical engineering. Group contribution methods are often applied in this context in order to calculate, for example, the vapour–liquid equilibrium of multicomponent mixtures for distillation processes. The adsorption from the liquid phase is seldom used for a separation process, owing to the fact that experimental data for the given mixture are often not available or no computational method exists. Therefore, the question is whether the group contribution principle in combination with a corresponding solution theory can be used in order to describe liquid-phase adsorption equilibria. The main advantage of this principle is the possibility of carrying out predictions for adsorption equilibria onto different activated carbons. The surface-excess isotherms of binary liquid systems onto zeolites were successfully described [1] by the use of the so-called adsorbate–solid solution theory (ASST) [2] in combination with the modified UNIFAC equation [3]. Activated carbons are much more heterogeneous than zeolites and therefore the questions of which properties of activated carbons have to be used and how it is possible to introduce these properties into the thermodynamic model arise.

Theory

The thermodynamic framework of the ASST is given in Ref. [2]. It is based on the equality of the chemical potential for a component, μ_i, in every phase. The adsorbed liquid phase is taken together with the active sites of the adsorbent in order to obtain a so-called reference phase. The chemical potential for a liquid component i has, therefore, to be equal in the bulk (index b) and in the reference phase (index *):

$$\mu_i^{\mathrm{b}} = \mu_i^* \ . \tag{1}$$

This leads to the formulation of the phase equilibrium:

$$x_i^{\mathrm{b}}\gamma_i^{\mathrm{b}} = x_i^*\gamma_i^* \exp\left(-\frac{\Phi^* - \Phi_{0i}^*}{RT\Gamma_{\mathrm{m},i}^{\mathrm{s}}}\right) \ , \tag{2}$$

where x_i denotes the molar fraction of this component, γ_i the activity coefficient, Φ the free enthalpy of immersion of the mixture, Φ_{0i} the free enthalpy of a pure liquid

Table 1 Characterisation of the activated carbons used

Activated carbon (from)	CWS (Lurgi)	ROTH (Roth)	L3S (Elf)	CPL (Elf)
Raw material	Hard coal	Peat	Pine wood	Pine wood
Activation procedure	Steam	Steam	Steam, acid washed	Phosphoric acid
Brunauer–Emmett–Teller surface area (m^2g^{-1})	500	750	1,150	1,700
PH	9–10	9–10	4	4–7
Acid solubility (%)	12.0	5.5	0.5	2.0
Total pore volume (N_2 isotherm) (cm^3g^{-1})	0.297	0.441	0.659	1.122
Acidic surface groups ($mmolg^{-1}$):				
Carboxylic	0.000	0.000	0.005	0.430
Lactonic	0.000	0.000	0.105	0.115
Phenolic	0.290	0.400	0.200	0.425
Basic surface groups	2.100	1.180	0.395	0.170

under the same conditions as the mixture, R the molar gas constant, T the temperature, and $\Gamma^s_{m,i}$ the saturation capacity of the pure liquid onto the activated carbon. This equation is similar to the often-used equation describing the phase equilibrium between a bulk liquid mixture and the adsorbed surface phase [3, 4].

For the calculations of the activity coefficient of a liquid component in the bulk phase γ_i^b the modified UNIFAC equation was taken [5]. For the introduction of the active sites as a component a molar mass has to be defined:

$$M_{\text{activated carbon}} = \frac{\bar{A}_{\text{C-atom}} N_A}{\text{BET}} \quad . \tag{3}$$

The calculation consists of the average area of a carbon atom $\bar{A}_{\text{C-atom}}$ within a graphitic structure, the Avogadro constant, N_A, and the Brunauer–Emmett–Teller (BET) number (Table 1) of the activated carbon.

The group classification for the active sites was subdivided into three main groups:the nonpolar group, the basic group, and the acidic group. The nonpolar group contains only a C atom of a graphitic structure (N_gC). The basic group contains one group (B_O), which covers all types of basic oxides on the surfaces and which can be determined by Boehm titrations altogether. The basic group is subdivided into the carboxylic group (A_COOH), the lactonic group (A_OCO), and the phenolic group (A_OH). These groups can also be determined by stepwise Boehm titrations. The number of each group on the surface is calculated by an algorithm which takes into account the number of C atoms per gram of adsorbent (calculated from the BET number and the average area per C atom), and from the results of the Boehm titrations, which give the number of basic and acidic groups per gram of adsorbent. From that, the corresponding parts for "one" group (a "unit cell"). is calculated (Table 2). For the investigations experimental data of four commercially available activated carbons were used [6, 7].

For every kind of group the relative van der Waals volumes (R_{vdW}) and surfaces (Q_{vdW}) have to be known. This is necessary owing to the fact that the group interactions in the modified UNIFAC model are weighed

		1	2	3	4	5
1	*Nonpolar*	0				
2	*Acidic*	0	0			
3	*Basic*	0	0	0		
4	CH_n				0	
5	cCH_n					0

Fig. 1 The interaction matrix of the chosen test system

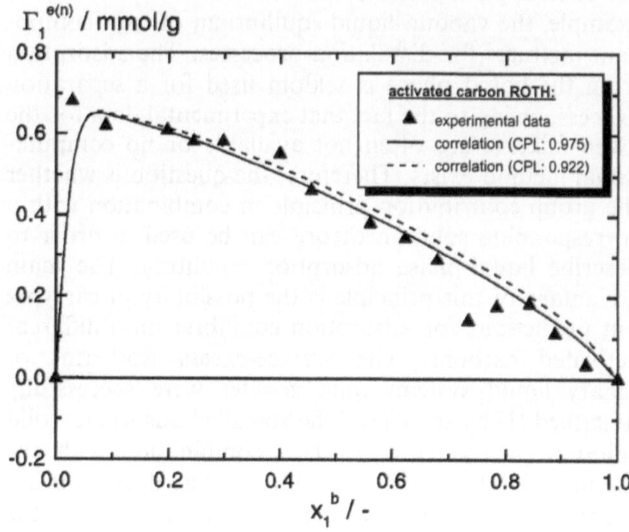

Fig. 2 Correlation of *n*-hexane (1)–2,2,4-trimethylpentane (2) onto ROTH

Table 2 Number of different groups (active sites) found on the activated carbons

Activated carbon	Carboxylic group, A_COOH	Lactonic group, A_OCO	Phenolic group, A_OH	Basic group, B_O	Nonpolar group, N_gC
CWS	0.000	0.000	0.008	0.062	0.930
ROTH	0.000	0.000	0.008	0.024	0.968
L3S	0.000	0.000	0.003	0.005	0.990
CPL	0.004	0.001	0.004	0.002	0.989

Table 3 Relative van der Waals properties of the active sites

Group	Relative volume, R_{vdW}	Relative surface, Q_{vdW}
N_gC	0.3652	0.120
B_O	0.7536	0.526
A_COOH	1.7647	1.396
A_OCO	2.6094	1.392
A_OH	0.7688	0.570

Table 4 Enthalpies of wetting and the factors for the saturation capacities

Activated carbon	Wetting enthalpy of *n*-octane (J/g)	Wetting enthalpy of 2,2,4-trimethylpentane (J/g)	Ratio of enthalpies of wetting	Chosen ratio
CWS	56.7	40.0	0.705	0.705
ROTH	100.1	71.1	0.710	0.710
L3S	128.0	117.1	0.915	0.915
CPL	147.8	136.3	0.922	0.975

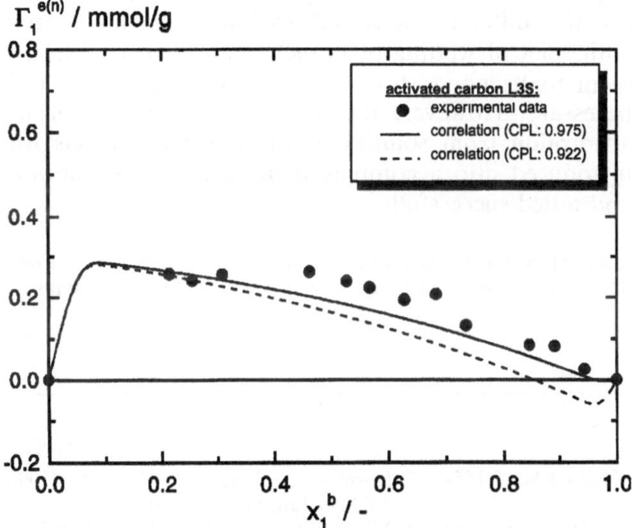

Fig. 3 Correlation of *n*-hexane (1)–2,2,4-trimethylpentane (2) onto L3S

by the volume and surface fractions. The values were calculated from the atom diameters and the bond lengths (Table 3).

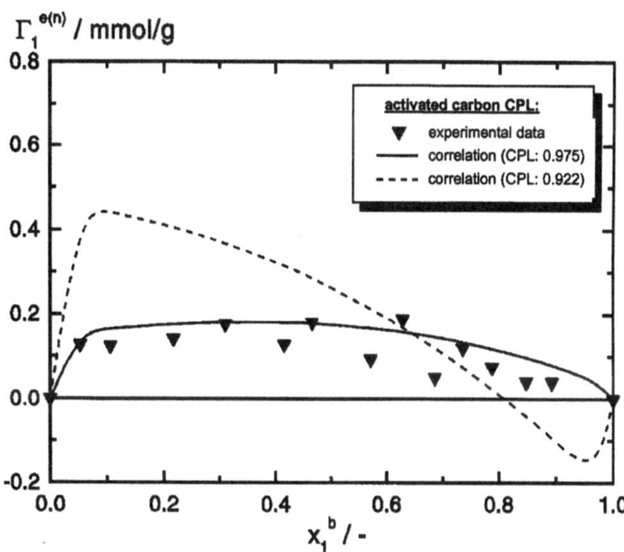

Fig. 4 Correlation of *n*-hexane (1)–2,2,4-trimethylpentane (2) onto CPL

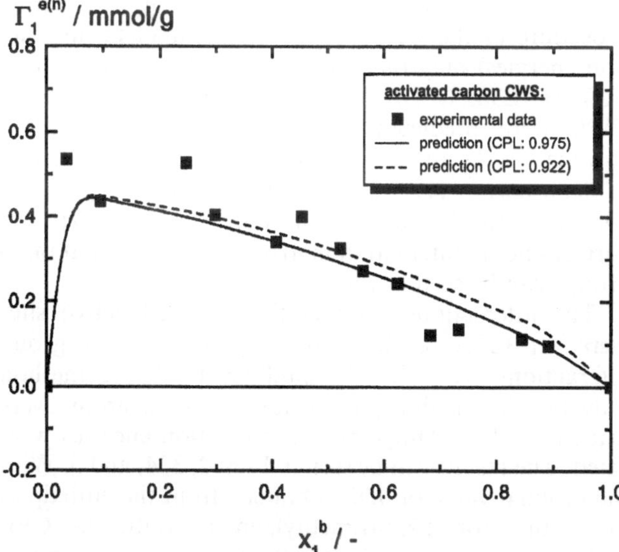

Fig. 5 Prediction of *n*-hexane (1)–2,2,4-trimethylpentane (2) onto CWS

One of the most important parameters is the saturation capacity, $\Gamma^s_{m,i}$, which describes the total amount of pure liquid within the pore system influenced by the solid. This quantity usually cannot be measured and therefore assumptions have to be made. A first assumption was taken for the maximum of the saturation capacity, $\Gamma^{s,max}_{m,i}$, the whole amount of liquid with its density, ρ_i, at 298.15 K which can be placed inside the pore system V_{pore}:

$$\Gamma^{s,max}_{m,i} = \rho_i V_{pore} \quad . \tag{4}$$

In wide pores only a part of the liquid is influenced by the solid and therefore the saturation capacity was treated as parameter which can be fitted [6]. Values of 60–75% of the maximum amount were obtained; therefore all saturation capacities were calculated as

$$\Gamma^{s}_{m,i} = 0.7\Gamma^{s,max}_{m,i} \ . \tag{5}$$

The values obtained were decreased additionally if a branched molecule had to be adsorbed. In the case of 2,2,4-trimethylpentane enthalpies of wetting, $\Delta_w H$, were known from previous investigations [8, 9]. The ratios of the enthalpies of wetting of 2,2,4-trimethylpentane to *n*-octane were taken as a factor in order to decrease the accessible pore volume (Table 4) and therefore the saturation capacity corresponding to

$$\Gamma^{s}_{m,i} = \frac{\Delta_w H_{\mathrm{branched}}}{\Delta_w H_{\mathrm{linear}}} 0.7\Gamma^{s,max}_{m,i} \ . \tag{6}$$

Results

For the investigations alkane–cycloalkane mixtures onto the four carbons were considered. For the fitting procedure of the UNIFAC model parameters (interaction energies) only the data of three carbons were used (ROTH, L3S, CPL). The surface-excess isotherms for CWS were predicted by making use of the fitted parameters. As an example, the surface-excess isotherms of the n-hexane (1)–2,2,4-trimethylpentane (2) system at 298.15 K on all four carbons were chosen [6, 7]. The corresponding interaction matrix of the functional main groups can be seen in Fig. 1.

The interactions between the localised active sites were set to zero; therefore only seven main group interactions exist. For the modified UNIFAC method only two interaction parameters per main group were fitted ($\Delta_{u_{ij}}$, $\Delta_{u_{ji}}$). Altogether 14 interaction energies were fitted. The results can be seen in Figs. 2, 3, 4, and 5. With the original value of 0.922 obtained from the enthalpies of wetting for 2,2,4-trimethylpentane onto the CPL carbon the results were not sufficient. By increasing this value up to 0.975 good results were obtained for the simultaneous correlation of the surface-excess isotherms of the test system onto three activated carbons. With the fitted parameters the surface-excess isotherm of the test system was predicted for the CWS activated carbon. As can be seen from Fig. 5 a good prediction was possible and it has to be emphasised that this prediction was obtained by only introducing the properties of this carbon into the model.

The results show that this theoretical framework can be used in order to correlate and predict the surface-excess isotherms for a whole class of adsorbents, in this investigation activated carbons; however, the carbon- and molecule-dependent parameters, especially the saturation capacities of the pure liquids, have to be known with only a small uncertainty. Nevertheless predictions for a completely different activated carbon were carried out with success for the first time. Additional investigations have to be carried out in the future in order to find an algorithm for calculating the saturation capacities of pure liquids as a function of the pore size distribution of the activated carbon.

Conclusion

The simultaneous correlation of surface-excess isotherms onto different activated carbons was performed by using the ASST, the modified UNIFAC model, and measured and calculated properties of the activated carbons. By using this whole framework, a prediction for a completely different activated carbon was carried out successfully. The limitations of the whole theory were apparent: the saturation capacities of the pure liquids on the different activated carbons calculated from enthalpies of wetting of branched and linear molecules seem to be good choice, but are not as precise as is necessary. However, the physical background of the adsorption from solution onto activated carbons was introduced into a complex thermodynamic framework and tested successfully.

Acknowledgement The authors thank the German Research Foundation (Deutsche Forschungsgemeinschaft) for financial support.

References

1. Berti C, Ulbig P, Schulz S (2000) Adsorption 6:79–92
2. Berti C, Ulbig P, Burdorf A, Schulz S (1999) Langmuir 15:6035
3. Everett DH (1965) Trans Faraday Soc 61:2478–2495
4. Schay G (1969) Surf Colloid Sci 2:155–211
5. Weidlich U, Gmehling J (1987) Ind Eng Chem Res 26:2274
6. Seippel J, Ulbig P, Klueppel M, Berti C, Schulz S (1999) Chem Tech 51:129–133
7. Seippel J, Ulbig P, Schulz S (2000) J Chem Eng Data 45:780–783
8. Ulbig P, Surya-Lukito, Schulz S, Seippel J (1998) J Therm Anal 54:333–342
9. Ulbig P, Friese T, Schulz S, Seippel J (1998) Thermochim Acta 310:217–222

Progr Colloid Polym Sci (2001) 117: 5–12

Krisztina László

Adsorption from aqueous phenol and 2,3,4-trichlorophenol solutions on nanoporous carbon prepared from poly(ethylene terephthalate)

K. László
Department of Physical Chemistry, Budapest University of Technology and Economics, 1521 Budapest, Hungary
e-mail: klaszlo.fkt@chem.bme.hu
Tel.: +36-1-4631893
Fax: +36-1-4633767

Abstract Highly microporous poly(ethylene terephthalate) based activated carbon exhibits a Brunauer–Emmett–Teller surface area of 1170 m^2/g and a total pore volume of 0.625 cm^3/g. It contains mesopores in a sufficient proportion of 42% of the total pore volume, which promotes the diffusion availability of the micropore region. The surface possesses an amphoteric nature owing to the oxygen functionalities present; however, the majority of the groups are basic. The adsorption properties from aqueous solution of weak acids, such as phenol and 2,3,4-trichlorophenol, depend both on the pH of the solutions and on the pK_a of the phenols, as the pH influences both the surface chemistry of the carbon and the dissociation of the weak acids used. In the case of phenol, a competitive adsorption takes place, as the interactions are weak in the three media investigated (pH 3, unbuffered and pH 11). The triple chlorine substitution in 2,3,4-trichlorophenol significantly enhances the surface interactions. Owing to the smaller pK_a value of trichlorophenol in an unbuffered medium, ionic interaction occurs as has been concluded from the outstandingly high value of the adsorption equilibrium constant. At pH 11 the adsorption of the phenols is hindered by electrostatic repulsion.

Key words Activated carbon · X-ray photoelectron spectroscopy · Amphoteric surface · Functional groups · pH effect

Introduction

Nanostructured materials can be manufactured either by building the system atom by atom or molecule by molecule or by dispersing a continuous matrix. One example is tailoring nanosized cavities into a solid phase. Thus, activated carbon might be considered as nanostructured material known and applied since ancient times.

Activated carbons, owing to their versatility, are the most frequently applied adsorbents of gases, vapors and liquids. Activated carbons of desired surface area and pore structure are now commercially available. It was in the late 1980s when the previously neglected importance of carbon surface chemistry was first analyzed in depth, as neither the surface area nor the pore structure was sufficient to explain many of the properties of the carbons. Their surface chemistry is governed by the presence of heteroatoms such as hydrogen, oxygen, nitrogen, phosphorus, etc., originating from the organic precursor and the method of carbon preparation, activation or further treatments [1–5]. The most frequent heteroatom in the carbon matrices is oxygen, which is generally bonded along the edges of the turbostratic graphite crystallites. Oxygen may be present in various forms, such as carboxyls, carbonyls, phenols, lactones, aldehydes, ketones, quinones, hydroquinones, anhydrides or ethereal structures [1, 4, 6–8]. Carbonyl, carboxyl, phenolic hydroxyl and lactonic groups are acidic, while pyrene, chromene and quinone are basic

[1, 2, 9–14]. These groups and the delocalized electrons of the graphitic structure determine the apparent acid/base character of the activated carbon surface [4]. Thus, the nature and distribution of surface functionalities is fundamental in activated-carbon-based water-treatment processes [15–18].

Several workers have exploited the potential of activated carbons for studying the adsorption behaviour of phenols and other organic compounds [19, 20]. The removal of phenol by activated carbon was first reported by Honig [21]. Phenol and its derivatives are antropogenic pollutants, existing widely in industrial effluents such as those from oil refineries and the coal tar, plastics, paint, pharmaceutical and steel industries. Since they are highly toxic and, in general, not easily biodegradable, methods of treatment are continuously being developed. Of all the methods, adsorption appears to provide the best prospect for overall treatment, especially in the lower concentration range. For phenol is one of the most frequent contaminants of industrial wastewaters, one of the important features of the activated carbons used in wastewater treatment is their capability to adsorb phenol [11, 16]. Besides being carcinogenic, in the chlorinating process, which is applied very often as a complementary purification step in wastewater treatment, it reacts with the chlorine and produces carcinogenic mono- and polychlorinated compounds [14, 17, 22]. In our experiments, the latter was represented by 2,3,4-trichlorophenol.

The objective of this work is to demonstrate the influence of pH on the surface chemistry and thus the adsorption mechanism and capacity of a highly microporous activated carbon derived from poly(ethylene terephthalate) (PET).

Experimental

Precursors and sample preparation

Granulated PET (Qualon) was obtained from Mitsubishi (Singapore). The polymer was carbonized at 750 °C for 30 min in a steel reactor flushed with nitrogen gas (50 dm^3/h). After cooling to room temperature, the pyrolyzed material was ground and sieved. Particles in the range 0.8–2.0 mm were activated at 900 °C, up to a burnoff of 50% in a flow of steam. Details have been published recently [23]. This carbon was studied without further treatment.

Characterization of the activated carbon

Nitrogen adsorption

The surface area and the pore size distribution were determined from nitrogen adsorption/desorption isotherms measured at 77 K using an AUTOSORB (Quantachrome, Syosset, N.Y., USA) computer-controlled surface analyzer. The sample was outgassed at 300 °C in a vacuum ($p < 3 \times 10^{-4}$ mbar). The apparent surface area was derived according to the Brunauer–Emmett–Teller (BET) model. The total pore volume, also called the Gurvitsch volume [24], the micropore volume derived from the t-method [25] and the pore size distribution calculated from the desorption branch of the isotherm, according to Barrett el al. [26], were used to characterize the pore structure.

Adsorption from completely miscible binary mixtures

Benzene–methanol binary liquid mixtures were used in order to characterize the chemical heterogeneity of the carbon surfaces. Isotherms were obtained by a batchwise method in the whole composition range, at ambient temperature. Dry methanol and benzene of high-performance liquid chromatography grade (Merck) were used. The solid/liquid ratio applied was 1:5. The 6-h contact time was established from preliminary experiments. The concentrations were derived from the refractive index [27]. Excess isotherms ($n_1^{\sigma(n)}$ versus x_1 functions) were computed from the primary experimental data, as

$$n_1^{\sigma(n)} \equiv n_0\left(x_{1,0} - x_1\right) \ , \qquad (1)$$

where $n_1^{\sigma(n)}$ is the specific surface-excess amount (millimoles per gram) of benzene in the interfacial layer, n_0 is the specific amount of the initial bulk liquid phase (millimoles per gram) and $x_{1,0}$ and x_1 are the initial and the equilibrium molar fractions of benzene, respectively, in the bulk phase. The intersection of the function with the abscissa (in the case of isotherms of S shape), the so-called adsorption azeotropic composition, $x_{1,a}$, can be used as a measure of the hydrophobic/hydrophilic character of the surface.

X-ray photoelectron spectroscopy

The surface chemical composition of the samples was determined by X-ray photoelectron spectroscopy (XPS) using an XR3E2 (VG Microtech) twin-anode X-ray source and a Clam2 hemispherical electron energy analyzer. The base pressure of the analysis chamber was 5×10^{-9} mbar. The Mg Kα radiation used (1253.6 eV) was nonmonochromatized. Wide-scan spectra in the 1000–0 eV binding-energy range were recorded with a pass energy of 50 eV for all samples. High-resolution spectra of the O 1*s* and C 1*s* signals were recorded in 0.05 eV steps with a pass energy of 20 eV. After the linear base line had been subtracted, curve-fitting was performed assuming a Gaussian peak shape.

pH

One gram of carbon sample was shaken with 35 ml bidistilled water in a sealed glass bottle for 72 h, at ambient temperature, to reach equilibrium. At the end of this period the pH of the aqueous phase was detected using an Orion 720 A (Inovata, Broma, Sweden) pH meter. Each determination was performed in triplicate, as well as that of the bidistilled water blank [17].

Boehm titration

The oxygenated surface groups were determined according to Boehm's method [7, 28–30]. One gram of carbon sample was immersed in 35 ml 0.05 M HCl, $NaHCO_3$, Na_2CO_3 and NaOH solutions, respectively. The vials were sealed and shaken for 72 h at ambient temperature. The filtrates were titrated with NaOH and HCl depending on the original titrant. The number of basic sites was calculated from the amount of HCl that reacted with the carbon. The various free acidic groups were derived using the assumption that NaOH neutralizes carboxyl, lactone and phenolic groups, Na_2CO_3 neutralizes carboxyl and lactone and $NaHCO_3$ neutralizes only carboxyl groups, respectively.

Sorption from dilute aqueous solutions of phenols

Solutions of phenol (Merck, 99.5%) and 2,3,4-trichlorophenol (Fluka, 98%) were prepared using bidistilled water or the appropriate buffer solutions (Titrisol pH 3 and Titrisol pH 11, Merck). The carbon (0.05 g) was shaken with 5–60 ml phenol (5 mmol/l) or 2,3,4-trichlorophenol (2 mmol/l) solutions for 24 h and 7 days, respectively, in sealed vials at ambient temperature. The contact times needed to reach equilibrium were concluded from preliminary kinetic measurements [31]. Initial and equilibrium concentrations were determined by detecting the UV absorption of phenol (265 nm) and 2,3,4-trichlorophenol ($\lambda = 290$ nm) using a UVIKON 930 UV–vis spectrophotometer (Kontron, Zurich, Switzerland). The amount of phenol adsorbed by 1 g activated carbon (n_a, millimoles per gram) was calculated according to the following equation:

$$n_a = \frac{V(c_0 - c_e)}{1000m} , \qquad (2)$$

where V is the volume of the liquid phase, m is the mass of the activated carbon and c_0 and c_e are the initial and equilibrium concentrations of the liquid phase, respectively. The curves were evaluated according to the linear Langmuir fit:

$$\frac{c_e}{n_a} = \frac{1}{Kn_m} + \frac{c_e}{n_m} , \qquad (3)$$

where n_a is the phenol or 2,3,4-trichlorophenol adsorbed by 1 g carbon, c_e is the equilibrium concentration in the aqueous phase, n_m is the monolayer capacity and K is the adsorption equilibrium constant. However, it should be noted that the surface homogeneity, which is one of the basic conditions applied in the Langmuir model, is hardly fulfilled in the case of these systems.

Selected physicochemical properties of the phenols are collected in Table 1. The electron-withdrawing effect of the chlorine atoms reduces the electron density of the aromatic ring, which is reflected in the lower pK_a value as well. The surface occupied by a single adsorbate molecule was applied to characterize the coverage of the carbon surface.

Results

Characterization of the carbon

Surface area and pore structure

The carbon obtained from the PET by the two-step physical activation applied exhibits high microporosity.

Table 1 Selected properties of phenol and 2,3,4-trichlorophenol

	Molecular weight	pK_a[a]	Solubility (g/l)[a]	Cross-sectional area (nm^2/molecule)	Ref.
Phenol	94.11	9.89	82	0.42	[41]
				0.30	[19]
				0.522	[16]
				0.437	[42]
				0.437	[20]
				0.43	[43]
2,3,4-Trichlorophenol	197.45	7.59	0.4	0.72	[41]
				0.631–0.649	[43]

[a] At ambient temperature

A BET surface area of 1,170 m^2/g and an average pore radius of 1.07 nm were derived from the nitrogen adsorption isotherm. According to the pore size distribution (Fig. 1a) it contains both micropores and mesopores; however, 0.425 cm^3/g of the 0.625 cm^3/g total pore volume is microporous (Fig. 1b). The high surface area and the pore structure designate this carbon for application in aqueous adsorption processes. It contains mesopores in a sufficient proportion of 42% of the total pore volume, which promotes the diffusion availability of the micropore region.

Chemistry of the surface

Elemental composition and surface heterogeneity

The bulk and surface compositions, obtained from elemental analysis and XPS, respectively, are compared in Table 2. The discrepancy of the surface and bulk regions is obvious from the deviation of the O/C proportions determined by the two methods, as the information obtained from the XPS analysis characterizes only the upper few nm layer of the sample.

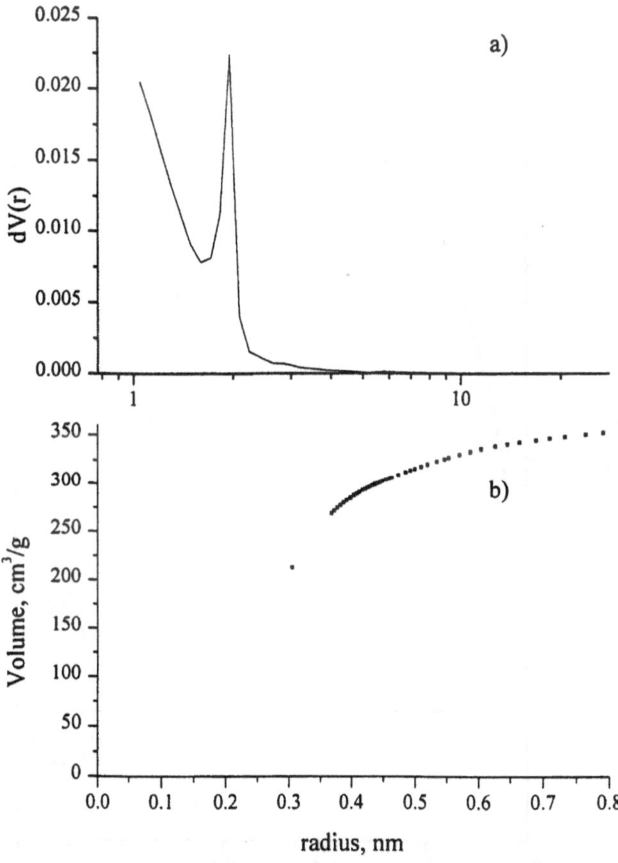

Fig. 1 **a** Differential pore size distribution derived from the desorption branch of the nitrogen adsorption/desorption isothem of the poly(ethylene terephthalate) (*PET*) carbon, 77 K; **b** t-plot derived from the same isotherm

Table 2 Elemental composition in atomic percent

	C	H	O	O/C
Bulk	90.7	4.6	4.7	0.052
Surface	95.7	–	4.3	0.045

The carbon displays an excess isotherm of type IV according to the Schay–Nagy classification (Fig. 2) [32]. The S shape means that the surface contains both hydrophilic and hydrophobic surface locations and, thus, the preferentially adsorbing component ($n_i^{\sigma(n)} > 0$) depends on the bulk concentration. The nonpolar benzene adsorbs preferentially below $x_{1,a} = 0.85$, while in the $0.85 < x_1 < 1$ range the adsorption of the polar methanol is preferred. Thus, the surface exhibits amphoteric character, as it contains both hydrophobic and hydrophilic surface sites. The adsorption of benzene is attributed mainly to the interaction between π electrons in the benzene ring and the condensed aromatic rings on the carbon surface [33] referred to as basic structural units [34]. The adsorption of methanol is preferred on hydrophilic functional groups as well as armchair and zigzag sites at the edges of the basal planes. The presence of hydrophilic centers, such as carboxylic or phenolic groups, renders the surface polar, enhancing the interaction with the polar methanol; however, interaction between the benzene and some of the oxygen-containing functional groups may take place as well [2, 35]. These surface functional groups were determined by the XPS method (Fig. 3).

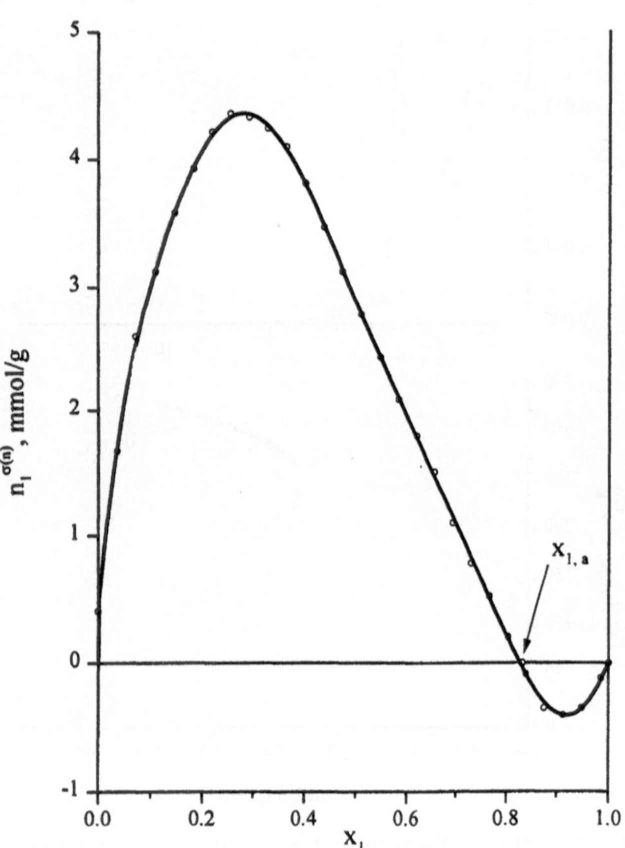

Fig. 2 Adsorption-excess isotherms on PET carbon from benzene (1)–methanol (2) binary liquid mixture

Five peaks, i.e. five different carbon bonds, were identified in the C 1*s* range, and two different oxygen bonds in the O 1*s* range. The results of the deconvolution and fitting processes are collected in Tables 3 and 4.

Acid/base properties

The pH of the aqueous slurry of the carbon was 6.2 (the pH of the bidistilled water blank was measured to be 5.5), i.e. this carbon can be classified as of type H because it

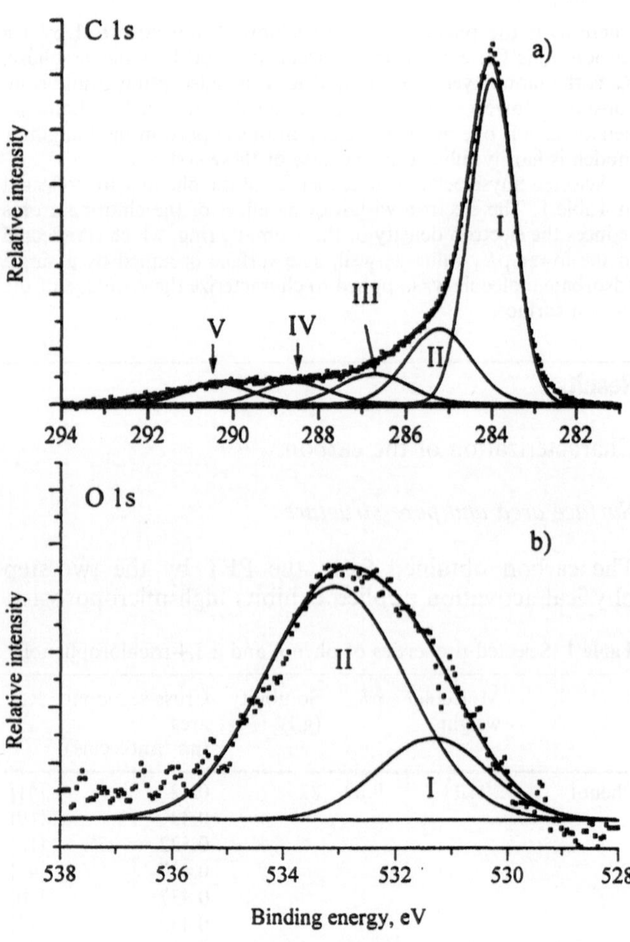

Fig. 3 **a** X-ray photoelectron spectroscopy (*XPS*) C 1*s* spectrum of the PET carbon. *I*: C–C, C–H; *II*: C–O; *III*: C=O, O–C–O; *IV*: O=C–O; *V*: shakeup satellite peaks owing to π-π* transitions in aromatic rings. **b** XPS O 1*s* spectrum of the PET carbon. *I*: C=O; *II*: C–O

Table 3 Distribution of carbon structures (atomic percent) from the X-ray photoelectron spectroscopy (*XPS*) C 1*s* spectrum [8, 44]

Peak	Binding energy (eV)	Carbon structure	Atomic %
I	284.0	C–C, C–H	60.0
II	285.2	C–O	18.1
III	286.7	C=O, O–C–O	8.1
IV	288.6	O=C–O	5.3
V	290.3	Shakeup satellite peaks owing to π–π^* transitions in aromatic rings	4.2

Table 4 Distribution of oxygen structures (atomic percent) from the XPS O 1*s* spectrum [8, 44]

Peak	Binding energy (eV)	Oxygen structure	Atomic %
I	531.3	C=O	1.2
II	533.1	C–O	3.1

exhibited a pH value higher than that of the blank [17]. The acid/base properties of the functional groups detected by XPS were determined by Boehm titration (Table 5).

In accordance with the pH shift experienced in the carbon suspension, 81.6% of the functional groups are basic. No carboxylic groups were detected within the experimental error of the Boehm titration [30]. The acidic character rises from the phenolic (81.1%) and lactonic (8.1%) groups. This is consistent with the ratio of the deconvoluted peaks of the C 1*s* spectrum, whereas, from among the carbon atoms related to oxygen (peaks II, III and IV), the ratio of peak II derived from carbons in phenolic, alcoholic or ethereal groups is 57% (Table 3). Similarly, in the O 1*s* spectrum, the ratio of the single C–O bonds is dominant (72%) (Table 4). The origin of the basic character can be partly traced back to the delocalized electrons of the graphitic structure. The C 1*s* spectrum also suggests that most of the surface carbon atoms are present in graphitic form.

Table 5 Results of Boehm titration of poly(ethylene terephthalate) (*PET*) carbon

Functional group	Concentration (μmol/g)	Group/100 nm^{2a}
Carboxylic	Below the detection limit	–
Lactonic	17.7	0.91
Phenolic OH	76.0	3.85
Total acidic groups	93.7	4.76
Total basic groups	416.8	21
Total functional groups	510.5	26

[a] Calculated as 100 × concentration × $\frac{N_A}{a_{S,BET}}$, applying the corresponding unit conversions; N_A is the Avogadro number

Adsorption of the phenols

The saturation capacities of the PET carbon from unbuffered phenol and 2,3,4-trichlorophenol solutions were found to be 2.58 and 3.16 mmol/g, respectively, i.e. significantly more adsorbed from the larger molecule. A nearly complete monolayer coverage, as can be deduced from the molecular areas, was achieved only in the case of trichlorophenol [6]. In order to understand this anomalous behavior, the pH dependence of the adsorption process was studied, as both the nature of the phenols and the chemistry of the surface sites are affected by pH. Solutions buffered to pH 3 and pH 11 were used as all the functional groups of the carbon surface are protonated below pH 3 and are deprotonated at pH 11 (Fig. 4).

The Langmuir parameters derived from the linearized isotherms (Figs. 5, 6, Table 6) confirm the expectations that the adsorption capacity and mechanism are influenced by the pH.

The sequence of the monolayer capacities is unbuffered > pH 3 > pH 11, and pH 3 ≈ unbuffered > pH 11 in the case of phenol and trichlorophenol, respectively. It is in accordance with former observations that in the pH range 9–12 the increasing pH results in a decreasing adsorption capacity [36, 37]. In the case of phenol, the values of the *K* parameter derived from the Langmuir fit are practically similar in the buffered solutions, but *K* exhibits a higher value when the system is not buffered. The sequence of the *K* values from the trichlorophenol isotherms is unbuffered ≫ pH 11 ≈ pH 3. The numerical values are 1 or 2 orders of magnitude higher than in phenolic systems. The enhanced interaction in the case of the trichlorophenol adsorbate is due to the electron-withdrawing phenomenon of the three chlorine substituents.

By comparing the number of functional groups per 100 nm^2 from the titration and the number of adsorbed

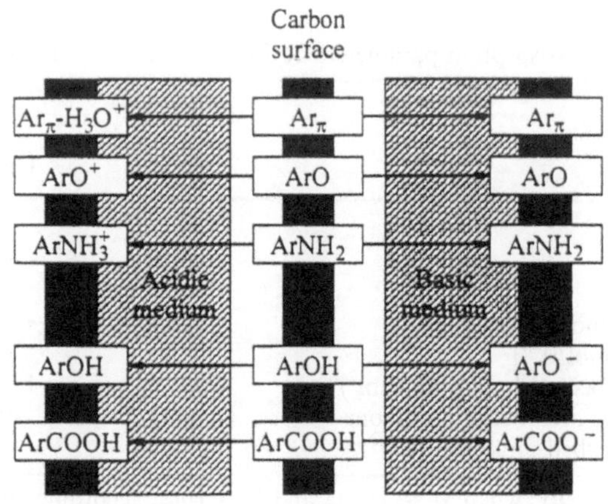

Fig. 4 Charge development on activated carbon surface in aqueous solution [45]

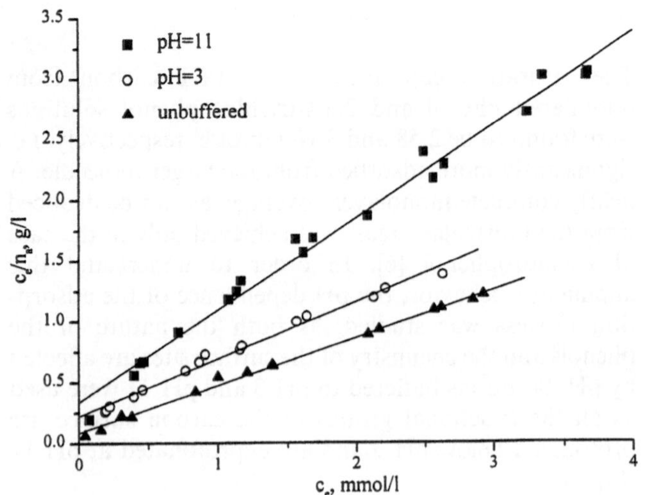

Fig. 5 Linearized adsorption isotherms (ambient temperature) from aqueous phenol solutions of various pH. The *symbols* represent measured values, the *solid lines* represent the fit by the Langmuir equation

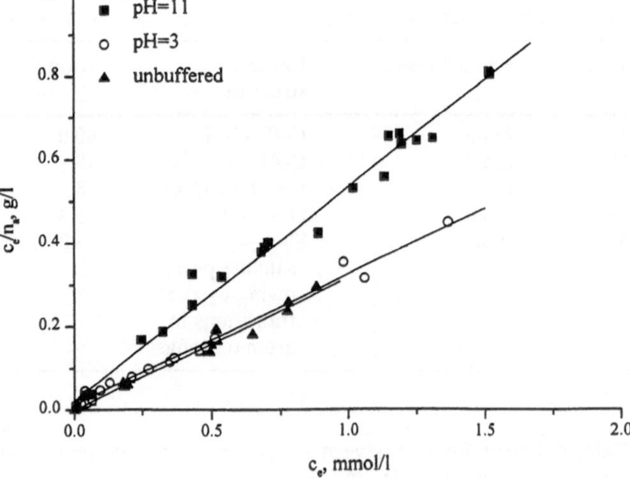

Fig. 6 Linearized adsorption isotherms (ambient temperature) from aqueous 2,3,4-trichlorophenol solutions of various pH. The *symbols* represent measured values, the *solid lines* represent the fit by the Langmuir equation

phenols per 100 nm^2 listed in Tables 5 and 6 respectively, it is clear that the adsorption does not takes place only on the titrated functionalities, as the number of the adsorbed molecules significantly exceeds that of the oxygen-containing surface sites. The highest surface population was developed from the unbuffered and acidic trichlorophenol solutions and the lowest when the media were basic.

The fundamental interactions between the carbon surface and phenols are a) dispersion effect between the aromatic ring and the π electrons of the graphitic structure and b) electrostatic attraction and repulsion if ions are present [16–19, 38–40]. The interaction between the water molecules and the surface sites cannot be excluded as well, and their competitive adsorption results in the depletion of the amount adsorbed [16]. Electron donor–acceptor interaction between the aromatic ring and the basic surface oxygens does not play a significant role in this case, as the concentration of the surface oxygen atoms is relatively low.

At pH 3, both the surface functional groups and the phenolic compounds are in nonionic form, that is, the surface groups are either neutral or positively charged. The interaction between the carbon surface and the phenolic compounds is weak and can be attributed to the dispersion effect (Fig. 7a).

This weak interaction in the case of the phenol results in the coadsorption of water molecules, yielding a reduced surface concentration in comparison with trichlorophenol. For the stronger interaction, trichlorophenol may complete a more or less complete monolayer. It cannot be excluded that some of the molecules interact via hydrogen bonds formed between the protonated surface functional groups along the edge of graphitic

Table 6 Adsorption parameters of aqueous phenol and 2,3,4-trichlorophenol solutions on PET carbon

	Phenol			2,3,4-Trichlorophenol		
	pH 3	Unbuffered	pH 11	pH 3	Unbuffered	pH 11
a^a	0.2244	0.0873	0.3367	0.0132	0.0007	0.0226
b^a	0.4684	0.3881	0.7574	0.3124	0.3167	0.5120
R	0.9936	0.9977	0.9938	0.9941	0.9952	0.9949
n_m (mmol/g)	2.14	2.58	1.32	3.20	3.16	1.95
K [(mmol/l)$^{-1}$]	2.09	4.45	2.25	23.67	479.85	22.66
Adsorbed molecule/100 (nm^2)[b]	108	130	67	161	159	99
Surface area occupied by one adsorbed molecule, (nm^2)[c]	0.93	0.77	1.50	0.62	0.63	1.01

[a] $y = a+bx$, where $a = 1/Kn_m$ and $b = 1/n_m$, according to Eq. (3)

[b] Calculated as $100 \times n_m \times \frac{N_A}{a_{S,BET}}$, applying the corresponding unit conversions

[c] Calculated as $\frac{a_{S,BET}}{n_m N_A}$, applying the corresponding unit conversions

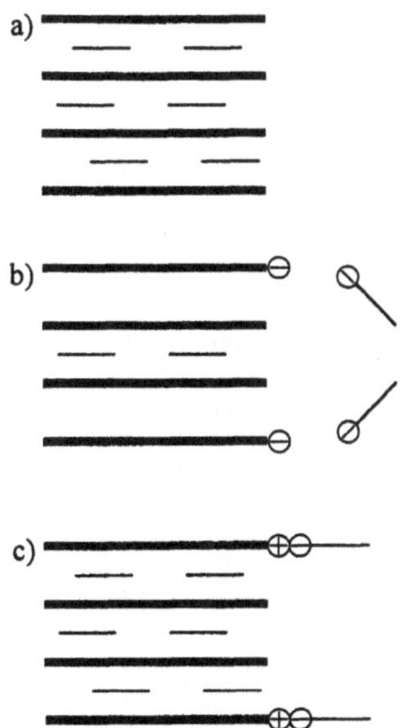

Fig. 7 Schematic representation of the adsorption of phenols on a carbon surface **a** by the dispersion effect, **b** when electrostatic repulsion takes place and **c** when electrostatic attraction is possible. (Water molecules are not shown.)

layers and the phenolic OH groups, which may reduce the availability of the graphitic region.

At pH 11 the phenols dissociate, forming phenolate anions, while the surface functional groups are either neutral or negatively charged. The electrostatic repulsion between the alike charges lowers the adsorption capacities in the case of both phenols (Fig. 7b). The deprotonated acidic groups along the graphitic layers repel the deprotonated phenolate ions, hindering the access of the phenolate ions into the space between the graphite sheets. Dispersion interaction may occur when the penetration of the phenol molecules into the graphitic region is not blocked. The competitive adsorption of water molecules occurs as well: the surface area related to a single adsorbed unit shows the highest value (i.e., the lowest coverage) at pH 11 in the case of both phenols.

When the adsorption takes place in unbuffered solutions (pH about 6.5), the carbon surface contains both protonated and deprotonated sites. Owing to the different pK_a values of the two adsorbate species, trichlorophenol produces a significantly higher number of ions than phenol. The ionic attraction acting between the trichlorophenolate anions and the positively charged surface sites (Fig. 7c) is reflected by the outstanding K value of the Langmuir fit. The surface coverage in this case is similar to that developing at pH 3, indicating also dispersion force driven adsorption within the graphene layers.

Conclusion

The microporous carbon produced from PET shows amphoteric surface nature, owing to the presence of oxygen-containing groups. 81.6% of the surface sites can be titrated by HCl, i.e., the majority of the functional groups exhibit basic character. The acidity derives from lactonic and phenolic oxygens. The adsorption mechanism fundamentally depends on the pH of the solution applied, as the phenols are weak acids and the surface is amphoteric, with acid/base character.

The relatively low K values and the high surface area available for a single phenol molecule compared to the theoretical cross-sectional area lead to the conclusion that in the three media applied phenol adsorbs mainly by dispersion interaction and, therefore, competitive adsorption of the solvent molecules may take place as well.

The introduction of the three chlorine substituents results in a significantly different electron structure in the case of 2,3,4-trichlorophenol and the dispersion interaction becomes stronger than in the case of phenol: this is reflected in the higher K values. In unbuffered medium the lower pK_a value of the trichlorophenol results in a significant number of trichlorophenolate anions in comparison with the simple phenol. This leads to ionic attraction between the trichlorophenolate ions and the positively charged surface sites. The ionic interaction is shown by the remarkably high K value. A practically complete monolayer is formed at pH 3 and from unbuffered solution as well.

At pH 11 the electrostatic repulsion existing between the negative surface sites and the phenolate or 2,3,4-trichlorophenolate ions reduces the adsorption capacities.

Acknowledgements This research was supported by the OTKA Fund (Hungary) nos. T 017019 and T 025581. The authors wish to express their appreciation to Katalin Josepovits for the XPS spectra. The technical support of Emese Fülöp and György Bosznai and András Szucs is gratefully acknowledged.

References

1. Bansal RC, Donnet JB, Stoeckli HF (1988) Active carbon. Dekker, New York
2. Puri BR (1970) In: Walker PL Jr (ed) Chemistry and physics of carbon, vol 6. Dekker, New York, pp 191–282
3. Boehm HP (1994) Carbon 32:759
4. Leon y Leon CA, Radovic LR (1994) In: Thrower PA (ed) Chemistry and physics of carbon, vol 24. Dekker, New York, pp 214–310
5. Benaddi H, Bandosz TJ, Jagiello J, Schwarz JA, Rouzaud JN, Legras D, Béguin F (2000) Carbon 38:669
6. Rodrígez-Reinoso F (1998) Carbon 36:159
7. Boehm HP, Diehl E, Heck WR Sappok (1964) Angew Chem Int Ed Engl 3:669
8. Zielke U, Hüttinger KJ, Hoffman WP (1996) Carbon 34:983

9. Lopez-Ramon MV, Stoeckli F, Moreno-Castilla C, Carrasco-Marin F (1999) Carbon 37:1215
10. Boehm HP, Voll M (1970) Carbon 8:227
11. Papirer E, Li S, Donnet JB (1987) Carbon 25:243
12. Papirer E, Dentzer J, Li S, Donnet JB (1991) Carbon 29:69
13. Bismarck A, Wuertz C, Springer J (1999) Carbon 37:1019
14. Suárez D, Menéndez JA, Fuente E, Montes-Morán MA (1999) Langmuir 15:3897
15. Salvador F, Merchán MD (1996) Carbon 34:1543
16. Nevskaia DM, Santianes A, Muñoz V, Guerrero-Ruíz A (1999) Carbon 37:1065
17. Tessmer CH, Vidic RD, Uranowski LJ (1997) Environ Sci Technol 31:1872
18. Brasquet C, Le Cloirec P (1997) Carbon 35:1307
19. Singh B, Madhusudhanan S, Dubey V, Nath R, Rao NBSN (1996) Carbon 34:327
20. Caturla F, Martín-Martínez JM, Molina-Sabio M, Rodriguez-Reinoso F, Torregrosa R (1988) J Colloid Interface Sci 124:528
21. Honig P (1926) Kolloid Chem Beih 22:345
22. Puri BR (1966) Carbon 4:39
23. László K, Bóta A, Nagy LG (2000) Carbon 38:1965
24. Unger K, Schadow E, Fischer H (1976) Z Phys Chem 99:245
25. Halsey GD (1948) J Chem Phys 16:931
26. Barrett EP, Joyner LG, Halenda PP (1951) J Am Chem Soc 73:373
27. Schay G, Nagy LG (1978) Adsorption from binary mixtures on S/L and S/G interfaces. Akadémiai, Budapest, p 106 (in Hungarian)
28. Tamon H, Okazaki M (1996) Carbon 34:741
29. Otowa T, Nojima Y, Miyazaki T (1997) Carbon 35:1315
30. Szucs A (1999) MSc thesis. Budapest University of Technology and Economics, Budapest, Hungary
31. Bóta A, László K, Nagy LG, Subklew G, Schlimper H, Schwuger MJ (1996) Adsorption 2:81
32. Schay G, Nagy LG (1961) J Chim Phys 149
33. Gasser G, Kipling JJ (1960) Proceedings of the Fourth Conference on Carbon. Pergamon, New York, p 55
34. Oberlin A (1989) In: Thrower PA (ed) Chemistry and physics of carbon, vol 22. Dekker, New York, pp 1–143
35. Kinoshita K (1988) Carbon. Electrochemical and physicochemical properties. Wiley, New York, p 144
36. Urano K, Kano H (1984) Bull Chem Soc Jpn 57:2501
37. Warch E (1983) Z Chem 23:427
38. Radovic LR, Ume JI, Scaroni W (1996) In: LeVan MD (ed) Fundamentals of adsorption. Kluwer, Boston, p 749
39. Radovic LR, Silva IF, Ume JI, Menendez JA, Leon y Leon CA, Scaroni W (1997) Carbon 35:1339
40. Ume JI, Scaroni W, Radovic LR (1993) Proceedings of the 21st Biennial Conference on Carbon, Buffalo. American Carbon Society, New York, p 468
41. Snyder LR (1968) Principles of adsorption chromatography. Dekker, New York, p 199
42. Haghseresht F, Lu GQ, Whittaker AK (1999) Carbon 37:1491
43. McClellan AL, Harnsberger HFJ (1967) J Colloid Sci 23:577
44. Biniak S, Szymanski G, Siedlewski J, Swiatkowski A (1997) Carbon 35:1799
45. Radovic LR (1999) In: Schwarz JA, Contescu CI (eds) Surfaces of nanoparticles and porous materials. Dekker, New York, pp 529–565

Progr Colloid Polym Sci (2001) 117: 13–17

Masashi Mizukami
Kazue Kurihara

Alcohol cluster formation on silica surfaces in cyclohexane

M. Mizukami · K. Kurihara (✉)
Institute for Chemical Reaction Science,
Tohoku University, Katahira 2-1-1,
Aoba-ku, Sendai 980-8577, Japan
e-mail: kurihara@icrs.tohoku.ac.jp
Tel.: +81-22-2175673
Fax: 81-22-2175673

Abstract Adsorption of ethanol onto silica surfaces from ethanol–cyclohexane binary liquids was investigated by a combination of colloidal probe atomic force microscopy and Fourier transform IR spectroscopy using the attenuated total reflection (ATR) mode. An unusually long range attraction was found between the silica surfaces in the presence of ethanol in the concentration range of 0.1–1.4 mol%. At 0.1 mol% ethanol, the attraction appeared at a distance of 35 ± 3 nm and turned into a repulsive force below 3.5 ± 1.5 nm upon compression. The ATR spectra demonstrated that ethanol adsorbed on the surfaces formed hydrogen-bonded clusters even in a low ethanol concentration range of 0.1–0.5 mol%, where the attractions were especially long-ranged and practically no ethanol cluster formed in the bulk solutions. The spectra also indicated that the cluster formation involved hydrogen-bonding interactions between surface silanol groups and ethanol hydroxyl groups in addition to those between ethanol molecules. We account for the observed long-range attraction by the bridging of opposed adsorption layers of ethanol on the silica surfaces.

Key words Adsorption from binary liquid · Alcohol cluster · Silica surface · Surface force · Fourier transform IR–attenuated total reflection

Introduction

Adsorption from solutions onto solid surfaces is a key in colloid and surface science. It plays important roles in many industrial processes, such as colloidal stabilization and catalysis. Elegant application of adsorption in advanced materials has been demonstrated recently in the formation of self-assembling monolayers and multilayers on various substrates [1, 2]. However, only a limited number of researches have been devoted to the study of adsorption in binary liquids. Adsorption isotherms and colloidal stability measurements have been the main tools for these studies, which limit our understanding of the phenomena to the macroscopic level. The molecular level of characterization is necessary to elucidate and further utilize the phenomenon in the future technology. We employed a combination of surface force measurement and Fourier transform IR spectroscopy in attenuated total reflection (FTIR–ATR) mode to shed light on the preferential (selective) adsorption of alcohol (methanol, ethanol and propanol) onto silica surfaces from their binary mixtures with cyclohexane.

Surface forces measurement is a unique tool for surface characterization [3]. It can directly monitor the distance dependence of surface forces, which reflect distance-dependent changes of surface properties from the surface to the bulk [4]. Such information is difficult to obtain by other techniques. One of the simplest examples is the case of the electric double layer force. The repulsion observed between charged surfaces follows the counterion distribution in the vicinity of surfaces and is called the electric double layer force (repulsion) [3]. It is easy to extend this idea to other surface properties, such as adsorption. Indeed, our studies have revealed the formation of alcohol clusters extending for about 10 nm on silica surfaces and a long-range attraction associated with this phenomenon [5, 6]. Among the alcohols studied, we report results obtained for ethanol adsorption on silica surfaces. Here, we use silica as a general word to call the substrates, glass (for colloidal probe atomic force microscopy, AFM) and oxidized silicon (for FTIR–ATR), chosen on the basis of the experimental requirement.

Experimental

Cyclohexane and ethanol were dried with sodium and magnesium, respectively, and distilled immediately prior to use.

The interaction force (F) between a glass sphere and a glass plate was measured as a function of the surface distance (D) in cyclohexane–ethanol mixtures using AFM (Seiko II, SPI3700-SPA300) [7]. Colloidal glass spheres (Polyscience) and glass plates (Matsunami, micro cover glass) were washed in a mixture of sulfuric acid and hydrogen peroxide (4: 1, v/v) and thoroughly rinsed with pure water. The colloidal glass sphere (4–5-μm radius) was then attached to the top of a cantilever (Olympus, RC-800PS-1) with epoxy resin (Shell, Epikote1004). The spheres and the plates were treated with water vapor plasma (Samco, BP-1) for 3 min. just prior to each experiment in order to ensure the existence of silanol groups on the glass surface [8]. The forces obtained were normalized by the radius (R) of the sphere using the Derjaguin approximation [1], $F/R = 2\pi G_f$. Here, G_f is the interaction free energy per unit area between two flat surfaces. R was measured using an optical microscope.

The adsorption-excess isotherm was measured using adsorbent glass spheres which were washed and water-vapor-plasma treated in the same manner as for the force measurements. The glass spheres (typically 1.0 g) dispersed in ethanol–cyclohexane mixtures (10 ml) precipitated after they were equilibrated for about 24 h at 20 ± 0.5 °C. The composition of the supernatant was determined using a differential refractometer (Otsuka Electronics, DRM-1021). The adsorption layer thickness (t) was estimated by assuming that only ethanol is present in the adsorption layer using the specific adsorption amount [9, 10].

IR spectra were recorded using a PerkinElmer FTIR 2000 system using a triglycine sulfate detector. For the ATR mode, the ATR attachment from Grasby Specac was used. A homemade stainless steel ATR flow cell was used and was sealed with a Teflon O-ring. Transmission IR spectra were obtained using a CaF_2 cell (Nihon Bunko) with a path length of 25 μm. An ATR prism made of silicon crystal (Nihon PASTEC, 60 × 16 × 4 mm trapezoid) was used as a solid adsorbent surface. It is known that on the silicon surface, the oxide layer gradually grows up to 3–5 nm thickness when the surface is exposed to air at room temperature [11], thus exhibiting similar properties as those of glass. In order to clean the oxide surface, the silicon crystal was immersed in a mixture of sulfuric acid and hydrogen peroxide (4:1, v/v) and thoroughly rinsed with pure water. The crystal was then treated with water vapor plasma for 20 min immediately prior to each experiment to ensure the formation of silanol groups on the silicon oxide surface [8]. It was kept in pure cyclohexane until assembled into the ATR cell.

Results and discussion

Typical force profiles measured between glass surfaces in ethanol–cyclohexane mixtures are shown in Fig. 1. In pure cyclohexane, the observed force agreed well with the

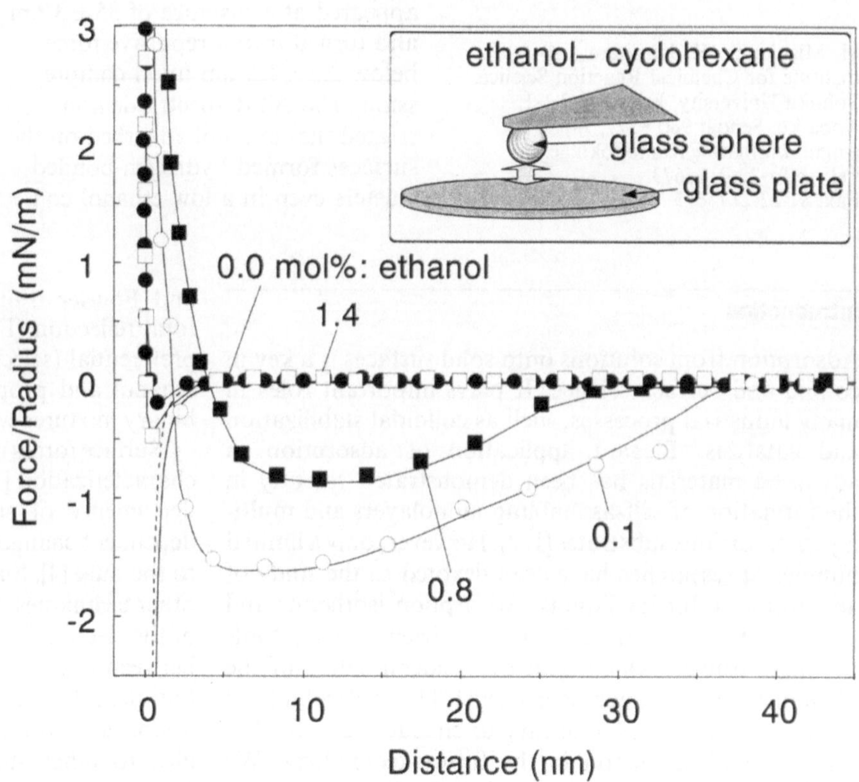

Fig. 1 Profiles of interaction forces between glass surfaces upon compression in ethanol–cyclohexane mixtures. The *dashed line* and the *solid line* represent the van der Waals force calculated using the non-retarded Hamaker constants of 3×10^{-21} J for glass/cyclohexane/glass and 6×10^{-21} J for glass/ethanol/glass, respectively [3]

conventional van der Waals attraction calculated with the nonretarded Hamaker constant for glass/cyclohexane/glass, 3.1×10^{-21} J. At an ethanol concentration of 0.1 mol%, the interaction changed remarkably: the long-range attraction appeared at a distance of 35 nm, showed a maximum around 10 nm, and turned into repulsion at distances shorter than 5 nm. The pull-off force of the contacting surfaces was 140 ± 19 mN/m, which was much higher than that in pure cyclohexane, 10 ± 7 mN/m. Similar force profiles were obtained at ethanol concentrations up to 0.4 mol%. A further increase in the concentration decreased the long-range attraction. At an ethanol concentration of 1.4 mol%, the interaction was identical to that in pure cyclohexane. The range where the long-range attraction extended changed parallel to the value of the pull-off force at various ethanol concentrations, indicating that both forces are associated with the same phenomenon, most likely the adsorption of ethanol. It is natural to ascribe the short-range repulsion to the steric force of structured ethanol molecules adjacent to the glass surfaces which is similar to the hydration force [3].

In order to understand the condition better, we determined the adsorption isotherm by measuring the concentration changes in the alcohol upon adsorption onto glass particles using a differential refractometer. Plots of the range of the attraction versus the ethanol concentration are shown in Fig. 2 together with the apparent adsorpion layer thickness estimated from the adsorption isotherm assuming that only ethanol is present in the adsorption layer [9]. For 0.1 mol% ethanol, half the distance where the long-range attraction appears, 18 ± 2 nm, is close to the apparent layer thickness of the adsorbed ethanol, 13 ± 1 nm. This supports our interpretation that the attraction is caused by contact of opposed ethanol adsorption layers. Half the attraction range is constant up to about 0.4 mol% ethanol and decreases with increasing ethanol concentration, while the apparent adsorption layer thickness remains constant for all the concentration ranges studied. The discrepancy between the two quantities indicates a change in the structure of the ethanol adsorption layer at concentrations higher than about 0.4 mol%, which we will discuss later.

The structures of the adsorbed ethanol were studied using FTIR–ATR spectroscopy. They turned out to be hydrogen-bonded clusters as described later. The spectra obtained were examined by referring to well-established, general spectral characteristics of hydrogen-bonded alcohols in the fundamental OH stretching region because ethanol is known to form hydrogen-bonded dimers and polymers (clusters) in nonpolar liquids [11]. We also examined hydrogen-bonded ethanol cluster formation in bulk ethanol–cyclohexane mixtures using transmission IR spectroscopy.

FTIR–ATR spectra of ethanol adsorbed onto silica surfaces in cyclohexane at various ethanol concentrations (0.0–2.0 mol%) are presented in Fig. 3. At 0.1 mol% ethanol, a narrow negative band at 3,680 cm^{-1}, a weak absorption at 3,640 cm^{-1} (free OH), and a broad strong absorption (3,600–3,000 cm^{-1}) with shoulders at 3,530 cm^{-1} (cyclic dimer

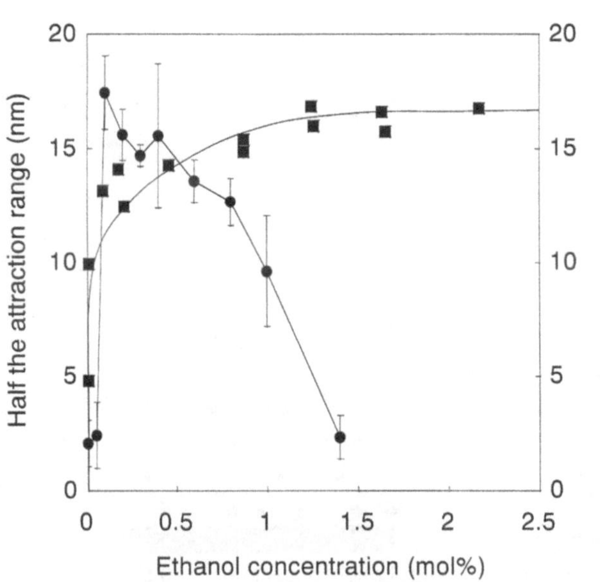

Fig. 2 A plot of half the attraction range between glass surfaces and that of the apparent adsorption layer thickness of ethanol on glass surfaces in ethanol–cyclohexane binary liquids

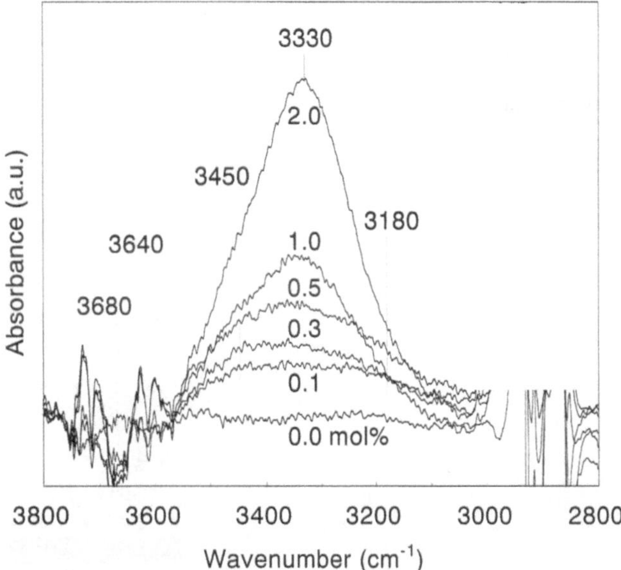

Fig. 3 Fourier transform IR–attenuated total reflection FTIR-AIR spectra of ethanol on an oxidized silicon surface in ethanol–cyclohexane binary liquids at various ethanol concentrations of 0.0–2.0 mol%

or donor end OH), 3,450 and 3,180 cm^{-1} were observed. The isolated silanol group is known to exhibit an absorption band at 3,675–3,690 cm^{-1} in a nonpolar liquid, for example, CCl_4 [12]. Therefore, the negative absorption at 3,680 cm^{-1} should correspond to the decrease in the isolated silanol groups. On the other hand, the positive shoulder at 3,450 cm^{-1} indicated the appearance of the silanol groups hydrogen bonded with the adsorbed ethanol, because the absorption band shifts to a lower wavenumber (3,425–3,440 cm^{-1}) when the silanol groups hydrogen bond with esters [12]. The strong broad band at 3,600–3,000 cm^{-1} (the polymer OH) demonstrated the hydrogen-bonded polymer cluster formation of ethanol adsorbed on the silicon oxide surface even at 0.1 mol% ethanol where no polymer peak appeared in the spectrum of the bulk solution at 0.1 mol% ethanol. With increasing ethanol concentration, the intensity of the polymer OH peak (3,330 cm^{-1}) increased.

At higher ethanol concentrations, the ATR spectra should contain a contribution from the bulk species, because of the long penetration depth of the evanescent wave, 250 nm. To distinguish the bulk contribution, the integrated peak intensities of polymer OH peaks of transmission (A_{TS}) and ATR (A_{ATR}) spectra are plotted as a function of the ethanol concentration in Fig. 4. The former followed cluster formation in the bulk liquid, whilst the latter contains contributions of clusters both on the surface and in the bulk. A sharp increase is seen in A_{ATR} even at 0.1 mol% ethanol, but there is no significant increase in A_{TS} at ethanol concentrations lower than 0.5 mol%. A comparison of A_{TS} and A_{ATR} clearly indicated that the ethanol cluster formed locally on the surface at ethanol concentrations lower than about 0.5 mol%, where practically only a negligible number of clusters existed in the bulk. In the adsorption layer, ethanol formed polymer clusters through hydrogen bonding of surface silanol groups and ethanol as well as those between ethanol molecules. A plausible structure of the ethanol adsorption layer is presented in Fig. 5.

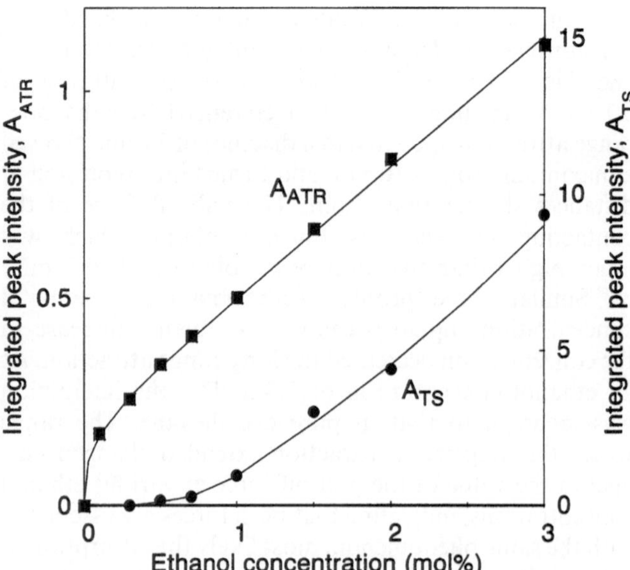

Fig. 4 Plots of integrated peak intensities of polymer OH as a function of the ethanol concentration. The *filled circles* represent the values obtained from transmission spectra (A_{TS}), while the *filled squares* represent those from ATR (A_{ATR})

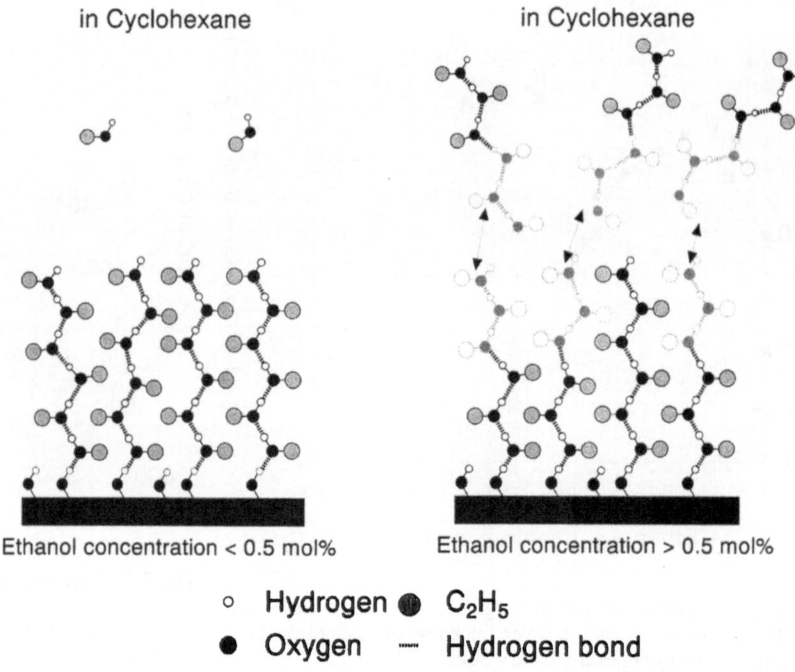

Fig. 5 Plausible structures of the adsorption layer composed of ethanol cluster

The contact of adsorbed ethanol layers should bring about the long-range attraction observed between glass surfaces in ethanol–cyclohexane mixtures. The attraction starts to decrease at about 0.5 mol% ethanol, where ethanol starts to form clusters in the bulk phase. It is conceivable that the cluster formation in the bulk influences the structure of the adsorbed alcohol cluster layer, thus modulating the attraction. We think that the decrease in the attraction is due to the exchange of alcohol molecules between the surface and the bulk clusters (Fig. 5).

Conclusion

This study, to the best of our knowledge, is the first attempt at employing the modern technique to understand the liquid adsorption at the molecular level. We found that ethanol formed hydrogen-bonded clusters on silica surfaces in cyclohexane, which probably brought about the long-range attraction between silica surfaces. A similar long-range attraction associated with cluster formation has been found for mixtures of other alcohols (methanol and propanol) and carboxylic acid with cyclohexane and is under active investigation in our laboratory. Such knowledge should be important for understanding surface treatment processes performed in solvent mixtures and for designing new materials with the use of molecular assemblies at solid–liquid interfaces. To show an example of utilizing these novel molecular assemblies, we have prepared polymer thin films by in situ polymerization of acrylic acid preferentially adsorbed on glass surfaces [14].

Acknowledgements This work was supported in part by a grant from the Ministry of Education, Science, Culture and Sports, Japan (11167204). K.K. is grateful to the Canon Foundation for a visiting professorship to Hungary which provided her with the opportunity for writing this article.

References

1. Adamson AW, Gast AP (1997) Physical chemistry of surfaces, 6th edn. Wiley, New York
2. Decher G (1997) Science 277:1232
3. Israelachivili JN (1992) Intermolecular and surface forces, 2nd edn. Academic, London
4. Kurihara K (2001) In: Rosoff M (ed) Nano-surface chemistry. Dekker, New York, pp 1–16
5. Mizukami M, Kurihara K (1999) Chem Lett 1005
6. Mizukami M, Kurihara K (2000) Chem Lett 248
7. Ducker WA, Sendan TJ, Pashley RM (1992) Langmuir 8:1831
8. Okusa H, Kurihara K, Kunitake T (1994) Langmuir 10:3577
9. Dékány I, Szántó F, Nagy LG (1978) Prog Colloid Polym Sci 65:125
10. Dékány I, Túri L, Tombácz E, Fendler JH (1995) Langmuir 11:2285
11. Sze SM (1985) Semiconductor devices:Physics and technology. Wiley, New York
12. Liddel U, Becker ED (1957) Spectrochim Acta 10:70
13. Cross SNW, Rochester CH (1979) J Chem Soc Faraday Trans 75:2865
14. Kurihara K, Mizukami M, Nakasone S, Miyahara T (2001) Trans MRS-J 26:(in press)

Progr Colloid Polym Sci (2001) 117: 18–26

E. Tombácz
M. Szekeres

Effects of impurity and solid-phase dissolution on surface charge titration of aluminium oxide

E. Tombácz (✉) · M. Szekeres
University of Szeged,
Department of Colloid Chemistry,
6720 Szeged, Aradi Vt. 1. Hungary

Abstract Potentiometic and calorimetric acid–base titration of aluminium oxide C (Degussa) was performed. This pyrogenic alumina contains chlorine impurity (about 1×10^{-4} mol/g). Its hydrolysis results in indefinite numbers of H^+ and Cl^- ions in aqueous suspension, which disturb the surface charge titration. The chlorine impurity could be removed by heat treatment of alumina powder at 1000 °C. The fitting of the titration data for heat-treated alumina in indifferent electrolyte (KNO_3) solutions between pH~5 and pH~9 is good enough assuming only surface charging reactions and any kind of surface complexation model. Outside these material-specific limits the dissolution of the amphoteric solid plays the governing role, especially below pH~4.5. A partial dissolution of the solid phase and the simultaneous equilibria of mononucleous aluminium-species formation have to be inserted into an appropriate model. Calorimetric data can be assigned to chemical reactions in the absence of impurities and dissolution and if the initial state of titration is fixed to the reference state of oxide suspensions. The partial molar enthalpy of the surface protonation process decreases in absolute value from −34 to −28 kJ/mol and that of deprotonation process increases from 33 to 41 kJ/mol with increasing ionic strength. The standard enthalpy of the surface protolysis reaction on alumina is 34.6 ± 0.6 kJ/mol.

Key words Oxides · Surface charging · Surface complexation model · Potentiometry · Calorimetry

Introduction

Aluminium oxides are of great practical importance and are frequently used as adsorbents in aqueous electrolyte solutions. The acid–base properties of amphoteric aluminium oxides have been studied by many authors with various techniques for many years. The popular theories for the surface chemistry of hydrous oxides based on the data mainly from acid–base titration and electrophoresis measurements appear to be inadequate when results from other types of investigations are considered [1]. Titration calorimetry is a useful and adequate technique to measure the heat changes accompanying the processes during acid–base titration.

Potentiometric acid–base titration, frequently called surface charge titration of oxides, is the most often used method; unfortunately, the interpretation of experimental data has been extended beyond the limitation of this method in some cases. All the information from potentiometric titration on the composition of the interfacial layer is indirect and is considerably affected by the assumptions applied during the data evaluation, since only the change in the activity of H^+ ions in bulk solution is obtained directly from the experiments. The effect of a trace amount of impurity and of additional acid–base reactions, however, is often neglected and the whole H^+/OH^- consumption is taken for the surface charge formation alone. In addition, the calculation of the surface charge density involves the specific surface area of the solid, obtained usually from nitrogen adsorption measurement, which may not be relevant to the available surface area in aqueous media; therefore, the calculation of charge potential functions from titration data without verifying the reality of the source data is questionable.

It is well known that different crystalline forms of aluminium oxides/hydroxides result from different processes and different raw materials. The products always contain more or less impurity. Pyrogenic alumina obtained from hydrolysis of aluminium chloride in a flame, such as the product of Degussa (type C), for example, contains chlorine impurity. Chlorine is bound to the active surface sites and influences their number and acid–base properties in a vacuum [2]. In aqueous suspension the hydrolysis of bound chlorine results in acidic species. As a consequence, the pH of dense suspensions is much lower (e.g. pH 5–5.5) than the pH of the point of zero charge (PZC) [3] for pure alumina suspensions. Heat treatment at high temperature is an effective tool to remove chlorine contamination [2]. The usual cleaning procedure, i.e. exhaustive washing with pure water, however, is ineffective in this particular case, since chloride ions are involved in surface reactions and can suppress the surface hydroxylation of alumina [4].

Dissolution of aluminium oxide in both acidic and alkaline solutions and its dependence of the crystal structure is well known [5]. Below pH$\sim$4 and above pH$\sim$10, the dissolution of this amphoteric solid becomes perceptible; therefore, the interfacial charging is often studied within these pHs [6, 7]. A schematic representation of charge formation on the aluminium oxide surface, the dissolution of amphoteric solid and some of the aqueous aluminium species formed is shown in Fig. 1.

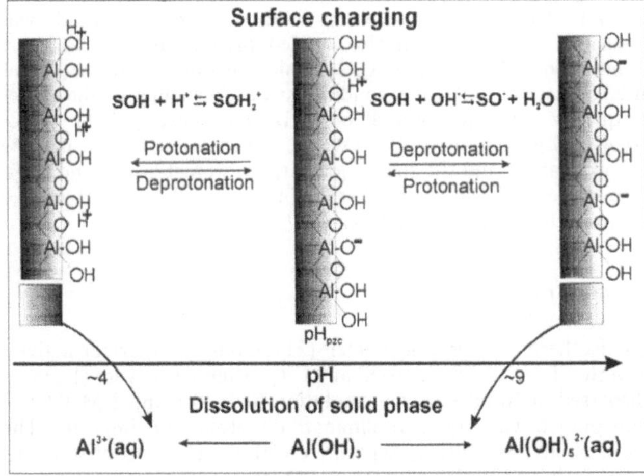

Fig. 1 A schematic representation of interfacial acid–base processes taking place spontaneously in aqueous suspensions of aluminium oxide with changing pH of the equilibrium solution phase

The study of the dissolution kinetics of aluminium oxide [8] shows that the dissolution rate considerably depends on the pH of the aqueous solution. In some work the possibility of alumina dissolution is excluded with reference to chemical equilibrium calculations [6] or it is not mentioned at all [9–11]. The fitting of experimental surface charging curves led to more and more complicated theoretical approaches with increasing numbers of layers for charge-compensating ions in the surface complexation models (diffuse double-layer model, triple-layer model, TLM, and four-layer model [10, 11]) and with the introduction of surface site heterogeneity parameters. The heterogeneity of the proton binding sites at the oxide/solution interface has been studied theoretically [12–14] and both theoretically and experimentally [6, 9–11, 15, 16].

Interfacial acid–base reactions in aqueous oxide suspensions can be characterized by direct calorimetric measurements. In the interpretation of the data, however, one meets the problem of distinguishing between the different contributions [17]. In simple cases, such as surface charging owing to the adsorption of potential-determining ions, calorimetric experiments may be designed in an appropriate way and the data may be interpreted [17]. An additional problem is how to account for the incorporated electrostatic effect. The enthalpy of the surface charging reaction (ΔH_r) can be separated into a "chemical" (i.e. the standard ΔH_r^0) and an electrostatic (ΔH_{elec}) contribution [18]:

$$\Delta H_r = \Delta H_r^0 + \Delta H_{elec} \ . \tag{1}$$

The electrostatic contribution is given by a Gibbs–Helmholtz relationship [1]:

$$\Delta H_{elec} = F\Delta z T(\partial\psi_0/\partial T)_p + F\Delta z\psi_0 \ , \tag{2}$$

where F is the Faraday constant, Δz is the change in charge of the surface owing to the adsorption reaction, T is the temperature; $(\partial\psi_0/\partial T)_p$ is the temperature coefficient of the surface potential (ψ_0) at constant pressure and $F\Delta z\psi_0$ is the electrostatic contribution to the Gibbs energy (ΔG_{elec}). The electrostatic enthalpy contribution can be larger or smaller than the electrostatic free energy, depending on the sign of the coefficient $(\partial\psi_0/\partial T)_p$.

The surface protolysis enthalpy [1], i.e. the difference between the standard enthalpies of protonation

$$SOH + H^+ \rightarrow SOH_2^+, \quad \Delta H_{pr}^0 \tag{3}$$

and deprotonation

$$SOH \rightarrow SO^- + H^+, \quad \Delta H_{depr}^0 \tag{4}$$

of surface groups [18], can be determined from the temperature dependence of the PZC of the amphoteric oxides. The plot pH_{PZC} versus $1/T$ is a linear function and the standard enthalpy of the surface protolysis process,

$$SOH_2^+ \rightarrow SO^- + 2H^+, \quad \Delta H^0 = \Delta H^0_{\mathrm{depr}} - \Delta H^0_{\mathrm{pr}} , \qquad (5)$$

calculated from the slope is 14.6 kJ/mol for TiO_2 [17] and 33.2 kJ/mol for hematite [19].

It has been proved [17, 18] that the electrostatic contribution to the enthalpy can be neglected in a smartly designed, so-called "symmetric" experiment, where the difference in the standard enthalpies ($\Delta H^0_{\mathrm{ch}} = \Delta H^0_{\mathrm{depr}} - \Delta H^0_{\mathrm{pr}}$) can be obtained from direct calorimetric measurements. The published values, 14.7 kJ/mol for TiO_2 [18] and 33.3 kJ/mol for hematite [19], agree well with that from the temperature dependence of $\mathrm{pH}_{\mathrm{PZC}}$.

Confusing experimental data and theoretical considerations are presented in some cases. Some metal oxide–aqueous solution systems have been investigated [1, 20–22] by titration calorimetry; however, the different enthalpy values for the interfacial processes scatter widely. An essential feature is that the enthalpies of the surface charging processes should exhibit an ionic strength dependence; however, there are published results [1, 20, 21] showing that the electrolyte concentration practically has no influence on the enthalpies observed. Although it is unquestionable that the proton adsorption is an exothermic process and that the proton desorption is endothermic, data are given [21] for surface deprotonation due to the adsorption of hydroxyl ions with an exothermic heat effect. The situation is even for the theoretical analysis of ion adsorption at oxide–electrolyte interfaces [23–25], since it is not determined whether exothermic or endothermic processes are involved.

A thermodynamic analysis for the ionization process on the surface of insoluble solids with fixed dissociable groups has recently been published [26]. It is stated that the partial molar enthalpy of surface group protonation, ΔH_{p}, is physically well defined and accessible via calorimetry.

Different aluminium oxides have been investigated in our laboratory [27–29] by potentiometric acid–base titration using a variety of electrolytes containing K^+, Na^+, NO_3^- or Cl^- ions. The purpose of this study is to show some results from potentiometric and calorimetric acid–base titration of a pyrogenic sample (aluminium oxide C, Degussa) with special attention to the effect of impurities and dissolution of alumina on the data evaluation.

Experimental

Materials

Aluminium oxide C (highly dispersed commercial products of Degussa) was investigated. This fine powder is produced by flame hydrolysis of $AlCl_3$ and is characterized by a nonporous δ-Al_2O_3 structure, a high specific surface area (100 ± 15 m^2/g) and a small particle size (13 nm). The density is 3.2 g/cm^3 and the HCl content is less than 0.5%. The pH of the aqueous suspension is 5–5.5. Data were taken from the brochure of Degussa [30]. In order to eliminate chlorine contamination, a powder sample was heated for 6 h at 1,000 °C. Only the freshly purified alumina powder was measured in the calorimetric experiments. Millipore water was used and all the chemicals were analytical reagent grade products of Merck.

Methods

Potentiometric titration

The pH-dependent surface charge state was determined from acid–base titration under CO_2-free conditions using different background electrolytes (KCl, KNO_3) to maintain the constant ionic strength over the region 0.005–0.5 M. Before titration, the solid samples were equilibrated with the electrolyte solution for 1 h, with gentle stirring under a continuous stream of purified, wet nitrogen. Equilibrium titration was performed by means of a self-developed titration system (GIMET1) with 665 Dosimat (Metrohm) burettes, nitrogen bubbling, a magnetic stirrer and a high-performance potentiometer. The whole system (millivolt measurement, stirring, bubbling, amount and frequency of titrant) was controlled by an IBM PS/1 computer using AUTOTITR software. A Radelkis OP-0808P (Hungary) combination pH electrode was calibrated for three buffer solutions to check the Nernstian response. The hydrogen ion activity versus concentration relationship was determined from reference solution titration so that the electrode output could be converted directly to the hydrogen ion concentration instead of activity. The net proton surface excess amount (Δq, moles per gram) has been defined as the difference of the H^+ and OH^- surface-excess amounts ($n^\sigma_{H^+}$, $n^\sigma_{OH^-}$) related to the unit mass of solid ($\Delta q = n^\sigma_{H^+} - n^\sigma_{OH^-}$). The surface-excess amount defined for adsorption [31] can be determined directly from the initial (c_i^0, moles per litre) and equilibrium (c_i^e, moles per litre) concentration of solute ($n_i^\sigma = (c_i^0 - c_i^e)V/m$, where V is the volume (litres) of the liquid phase and m is the mass of adsorbent) for adsorption from dilute solution.

Determination of dissolved aluminium species

Several alumina suspensions with the same composition as that in the acid–base titration were prepared between pH~3 and pH~10. Samples from the equilibrium liquid phase at each pH were taken after standing for 1 day by means of centrifugation at 13,000 rpm for 2 h. The aluminium content of the supernatant liquids was measured by the inductively coupled plasma (ICP) method. A Jobin Yvon 24 sequential ICP–atomic emission spectrometer was used under the following conditions: the frequency of the radio-frequency generator was 40.68 MHz; the power of the radio-frequency generator was 0.8 kW; the flow rates of the plasma, the aerosol carrier and the sheath gas (argon) were 12, 0.37 and 0.2 l/min, respectively; the flow rate of nebulization (Babington) was 1.4 ml/min; the wavelength (Al) was 237.324 nm.

Titration calorimetry

An isothermal microcalorimeter (Thermometric thermal activity monitor TAM 2277) was used at 25 °C. Alumina powder (1 g) was dispersed in 10 ml electrolyte solution (0.01, 0.1 and 1 M KNO_3) and purged with argon to eliminate dissolved CO_2 impurity. The suspensions were titrated separately with the portions of standard acid or base solutions under a CO_2-free argon atmosphere. In parallel, blank titrations were performed in the absence of the alumina at the same concentrations of indifferent electrolyte KNO_3. The heat flow across the titration cell of the TAM was continuously recorded. The accuracy of the measured reaction heats was tested by measuring the standard heat of reaction between tris(hydroxymethyl)aminomethane [$(HO-CH_2)_3-C-NH_2$] and HCl ($\Delta H = -55 \pm 1$ kJ/mol).

The measured heat (Q_{meas}) is the overall heat flow of the simultaneous chemical reactions and the mixing processes in the calorimeter cell:

$$Q_{meas} = Q_{ri} + Q_{mix} = \sum_i \Delta H_{ri} \Delta \xi_i + Q_{mix} , \qquad (6)$$

where ΔH_{ri} is the enthalpy change of reaction i, $\Delta \xi_i$ is the change of the extent of reaction i and Q_{mix} is the mixing heat of the titrant portion. Reactions with added acid (ΔH_a), reactions with added base (ΔH_b) and the water formation (neutralization) reaction (ΔH_n) can be distinguished. The heat of mixing was determined in blank experiments. Uncontrolled acid–base reactions with significant heat effects were eliminated.

Results and discussion

Potentiometric titration

The aluminium oxide C (δ-Al_2O_3) contains chlorine contamination originating from the production procedure (flame hydrolysis of $AlCl_3$). The experimental net proton consumption versus pH functions of the original and purified alumina samples are obviously different as shown in Fig. 2. The most conspicuous feature is that the pH of the suspensions from heat-treated alumina is almost independent (pH between 8.2 and 8.4) of the ionic strength, while in the suspensions of the original sample, it significantly increases from pH 6.1 to 7.5 with increasing KCl concentration. Bonded chlorine atoms hydrolyse in aqueous media releasing acidic species into the bulk liquid phase and so the pH of the suspension is acidic. The pH of the unpurified alumina suspensions has to increase with the KCl addition, since the increasing concentration of chloride ions in the equilibrium liquid phase hinders the propagation of $AlCl_3$ hydrolysis.

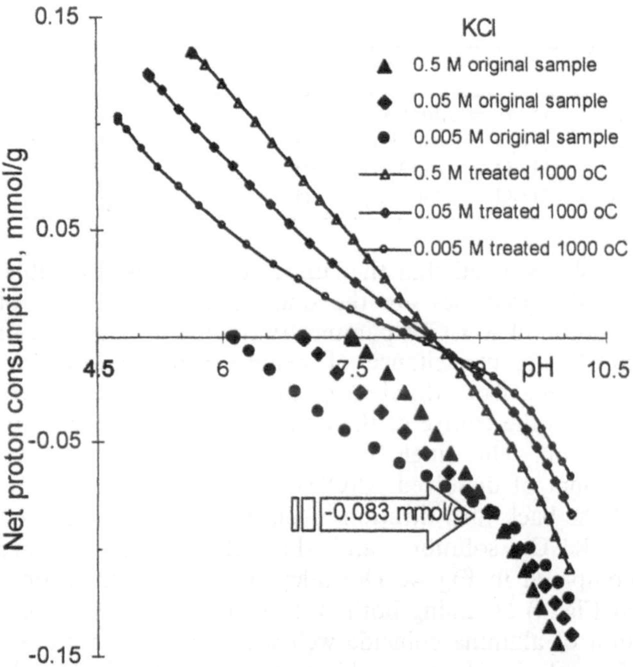

Fig. 2 Experimental net proton consumption curves for original and purified (treated at 1,000 °C) δ-Al_2O_3, dispersed in KCl solutions at room temperature. The points were calculated from the data of a forward equilibrium titration with 0.1 M KOH solution. The initial suspension were prepared with (purified samples) and without (original samples) adding a given amount of 0.1 M HCl solution

$$AlCl_3 + 3H_2O \rightarrow Al(OH)_3 + 3H^+ + 3Cl^- \qquad (7)$$

The common intersection point (CIP) [32] of curves measured at different ionic strengths, assigned as the point of zero salt effect (PZSE) by Sposito [33, 34] shifts up after removal of $AlCl_3$ contamination, showing the absence of base-consuming hydrolysis product (H^+). The pH of the PZSE for the original sample is about 9.2 and it slightly shifts down ($pH_{PZSE} \sim 8.6$) for the purified sample. It has to be stated that a simple additive correction of base-consuming impurities cannot be the correct way of experimental data evaluation.

Since the position of pH_{PZSE} for the heat-treated alumina sample dispersed in KCl solution (Fig. 2) is indicative of a weak specific anion adsorption [35], it was also titrated in an indifferent electrolyte (KNO_3). The pH dependence of the net proton consumption for δ-Al_2O_3 heat treated at 1000 °C, dispersed in KNO_3 solutions at room temperature, is demonstrated in Fig. 3. The data calculated can be identified with the net proton surface excess amount ($\Delta q = n^\sigma_{H^+} - n^\sigma_{OH^-}$), which would be proportional to the surface charge density [$\sigma_0 = F(n^\sigma_{H^+} - n^\sigma_{OH^-})/a^S$ where a^S is the specific surface area] of aluminium oxide, since this solid can develop charges only conditionally [17, 27, 36, 37] and electrolyte KNO_3 is indifferent [27].

The forward and backward curves in Fig. 3 are reversible within the experimental error of this method down to pH~5. Below pH~4.5, however, the backward curves show a sharp increase in proton consumption with decreasing pH and gradually become independent of ionic strength. A similar feature of the curves was explained [15] as the protonation of the most acidic triply coordinated groups of type III on the alumina surface neglecting the dissolution of the solid phase, although beyond the roughly defined limits of the pH values the dissolution of amphoteric aluminium oxides becomes perceptible [38, 39].

Evaluation of titration results

The evaluation of the experimental data from acid–base titration of amphoteric solid material demands cautious work. Other acid- or base-consuming reactions (e.g.

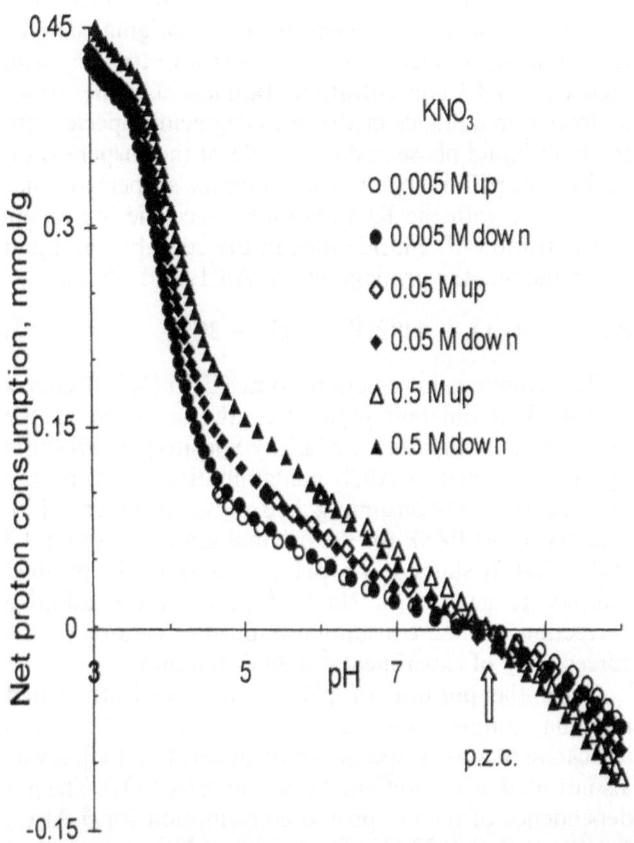

Fig. 3 Experimental net proton consumption curves for purified δ-Al_2O_3 (treated at 1,000 °C for 6 h) dispersed in KNO_3 solutions at room temperature. The points were calculated from the data of a forward (*open symbols*) and a backward (*closed symbols*) equilibrium titration with 0.1 M KOH and 0.1 M HNO_3 solutions, respectively

dissolution of solid at lower or higher pHs) can take place in parallel with the surface charging processes, which cannot be separated experimentally. Problems in modelling the surface charging of oxides have been discussed in several articles [17, 40, 41]. The pH- and ionic-strength-dependent surface charge formation process can be described by various model approximations, the most widely accepted models are the site-binding electrostatic models [7, 9–11, 27, 42, 43]. The surface charge development can be affected by the solubility of the solid [44], which is not incorporated in the models.

The experimental data shown in Fig. 3 were evaluated using a numerical data-fitting program [45]. In the optimization procedure of FITEQL the experimental data and the corresponding theoretical values calculated from a multicomponent chemical equilibrium model incorporating interfacial charging equilibria are compared. The parameters of the model are adjusted to minimize the difference between the measured and calculated data. The choice of different surface complexation [constant-capacitance model (CCM), diffuse-layer model (DLM), Stern and TLM] models is optional.

The measured data of the forward titration in indifferent electrolyte (KNO_3) solutions might be well fitted by choosing any surface complexation model. In the presence of KCl electrolyte only the choice of the TLM option led to an acceptable level of the optimization process by FITEQL [45], presumably because of specific adsorption of Cl^- ions. The calculated intrinsic equilibrium constants for charge- and ion-pair-formation reactions occurring on the surface hydroxyl groups of alumina in different electrolyte solutions are summarized in Table 1. The pK values obtained in our previous work [27] and this study lie within the values for different Al_2O_3 reported in the literature [7, 9, 16, 37]. The data in Table 1 clearly show that there is more than 1 order of magnitude difference between the log$K$ values of the protonation and deprotonation reactions on surface site (SOH) depending on the evaluation methods.

The fitting of the backward titration data was impossible when only the surface charge formation processes were postulated as H^+/OH^- consuming reactions in alumina suspensions; therefore we assumed a partial dissolution of alumina particles during the long period of equilibrium titration. A component identified as a solid compound $Al(OH)_3(s)$ and its solubility product $[Al(OH)_3(s) = Al^{3+} + 3OH^-]$ as well as the formation of different mononuclear aluminium species were inserted in the stoichiometry matrix of FITEQL [45]. The following reactions and log$K$ values [45] were assumed:

	logK
$Al^{3+} + H_2O \rightleftharpoons Al(OH)^{2+} + H^+$	−4.99
$Al^{3+} + 2H_2O \rightleftharpoons Al(OH)_2^- + 2H^+$	−9.30
$Al^{3+} + 3H_2O \rightleftharpoons Al(OH)_3 + 3H^+$	−15.00
$Al^{3+} + 4H_2O \rightleftharpoons Al(OH)_4^- + 4H^+$	−23.00

We assumed that the surface charging is reversible, i.e. the processes are the same in both directions of titration. The CCM parameters (total concentration of SOH site, capacitance of oxide–electrolyte interface) calculated from the data of the forward titration were fixed in the course of the data fitting for the backward titration. The single fitting parameter was the total amount of dissolved $Al(OH)_3(s)$. The measured points of the backward titration of purified alumina dispersed in KNO_3 solutions and the calculated curves are compared in Fig. 4. The calculated curves (thick lines in Fig. 4) assuming both surface charging and dissolution of alumina coincide well with the measured data. The calculated curves (thin lines in Fig. 4) related only to the H^+/OH^- consumption of interfacial protonation/deprotonation reactions resulting in surface charging are also plotted. It can be seen that the net proton consumption from the dissolution of amphoteric solid becomes significant where the calculated curves start to

Table 1 Intrinsic equilibrium constants for charge- and ion-pair-formation reactions occurring on the surface hydroxyl groups of aluminium oxide C (Degussa) in electrolyte solutions

	Electrolyte, KNO_3 0.005 M	0.05 M	0.5 M	Average
Graphic extrapolation [27]				
$LogK_{a1}^{int}(SOH_2^+)$				5.8 ± 0.2
$LogK_{a2}^{int}(SO^-)$				−10.2 ± 0.2
Point of zero charge				8.0 ± 0.1
FITEQL [45] optimization	Constant-capacitance model			
$LogK_{a1}^{int}(SOH_2^+)$	6.56	7.04	7.61	7.07 ± 0.51
$LogK_{a2}^{int}(SO^-)$	−9.39	−9.13	−8.79	−9.10 ± 0.31
Point of zero charge	7.97	8.09	8.20	8.09 ± 0.12
	Diffuse-layer model			
$LogK_{a1}^{int}(SOH_2^+)$	6.95	7.47	7.63	7.35 ± 0.30
$LogK_{a2}^{int}(SO^-)$	−9.00	−8.71	−8.77	−8.82 ± 0.18
Point of zero charge	7.98	8.09	8.20	8.09 ± 0.11
	Triple-layer model			
$LogK_{a1}^{int}(SOH_2^+)$ fixed				6.8
$LogK_{a2}^{int}(SO^-)$ fixed				−9.2
$LogK_{aA}^{int}(SOH_2^+A^-)$	8.77	8.35	8.26	8.46 ± 0.31
$LogK_{ak}^{int}(SO^-K^+)$	−7.15	−7.85	−8.15	−7.72 ± 0.57
	Electrolyte, KCl, triple-layer model			
$LogK_{aA}^{int}(SOH_2^+A^-)$	8.95	8.49	8.06	8.50 ± 0.45
$LogK_{aK}^{int}(SO^-K^+)$	−7.35	−8.25	−8.81	−8.13 ± 0.78

diverge from each other, i.e. below pH∼5 and above pH∼9.5. On the basis of the experimental curves showing a sharp increase in the net proton consumption values at pH∼4 (Fig. 3), the assumption of alumina dissolution at low pHs seems to be evident.

It is a remarkable result of our model calculation that there is a probable interference of dissolution under alkaline conditions during surface charge titration above pH∼9.5, where alkaline dissolution of alumina starts to contribute to the measured OH^- consumption. The calculated curves also coincide well with the measured points above pH∼9.5, if the dissolution of alumina is inserted into the chemical model of the data-fitting procedure.

To verify the assumption of alumina dissolution, we measured the total aluminium concentration in equilibrium supernatants separated from alumina suspensions with different pHs between pH∼4 and pH∼10 by means of the ICP method. The measured values together with the calculated distribution of the dissolved aluminium species from the FITEQL output files and their total amounts are plotted as a function of pH in Fig. 5. The measured values coincide very well, especially in the alkaline region, with the calculated total amount of dissolved aluiminum species.

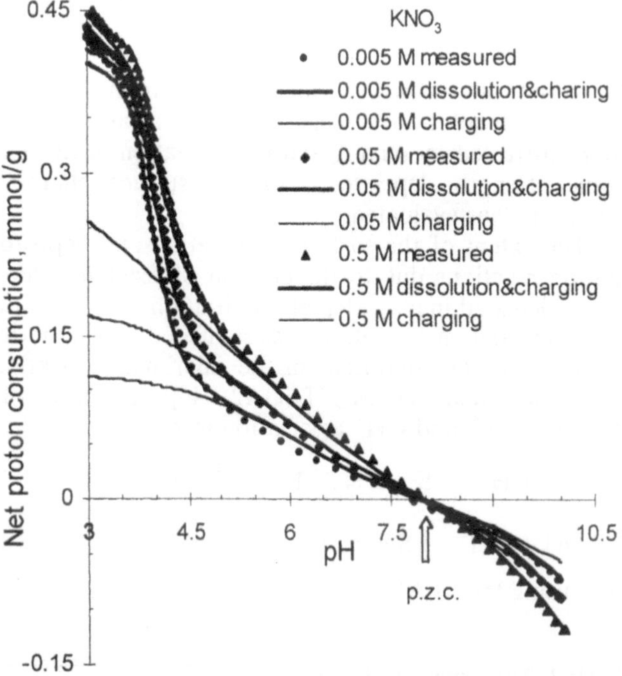

Fig. 4 Experimental points of net proton consumption from backward titration with 0.1 M HNO_3 for purified δ-Al_2O_3 (treated at 1000 °C for 6 h) dispersed in KNO_3 solutions at room temperature. The *continuous thick lines* are numerically fitted (FITEQL) using a combined surface complexation and chemical dissolution model: the constant capacitance model ($C=1.2F/m^2$) for surface charging, partial dissolution of solid alumina and formation of mononuclear aluminium species with log$K$ values from the literature [45]. The *continuous thin lines* show that part of the net proton consumption which belongs to the surface charge formation

Calorimetric titration

The measurements were performed under delicate experimental conditions. Freshly purified aluminium oxide samples were measured in the presence of only indifferent

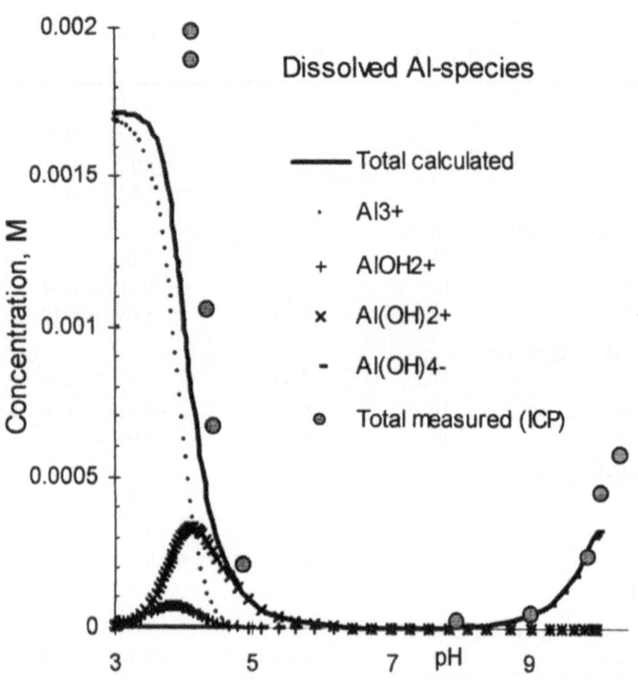

Fig. 5 Comparison between the calculated distribution of dissolved aluminium species from FITEQL output files (same as shown in Fig. 4) and the experimental values measured by means of the inductively coupled plasma method

electrolyte and within a region of pH where the extent of alumina dissolution is negligible. The measured data were corrected by mixing and neutralization heats. The ionic strength dependence of the surface charging processes was evaluated.

The extent of the surface protonation and deprotonation reactions due to the amount of reactant added was calculated in each step of the titration. The titration with acid and base solutions was simulated. The material balance for the conditions in the calorimetric titration was calculated by MINTEQA2 [46] assuming the following H^+ and OH^- association reactions

$$S{-}OH + H^+ \rightarrow S{-}OH_2^+, \quad \log K_1^{int} = 6.8 \;, \tag{7}$$

$$S{-}OH + OH^- \rightarrow S{-}O^- + H_2O, \quad \log K_2^{int} = 9.2 \tag{8}$$

and using the CCM.

The reaction in Eq. (7) is the surface protonation process, while the OH^- association reaction in Eq. (8) differs from the surface deprotonation reaction

$$SOH \rightarrow S{-}O^- + H^+ \tag{9}$$

since Eq. (8) involves water formation

$$H^+ + OH^- \rightarrow H_2O, \quad \Delta H^0 = -56.5\,\text{kJ/mol} \tag{10}$$

Therefore, if we wish to identify the reaction heat of the negative charge formation on the oxide surface, i.e. of the surface deprotonation reaction (Eq. 9), the reaction heat of the formation of a given amount of water in each step must be subtracted from the measured heat effects of titration with base solution (the OH^- association reaction, Eq. 8).

The cumulative heats of the surface protonation and deprotonation reactions during titration, measured directly or calculated with the correction for water formation, are plotted against the calculated amount of the surface charges formed in the reactions in Eqs. (7) and (9). This presentation of experimental data shows a satisfactory linear relationship for both the acidic and alkaline sides of the titration. This linearity means that the molar enthalpy changes of the surface charging processes (ΔH_{depr} and ΔH_{pr}) are independent of the surface charge density within experimental accuracy. Therefore, the slope of the cumulative heat of surface charging versus the charged sites' amount functions would be identified with the partial molar enthalpies of the surface protonation and deprotonation reactions (ΔH_{pr} and ΔH_{depr}). The calculated values are given in Table 2.

While the partial molar enthalpy data for the surface protonation and deprotonation reactions (second and third columns of Table 2) show a systematic change with increasing concentration of indifferent electrolyte, the difference between them results in a molar enthalpy which is practically independent of ionic strength. Applying the basic thermochemistry rule, subtracting one reaction (Eq. 7) from another (Eq. 9), we get

$$SOH_2^+ \rightarrow SO^- + 2H^+, \quad \Delta H_{depr} - \Delta H_{pr} \tag{11}$$

or written for one proton

$$SOH_2^{1/2+} \rightarrow SOH^{1/2-} + H^+, (\Delta H_{depr} - \Delta H_{pr})/2 \tag{12}$$

Table 2 Partial molar enthalpy values of surface charging process in indifferent electrolyte solutions for aluminium oxide at 25 °C

Partial molar enthalpies of surface charging process (kJ/mol)			
Surface reaction	$SOH + H^+ \rightarrow SOH_2^+$ Exothermic	$SOH \rightarrow SO^- + H^+$ Endothermic	$SOH_2^{1/2+} \rightarrow SOH^{1/2-} + H^+$ Endothermic
c_{KNO_3},(M)	ΔH_{pr}	ΔH_{depr}	$(\Delta H_{depr}-\Delta H_{pr})/2$
0.01	−34.16	33.61	33.88
0.1	−31.11	39.19	35.15
1	−28.28	41.34	34.81
Average			34.6 ± 0.6

and the enthalpy change of the resulting reaction can be calculated.

It can be stated that the calculated, ionic-strength-independent enthalpy value of 34.6 ± 0.6 kJ/mol is the same standard reaction enthalpy for surface charging of alumina ($SOH_2^{1/2+} \rightarrow SOH^{1/2-} + H^+$) in a thermodynamic sense as that given for titania (14.6 kJ/mol) and hematite (33.2 kJ/mol) from the temperature dependence of the PZC and from calorimetric measurements performed under a "symmetric" condition [18, 19]. These results can be hardly compared with the uncertain data published for alumina in the literature [20] apart from the sign and the magnitude of the proton adsorption enthalpies. The proton adsorption enthalpy value of −42 kJ/mol is comparable with our data between −34 and −28 kJ/mol given in the second column of Table 2, which, in general, are between −20 and −50 kJ/mol for metal oxides [20].

It has to be emphasized that the evaluation of the titration calorimetric measurements results in valuable thermodynamic quantities only if the experiments are performed carefully, with controlled systems without any side reactions from impurities and dissolution, since only these can give us a chance to assign the measured heats to the corresponding reactions.

Conclusions

The results of the potentiometric acid–base titration of aluminium oxide suspensions allow us to conclude that the effect of a trace amount of impurity and dissolution of solid cannot be separated experimentally and the net H^+/OH^- consumption is considerably affected in their presence. The position of the CIP may refer to them, for example, an acidic impurity shifts the CIP to the negative region of the surface charge density. In the presence of additional acid–base consuming reactions the calculation of charge potential functions is meaningless. The goodness of the fit between the experimental points and the calculated curves does not prove the reality of the theoretical model, the assumed structure of the surface layer and the presumed interfacial reactions. The assumed chemical model has to be supported by independent measurements (e.g. spectroscopy, analysis of the bulk phase) to prove its reality.

On the basis of the calorimetric titration of alumina, it can be stated that the reference state of aqueous oxide suspensions, the PZC, is a correct initial state of the calorimetric titration, where the electrostatic contribution can be neglected. The partial molar enthalpies of the surface ionization processes ($S{-}OH + H^+ \rightarrow S{-}OH_2^+$, ΔH_{pr} and $S{-}OH \rightarrow S{-}O^- + H^+$, ΔH_{depr}) and the standard reaction enthalpy for surface charging ($SOH_2^{1/2+} \rightarrow SOH^{1/2-} + H^+$, ΔH^0_{ch}) can be determined. Calorimetric titration at any single concentration of indifferent electrolyte is sufficient to determine the standard reaction enthalpy (ΔH^0_{ch}). The definite ionic strength dependence of the partial molar enthalpy changes of the surface charging processes ($\Delta H_{depr,I}$ and $\Delta H_{pr,I}$) was determined; it is a small, uniform endothermic effect of increasing salt concentration for both sides of surface charging.

Acknowledgements Calorimetric measurements were performed at the Department of Applied Physical Chemistry, Research Center Jülich, Germany. Dissolved aluminium was determined by ICP at the Department of Inorganic and Analytical Chemistry, University of Szeged. The authors are grateful for the measurements. This work was supported by grants OTKA T034755 and FKFP 0587/1999.

References

1. Machesky ML, Anderson MA (1986) Langmuir 2:582
2. Sharanda LF, Shimansky AP, Kulik TV, Chuiko AA (1995) Colloids Surf 105:167
3. Zalac S, Kallay N (1992) J Colloid Interface Sci 149:233
4. Klug O, Forsling W (1999) Langmuir 15:6961
5. Lindsay WL, Walthall PM (1989) In: Sposito G (ed) The environmental chemistry of aluminum. CRC, Boca Raton, pp 221–239
6. Hiemstra T, Yong H, Van Riemsdijk WH (1999) Langmuir 15:5942
7. Wood R, Fornasiero D, Ralston J (1990) Colloids Surf 51:389
8. Kraemer SM, Chiu VQ, Hering JG (1998) Environ Sci Technol 32:2876
9. Rudzinski W, Charmas R, Partyka S, Thomas F, Bottero JY (1992) Langmuir 8:1154
10. Charmas R, Piasecki W (1996) Langmuir 12:5458
11. Charmas R (1999) Langmuir 15:5635
12. Cernik M, Borkovec M, Westall JC (1996) Langmuir 12:6127
13. Borkovec M (1997) Langmuir 13: 2608
14. Rustad JR, Wasserman E, Felmy AR, Wilke C (1998) J Colloid Interface Sci 198:119
15. Contescu C, Jagiello J, Schwarz JA (1993) Langmuir 9:1754
16. Rudzinski W, Charmas R, Borowiecki T (1996) In: Dabrowski A, Tertykh VA (eds) Adsorption on new and modified inorganic sorbents. Studies in surface science and catalysis. Elsevier, Amsterdam, pp 357–409
17. Kallay N, Zalac S, Kobal I (1996) In: Dabrowski A, Tertykh VA (eds) Adsorption on new and modified inorganic sorbents. Studies in surface science and catalysis. Elsevier, Amsterdam, pp 857–877
18. Kallay N, Zalac S, Stefanic G (1993) Langmuir 9:3457
19. Kallay N, Zalac S, Culin J, Beiger U, Pohlmeier A, Narres HD (1994) Prog Colloid Polym Sci 95:108
20. (a) Machesky ML, Jacobs PF (1991) Colloids Surf 53:297; (b) Machesky ML, Jacobs PF (1991) Colloids Surf 53:315
21. De Keizer A, Fokkink LGJ (1990) Colloids Surf 49:149
22. Casey WH (1994) J Colloid Interface Sci 163:407
23. Rudzinski W, Charmas R, Partyka S (1991) Langmuir 7:354
24. Rudzinski W, Charmas R, Cases JM, Francois M, Villieras, Michot LJ (1997) Langmuir 13:483
25. Rudzinski W, Charmas R, Piasecki W, Groszek AJ (1999) Langmuir 15:5921

26. Hall DG (1997) Langmuir 13:91
27. Tombácz E, Szekeres M, Kertész I, Turi L (1995) Prog Colloid Polym Sci 98:160
28. Szekeres M, Tombácz E, Ferencz K, Dékány I (1998) Colloids Surf 141:319
29. Tombácz E, Filipcsei G, Szekeres M, Gingl Z (1999) Colloid Surf 151:233
30. Technical Bulletin Pigments, no 56, Degussa AG, Frankfurt
31. Everett DH (1986) Pure Appl Chem 58:967
32. Lyklema J (1991) Pure Appl Chem 63:895
33. Sposito G (1992) In: Buffle J, Van Leeuwen HP (eds) Environmental particles, vol 1. Lewis, Boca Raton, pp 291–314
34. Sposito G (1998) Environ Sci Technol 32:2815
35. Lyklema J (1995) Fundamentals of interface and colloid science. Solid–liquid interfaces, vol 2. Academic, London, chapter 3.6, pp 3.44–3.84
36. Davis JA, Hem JD (1989) In: Sposito G (ed) The environmental chemistry of aluminum, CRC, Boca Raton, pp 185–219
37. James RO, Parks GA (1982) In: Matijevic E (ed) Surface and colloid science, vol 12. Plenum, New York, pp 119–216
38. Bartoli F, Philippy R (1987) Clay Miner 22:93
39. Tombácz E, Dobos Á, Szekeres M, Narres HD, Klumpp E, Dékány I (2000) Colloid Polym Sci 278:337
40. Sposito G (1983) J Colloid Interface Sci 91:329
41. Zhang ZZ, Sparks DL (1994) J Colloid Interface Sci 162:244
42. Westall J, Hohl H (1980) Adv Colloid Interface Sci 12:265
43. Rakotonarivo E, Bottero JY, Thomas F, Poirier JE, Cases JM (1988) Colloids Surf 33:191
44. Ludwig C, Casey WH (1996) J Colloid Interface Sci 178:176
45. Herbelin AL, Westall JC (1996) FITEQL version 3.2. Oregon State University, Corvallis, Ore, USA
46. Allison JD, Brown DS, Novo-Dradac KJ (1991) MINTEQA2/PRODEFA2, version 3.0. US EPA, Athens, Ga, USA

Progr Colloid Polym Sci (2001) 117: 27–31

G. Horányi
P. Joó

Radiotracer study of the specific adsorption of anions on oxides

G. Horányi (✉)
Institute of Chemistry,
Chemical Research Center,
Hungarian Academy of Sciences,
P.O. Box 17, 1525 Budapest, Hungary
e-mail: hor34@ludens.elte.hu
Tel.: +36-1-2670820
Fax: +36-1-2663899

P. Joó
Department of Colloid Chemistry,
University of Debrecen, P.O. Box 31,
4010 Debrecen, Hungary

Abstract The possibility of the extension of the radiotracer technique to the study of specific adsorption of anions on oxides is discussed. It is shown that the indirect version of the so-called "thin foil" method could be a useful tool for the clarification of adsorption behavior of certain species. The main tendencies characterizing the adsorption of phosphate species on hematite and γ-Al_2O_3 are demonstrated using ^{35}S-labeled sulfate ions as indicator species.

Key words Specific adsorption · Anions · Hematite · γ-Al_2O_3 · Radiotracer technique

Introduction

The in situ study of the specific adsorption on electrodes using the radiotracer technique is a well-known method in electrochemistry that furnished in the second half of the past century important information on the state of the double layer formed at the metal/solution interface.

Several review articles published during the last two decades have given a survey on both the technical achievements and the phenomena studied in this field [1–10]. On the other hand, although the radiotracer technique found its application in colloid chemistry the methods used are far from being able to follow in situ the adsorption phenomena and to study the effect of various parameters on the adsorption without disturbing the original experimental conditions.

In our previous communications [11–14] it was demonstrated that some of the radiotracer methods elaborated for the investigation of electrosorption phenomena occurring at electrodes can be used for the study of adsorption phenomena at oxide surfaces.

In the present article an attempt is made to show that a version of the so-called "thin foil" method, the indirect method, could furnish important information concerning the adsorption phenomena occurring at oxide surfaces.

As to the role of the labeled species in the radiotracer study of adsorption phenomena, two different versions of the method may be distinguished. The first one, reported in our previous communications [11–14], is the direct method. In this case the species to be studied is labeled and the radiation measured gives direct information on the adsorption of this species. However, this method cannot be used in several cases owing to the technical restrictions connected with the very nature of the radiotracer method (limitations in the available concentration range; there is no possibility of distinguishing between the adsorption of the labeled compound studied and that of a product formed from it; the number of commercial easily available labeled compounds is restricted).

Considering all these problems the use of the so-called indirect radiotracer methods was suggested. Instead of labeling the species to be studied another adequately chosen labeled species (indicator species) is added to the system and the adsorption of this component is followed by the usual radiotracer measuring technique. Evidently, the sorption of the indicator species should be in relation to that of the species to be studied. The nature of this link could be different in different systems. For instance, in some cases the competitive adsorption with the labeled species, while in other cases induced adsorption of the labeled species may furnish information about the adsorption behavior of a given molecule. The principle of the study in the former case can be demonstrated by Scheme 1.

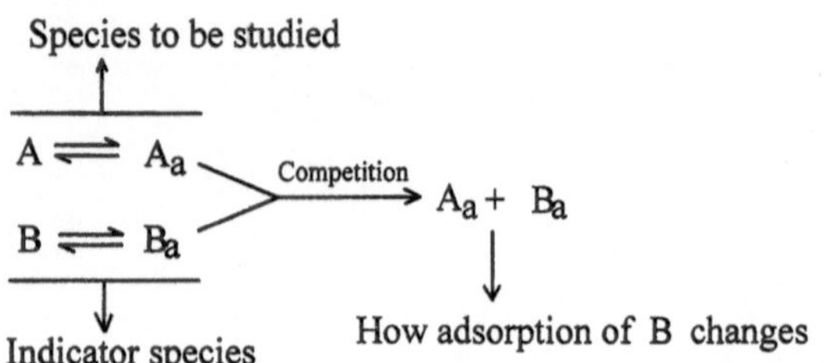

In the following, the results obtained from the indirect study of PO_4^{3-} adsorption on hematite and Al_2O_3 are reported. As the adsorption of labeled sulfate ions at hematite and Al_2O_3 was studied in detail in our previous communications [13, 14] and a reversible adsorption was found ^{35}S-labeled sulfate ions could be used as an ideal indicator species.

On the other hand, preliminary results with phosphate species allow us to assume that the adsorption of these ions is also reversible; thus, the validity of the principle of the indirect study characterized by Scheme 1 can be considered as a basis for the evaluation of the experimental results.

Experimental

The experimental technique and setup for the radiotracer study of sorption processes (involving powdered sorbents) are described and surveyed in Refs. [11–14].

The principle of the method used in the present study is the measurement of radiation intensity originating from adsorbed species on a powdered oxide layer sprinkled on a thin plastic foil that serves simultaneously as the window for the radiation measurement. The measurements were carried out at ambient temperature in an Ar atmosphere, bubbling Ar in the solution phase and letting it through the gap between the bottom of the measuring cell and the detector. Labeled H_2SO_4 was used (Amershem, specific activity: 200 MBqmmol^{-1}) for the preparation of the solutions.

The concentration of the labeled species was several orders of magnitude lower than that of the supporting electrolyte (0.5 moldm^{-3} $NaClO_4$) in order to get information concerning the specific interaction of the former species with the oxide surface. Thus, the nonspecific electrostatic forces and the formation of a double layer could not play a significant role in the sorption process of the sulfate species.

A commercial neutral Al_2O_3 powder was used (Merck, Art 1077, particle size: 0.063–0.200 mm). The specific surface area of the oxide powder, determined by the N_2–Brunauer–Emmett–Teller method, was 156 ± 9 m^2/g, while by the methylene blue adsorption method 17 ± 3 m^2/g was obtained.

A commercial hematite powder (Reanal, analytical grade) characterized by its X-ray diffraction pattern was used (ATOMKI/Siemens horizontal θ–2θ X-ray diffractometer with Cu Kα radiation).

The specific surface area of the oxide powder, 0.15 ± 0.05 m^2/g, was estimated from the particle size distribution according to the microscopic method based on the measurements with a Bürker chamber.

In the case of thin oxide layers, the radiation intensity measured should be proportional to the amount of the powder sprinkled on the bottom of the measuring cell. With increasing thickness the radiation intensity tends to a limiting value as a consequence of the self-adsorption of the soft β radiation of ^{35}S in the layer. In the case of powders with relatively low specific surface area (as in the present case of hematite) it is preferable to carry out the experiments at this limiting value in order to attain the highest count rates in the system. In contrast to this, thin layers were considered in the case of Al_2O_3 (high specific surface area).

The time required for reliable equilibrium measurements varied between 20 and 100 min, depending on the concentrations and on the amount of the powder used. The attainment of the equilibrium was controlled by the observation of the response of the system following its stirring. As a result of stirring, the oxide particles with adsorbed labeled species are lifted up from the bottom of the cell and the count rate decreases: the detector does not "see" these particles. Following the deposition of the powdered adsorbent the count rate increases and in the case of equilibrium attains its original value.

Results and discussion

The fundamental relationships for the evaluation of experimental data

Denoting by A the phosphate species and by B the indicator sulfate species and by taking into account that according to the results reported in Refs. [13, 14] the adsorption of sulfate ions can be described by a Langmuir isotherm in the case of both adsorbents studied the following equations can be used:

$$\theta_A = \frac{b_A c_A}{1 + b_A c_A + b_B c_B} , \tag{1}$$

$$\theta_B = \frac{b_B c_B}{1 + b_A c_A + b_B c_B} . \tag{2}$$

Transformation of Eq. (2) leads to the expression

$$\frac{1}{\theta_B} = 1 + \frac{1}{b_B c_B} + \frac{b_A c_A}{b_B c_B} . \tag{3}$$

At a fixed concentration of the indicator species (B) and taking into account that θ_B is proportional to the radiation intensity measured (I)

$$\theta_B = qI , \tag{4}$$

where q is a proportionality factor, we obtain

$$\frac{1}{I} = D + Fc_A . \tag{5}$$

From Eqs. (1) and (2)

$$\frac{\theta_A}{\theta_B} = \frac{b_A}{b_B} \cdot \frac{c_A}{c_B} \quad (6)$$

and

$$\theta_A = F' I c_A \ . \quad (7)$$

The problem of specific adsorption.

It was emphasized in our previous communications [13, 14] that one of the main and characteristic advantages of the radiotracer technique is that this method, using a great excess of supporting electrolyte, furnishes direct information on the specific adsorption of the labeled species present in very low concentrations in the solution phase (in comparison with that of the supporting electrolyte).

Our approach was based on the classical double layer theory in accordance with the well-known definition: "specific adsorption of ions is their adsorption by non-electrostatic forces". It was clearly demonstrated that owing to the presence of the great excess of supporting electrolyte the measurable adsorption values found with respect to the labeled species cannot be ascribed to electrostatic forces.

In the case of the indirect measuring technique the same statement refers to the nonlabeled species to be studied if its concentration remains 1–2 orders of magnitude lower that that of the supporting electrolyte.

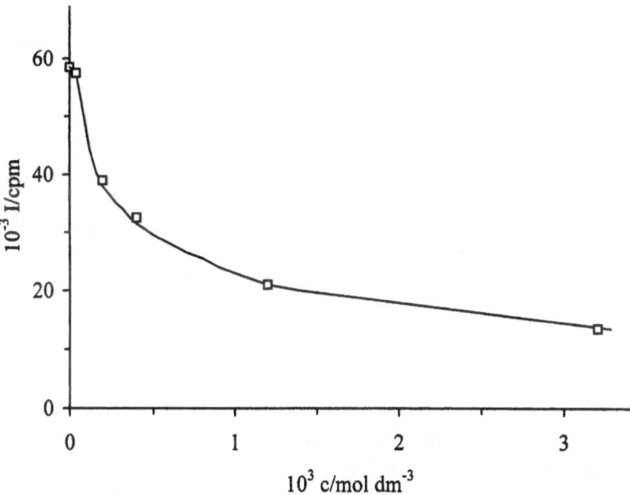

Fig. 1 The effect of H_3PO_4 concentration on the radiation intensity measured in the case of γ-Al_2O_3 (1 mg/cm^2 referred to the geometric surface area) in the presence of 0.5 moldm^{-3} $NaClO_4$, pH 2.5. The concentration of labeled H_2SO_4 was 2×10^{-4} moldm^{-3}

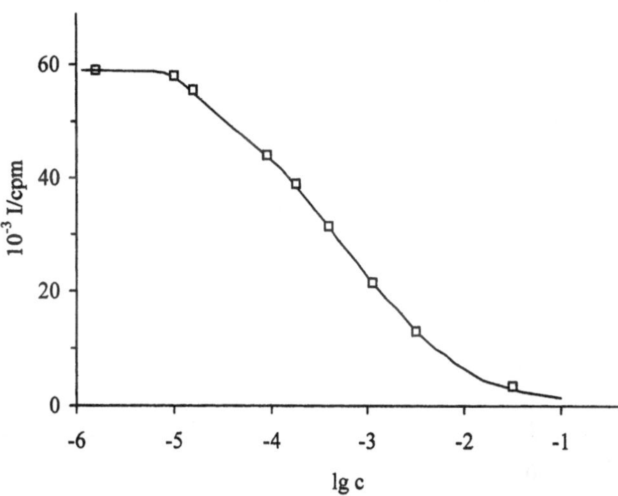

Fig. 3 The experimental data in Fig. 1 expressed as count-rate versus log$c$

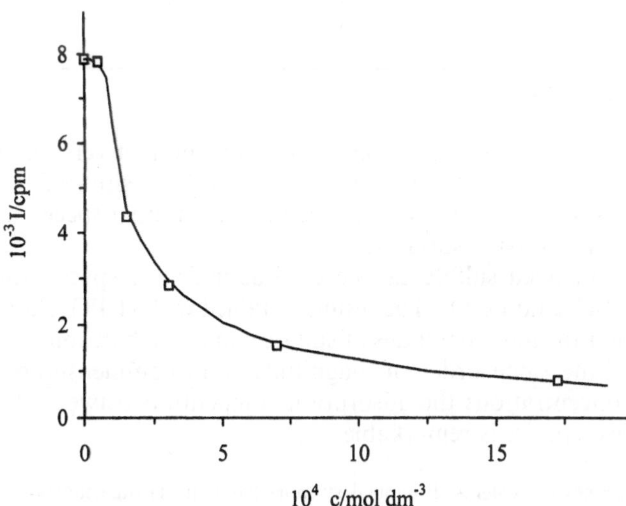

Fig. 2 The effect of H_3PO_4 concentration on the radiation intensity measured in the case of hematite (20 mg/cm^2). Other data as in Fig. 1

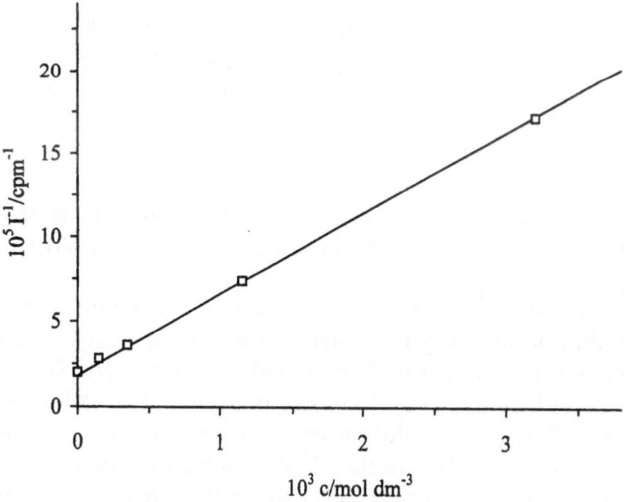

Fig. 4 The experimental data presented in Fig. 1 in a representation corresponding to Eq. (5)

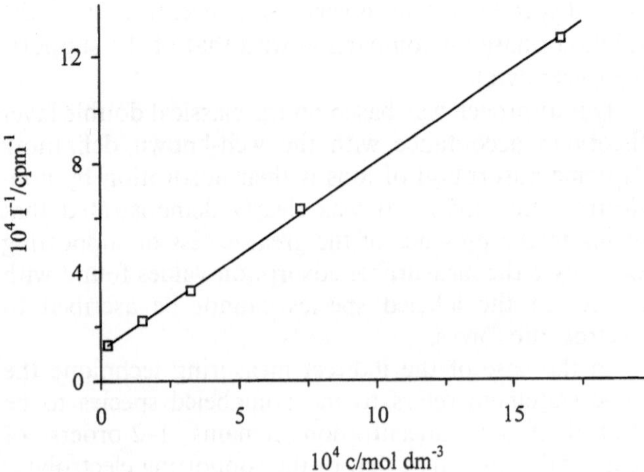

Fig. 5 The experimental data presented in Fig. 2 in a representation corresponding to Eq. (5)

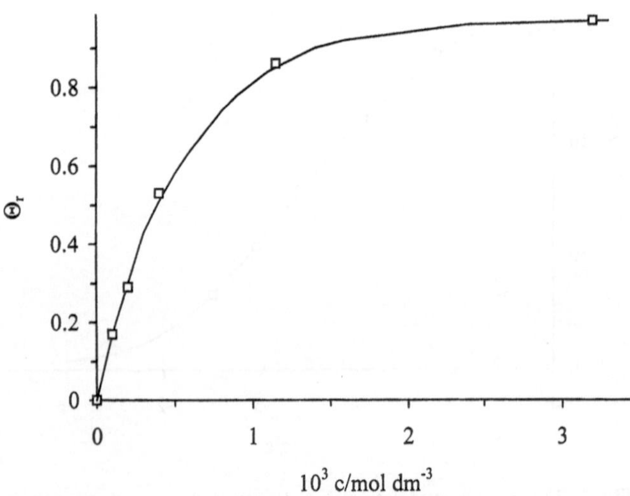

Fig. 6 The adsorption isotherm at phosphate species on Al_2O_3 calculated by Eq. (7) (coverage in arbitrary units)

On the basis of the approach outlined earlier 0.5 moldm^{-3} $NaClO_4$ supporting electrolyte was used to shift the pH of the solution to pH 2.5 by addition of $HClO_4$. The concentration of the indicator species, labeled H_2SO_4, was 2×10^{-4} moldm^{-3}. pH 2.5 was chosen in order to attain the limiting value of the sulfate adsorption determined by the complete protonation of the surface in both cases (Fe_2O_3 and Al_2O_3) [13, 14].

The direct effect of the addition of phosphate on the radiation intensity measured in the case of γ-Al_2O_3 and hematite is shown in Figs. 1 and 2, respectively. In the case of hematite (Fig. 2) it may be seen that the addition of H_3PO_4 to the solution phase in very low concentrations, $c < 1 \times 10^{-4}$ moldm^{-3}, does not exert influence on the sulfate adsorption. This phenomenon can be well demonstrated for Al_2O_3 as well by plotting the radiation intensity values against log$c$ (Fig. 3).

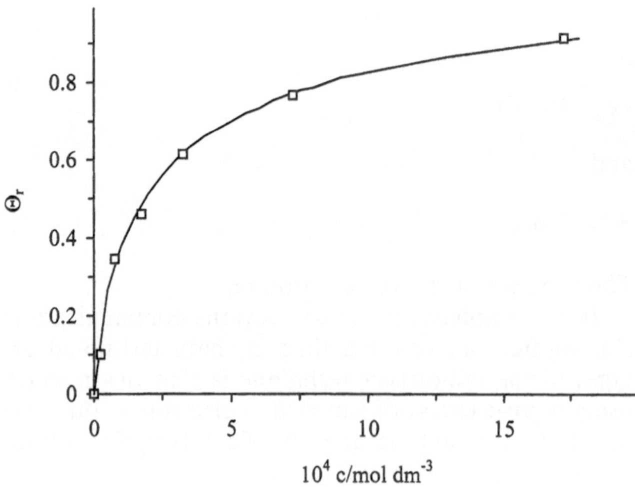

Fig. 7 The adsorption isotherm at phosphate species on hematite calculated by Eq. (7) (coverage in arbitrary units)

With respect to Eq. (5), by plotting the $1/I$ values against the concentration of phosphate species linear relationships are obtained as shown by Figs. 4 and 5 (the regressions are 0.9995 and 0.998, respectively). The existence of this linear relationship confirms the initial assumption of the applicability of the Langmuir isotherm for the specific adsorption of both anions.

The shape of the adsorption isotherm of the phosphate species can be obtained for both adsorbents using Eq. (7). The results are shown in Figs. 6 and 7. Only the shape of the curves in terms of arbitrary coverage values can be given at the present stage of the studies; nevertheless, the results obtained clearly demonstrate that the main tendencies of phosphate adsorption can be followed easily by the indirect radiotracer technique. Further results of the quantitative analysis of the results will be the subject of a further study.

Conclusions

The results obtained confirm that the indirect version of the thin foil method can be applied for the in situ study of adsorption of anions (and presumably of other species as well) on oxide surfaces.

Labeled sulfate can be used as indicator species for Al_2O_3 and Fe_2O_3. The results obtained with H_3PO_4 show that the adsorbabilities of sulfate and phosphate ions are of the same order of magnitude as in commensurable concentrations the adsorption competition between the two species is remarkable.

Acknowledgements Financial support from the Hungarian Scientific Research Fund and the Ministry of Environmental Protection is acknowledged (OTKA grants T 023056, T 031703, OKTKP grant 021856-01/2000).

References

1. Balashova NA, Kazarinov VN (1969) In: Bard AJ (ed) Electroanalytical chemistry, vol 3. Dekker, New York, pp 135–197
2. Horányi G (1980) Electrochim Acta 25:45
3. Kazarinov VE, Andreev VN (1984) In: Yeager E, Bockris JO'M, Conway BE (eds) Comprehensive treatise on electrochemistry, vol 9. Plenum, New York, pp 393–443
4. Wieckowski A (1990) In: Bockris JO'M, Conway BE, White RE (eds) Modern aspects of electrochemistry, vol 21. Plenum, New York, pp 65–119
5. Zelenay P, Wieckowski A (1991) In: Abruna HD (ed) Electrochemical interfaces: modern techniques for in situ surface characterization, VCH, New York, pp 479–527
6. Krauskopf EK (1992) In: Lipkowski J, Ross PN (eds) Adsorption of molecules at metal electrodes. VCH, New York, pp 119–169
7. Horányi G (1989) B Electrochem 5:235
8. Horányi G (1995) Rev Anal Chem 14:1
9. Horányi G (1996) In: Spivey JJ (ed) Catalysis. Specialist periodical report, vol 12. The Royal Society of Chemistry, Cambridge, pp 254–301
10. Horányi G (1999) In:Wieckowski A (ed) Interfacial electrochemistry, Theory, experiment, and applications. Dekker, New York, pp 477–491
11. Joó P, Horányi G (2000) J Colloid Interface Sci 223:308
12. Horányi G, Joó P (2000) Russ J Electrochem 36:1189
13. Horányi G, Joó P (2000) J Colloid Interface Sci 227:206
14. Horányi G, Joó P (2000) J Colloid Interface Sci 231:373

Progr Colloid Polym Sci (2001) 117: 32–36

D. Kovačević
A. Čop
A. Bradetic
N. Kallay
A. Pohlmeier
H.-D. Narres
H. Lewandowski

Interfacial equilibria at a goethite aqueous interface in the presence of amino acids

D. Kovačević · A. Čop · A. Bradetic
N. Kallay (✉)
Laboratory of Physical Chemistry,
Faculty of Science, University of Zagreb,
Marulićev trg 19, P.O. Box 163,
10001 Zagreb, Croatia
e-mail: nkallay@prelog.chem.pmf.hr
Tel.: +385-1-4895502
Fax: +385-1-4829958

A. Pohlmeier · H.-D. Narres
H. Lewandowski
Institute for Applied Physical Chemistry,
Research Centre Jülich, 52425 Jülich,
Germany

Abstract Adsorption of *o*-, *m*- and *p*-aminobenzoic acid on goethite was investigated by means of adsorption and electrokinetic measurements and was interpreted on the basis of the surface complexation model (SCM). All three acids examined showed similar behavior, with adsorption maximum at pH 4.5, while the adsorption affinity decreased in the order *p*-, *o*-, *m*-aminobenzoic acid. Electrokinetic measurements indicated that isoelectric points of goethite with adsorbed amino acids were shifted to lower pH values with respect to goethite in absence of amino acids, suggesting adsorption of negatively charged species (−1). This conclusion was confirmed by calculations based on the SCM. The equilibrium constant for exchange of Cl^- ions with amino acid ions (−1) were determined as $\log K_{A/L} = 2.62 \pm 0.13$ for *o*-aminobenzoic acid, $\log K_{A/L} = 2.37 \pm 0.11$ for *m*-aminobenzoic acid and $\log K_{A/L} = 2.85 \pm 0.14$ for *p*-aminobenzoic acid.

Key words Adsorption · Electrokinetics · Goethite · Aminobenzoic acid

Introduction

In the last decades, with the development of ecological conscience, the importance of understanding processes in natural systems is permanently growing. The binding of different pollutants on soil particles is, in laboratory conditions, examined through model systems, for example, specific adsorption of heavy metals, organic acids or amino acids on metal oxides. The simultaneous interpretation of adsorption and electrokinetic data at the metal oxide aqueous interface, based on the surface complexation model (SCM) [1], was introduced. This approach was successfully applied to lead ions/goethite [2], cadmium ions/goethite [3], salicylic acid/hematite [4, 5] and salicylic acid/titania [6] systems. In the study presented the adsorption of a more complex system, i.e. adsorption of amino acids (*o*-, *m*- and *p*-aminobenzoic) on metal oxide (goethite), is examined. In the interpretation, the effect of amino acid dissociation, i.e. the presence of negatively and positively charged ionic species, as well as zwitterionic species, in the bulk of the solution should be taken into account. Better understanding of adsorption of amino acids would be one of the crucial steps towards the solution of a more complex problem, which is the adsorption of proteins.

Theory

The interpretation of surface equilibria based on the SCM [1] takes into account interaction of ions from the bulk of the solution with active surface groups. The classical 2-p*K* model [7] assumes protonation (p) and deprotonation (d) of amphoteric MOH groups, resulting in positive and negative surface groups

$$MOH + H^+ \rightarrow MOH_2^+$$

$$K_p = \exp(\phi_0 F/RT)\frac{\Gamma(MOH_2^+)}{a(H^+)\Gamma(MOH)} , \tag{1}$$

$$MOH \rightarrow MO^- + H^+$$

$$K_d = \exp(-\phi_0 F/RT)\frac{a(H^+)\Gamma(MO^-)}{\Gamma(MOH)} , \quad (2)$$

where K_p and K_d are the corresponding equilibrium constants, Γ denotes surface concentration, while F, R and T have their usual meaning. Both positively (MOH_2^+) and negatively (MO^-) charged surface species are exposed to the same electrostatic potential, denoted as ϕ_0.

Counterion association, i.e. the binding of ions of the opposite charge (counterions) can be described as

$$MOH_2^+ + A^- \rightarrow MOH_2^+ \cdot A^-$$

$$K_A = \exp(-\phi_\beta F/RT)\frac{\Gamma(MOH_2^+ \cdot A^-)}{a(A^-)\Gamma(MOH_2^+)} , \quad (3)$$

$$MO^- + C^+ \rightarrow MO^- \cdot C^+$$

$$K_C = \exp(\phi_\beta F/RT)\frac{\Gamma(MO^- \cdot C^+)}{a(C^+)\Gamma(MO^-)} , \quad (4)$$

where A^- and C^+ denote anions and cations, and K_A and K_C are respective equilibrium constants. The state of the associated counterions is affected by electrostatic potential, ϕ_β.

The total surface concentration of the active surface sites in the interfacial layer (Γ_{tot}) is the sum of all contributions:

$$\Gamma_{tot} = \Gamma(MOH) + \Gamma(MOH_2^+) + \Gamma(MO^-) + \Gamma(MO^-C^+) + \Gamma(MOH_2^+A^-) . \quad (5)$$

For a metal oxide aqueous interface several versions of the electrical interfacial layer are described in the literature and all of them may be considered as simplifications of a general model [8]. That general model includes four equipotential planes: a 0 plane, in which surface charge groups formed by interactions with potential-determining ions are located, a β plane, in which centers of associated counterions are located, a d plane, which is the onset of the diffuse layer (characterized by a ϕ_d potential), and the electrokinetic slipping or shear plane, e, characterized by a ζ-potential and separated by a distance, s, from the d plane.

The surface charge densities in the 0 plane (σ_0) and in the β plane (σ_β) are given by

$$\sigma_0 = F[\Gamma(MOH_2^+) + \Gamma(MOH_2^+A^-) - \Gamma(MO^-) - \Gamma(MO^-C^+)] , \quad (6)$$

$$\sigma_\beta = F[\Gamma(MO^-C^+) - \Gamma(MOH_2^+A^-)] . \quad (7)$$

The surface charge density in the diffuse layer (σ_d) is equal in magnitude, but different in sign, to the net charge bound to the surface (σ_s)

$$\sigma_s = -\sigma_d = \sigma_0 + \sigma_\beta = F[\Gamma(MOH_2^+) - \Gamma(MO^-)] . \quad (8)$$

The potential drop between the 0 and the β planes depends on the constant capacitance (C_1) of the Helmholtz layer,

$$\phi_0 - \phi_\beta = \frac{\sigma_0}{C_1} , \quad (9)$$

while the potential drop in the region between the β and the d planes depends on the capacitance of the second capacitor (C_2)

$$\phi_\beta - \phi_d = \frac{\sigma_\beta}{C_2} . \quad (10)$$

The surface charge density in the diffuse layer (σ_d) is, according to the Gouy–Chapman theory, related to the ϕ_d potential as

$$\sigma_d = -\sqrt{8RT\varepsilon I_c}\sinh\frac{\phi_d F}{2RT} , \quad (11)$$

where I_c is the ionic strength and ε is the medium permittivity ($\varepsilon = \varepsilon_0\varepsilon_r$).

The relationship between the potential at the onset of the diffuse layer (ϕ_d) and the ζ-potential at a distance s from the onset of the diffuse layer is given by the Gouy–Chapman theory

$$\phi_d = 2RTF^{-1}\ln\left(\frac{\exp(-s\kappa) + \tanh(F\zeta/4RT)}{\exp(-s\kappa) - \tanh(F\zeta/4RT)}\right) , \quad (12)$$

with κ being the Debye–Hückel reciprocal length,

$$\kappa = \sqrt{\frac{2FI_c}{\varepsilon RT}} . \quad (13)$$

The zero-charge conditions at the surface of a metal oxide are characterized by two values: the point of zero charge (PZC), corresponding to the pH value at which $\sigma_0 = 0$, and the isoelectric point (IEP), characterized by $\zeta = 0$.

Specific adsorption

The previous considerations deal with metal oxides in aqueous solutions of "indifferent" electrolytes. Potential-determining ions (H^+ and OH^- in the case of metal oxides) are responsible for the formation of surface charge, while counterions do not chemically react with surface groups, but become associated with oppositely charged surface groups owing to electrostatic (Coulombic) interactions. However, several molecules and ionic groups may be bound "chemically" to the surface. Such processes are commonly called specific adsorption. Specific adsorption influences significantly the surface equilibria and should not be neglected in the interpretation. In the case of adsorption of an organic acid or an amino acid several possibilities regarding the species that actually adsorb exist, which is due to the acid dissociation in the bulk of the solution. Additionally, one has to take into account protonation and deprotonation of the surface, i.e. the types of surface sites (positive, negative, neutral) which participate in the specific adsorption.

Experimental

All the chemicals (*o*-, *m*-, *p*-aminobenzoic acid, HCl, NaOH, NaCl) used in this study were of analytical purity grade. The goethite was prepared by precipitation from $Fe(NO_3)_3$ and NaOH according to the procedure described by Cornell and Schwertmann [9]. The specific surface area, as determined by the Brunauer–Emmett–Teller method, is 69 $m^2\ g^{-1}$. The PZC obtained by titration at different ionic strengths is at pH 8.5, which coincides with the IEP of goethite in the absence of amino acid, as determined electrokinetically.

Adsorption experiments were performed at 20 °C. Suspensions of different pH ($2 < pH < 9$), containing a constant mass concentration of goethite (2.6 g cm^{-3}) and a constant initial concentration of amino acid ($c = 1 \times 10^{-3}$ mol dm^{-3}) were prepared. HCl and NaOH were used to adjust the pH. The samples were shaken for 16 h. Goethite was separated by centrifugation, and the concentration of amino acid in the supernatant was determined spectrophotometrically. The absorbance was measured in a basic medium at 310, 300 and 265 nm, which are characteristic bands of *o*-, *m*- and *p*-aminobenzoic acid, respectively.

Electrokinetic measurements, performed at 20 °C, were carried out with an Otsuka ELS-800 electrophoretic light scattering instrument. In these experiments the mass concentration of goethite was kept sufficiently low, while the bulk concentration of amino acid, pH and ionic strength were the same as in the adsorption experiments.

Results and discussion

The effect of pH on the adsorption of *o*-, *m*- and *p*-aminobenzoic acid on goethite is presented in Fig. 1. The adsorption behavior of all three amino acids is similar, with the maximum at pH ≈ 4.5, with the adsorption affinity decreasing in the order *p*-, *o*-, *m*-aminobenzoic acid.

The effect of pH on the electrokinetic potential of goethite particles after adsorption of *o*-, *m*- and *p*-aminobenzoic acid in comparison to the electrokinetic potential of goethite particles in the absence of amino acid is presented in Fig. 2. The shift of the IEP of goethite with adsorbed amino acid to lower pH values, with respect to the IEP of goethite in the absence of amino acid, confirms the assumption that the adsorption of negatively charged ionic species takes place.

Additional confirmation of the assumed reaction can be obtained from the comparison of the adsorption and electrokinetic data with the so-called "speciation diagram" describing fractions of species in the bulk of the solution as a function of pH. The "speciation diagram" of *o*-aminobenzoic acid is shown in Fig. 3. *m*- and *p*-aminobenzoic acid show similar behavior.

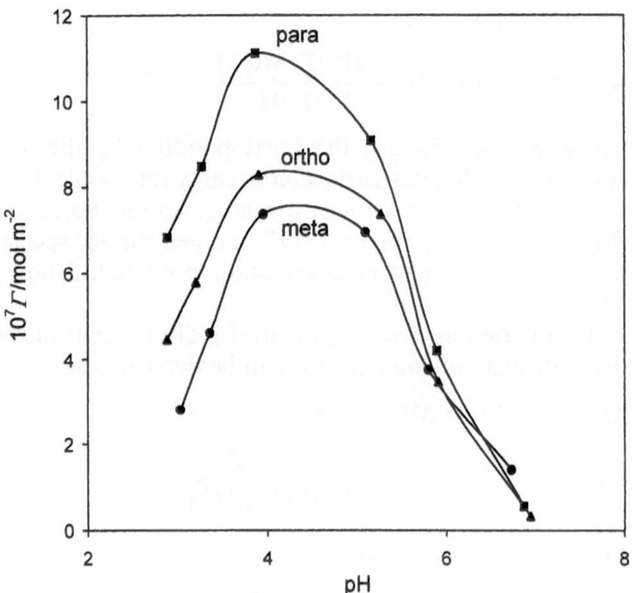

Fig. 1 Adsorption of *o*-, *m*- and *p*-aminobenzoic acid at the goethite interface. Dependency of the surface concentration of adsorbed amino acid species on pH at 20 °C and $I_c = 1 \times 10^{-2}$ mol dm^{-3}. The initial concentration of amino acid was 1×10^{-3} mol dm^{-3} and the mass concentration of goethite was 2.6 g cm^{-3}

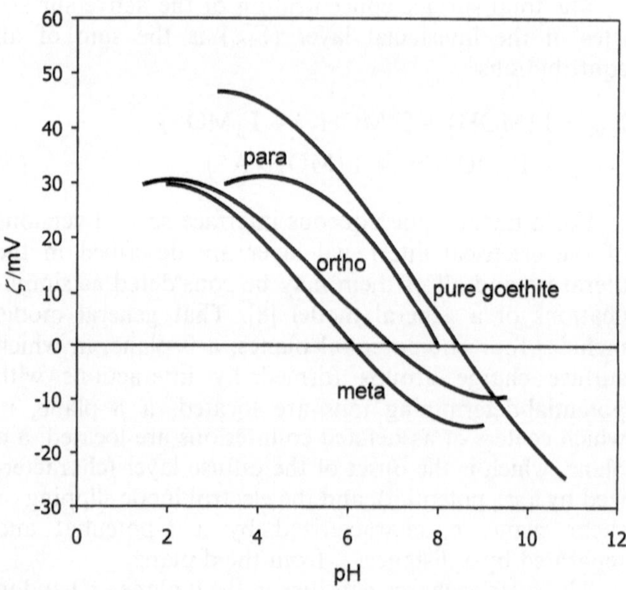

Fig. 2 Adsorption of *o*-, *m*- and *p*-aminobenzoic acid at the goethite interface. Dependency of the ζ-potential of goethite particles after adsorption of amino acids and goethite in the absence of amino acids (pure goethite) on pH at 20 °C and $I_c = 1 \times 10^{-2}$ mol dm^{-3}. The mass concentration of goethite was kept sufficiently low

In the pH region $2 < pH < 5$ the adsorption isotherm (Fig. 1) follows the fraction of negatively charged aminobenzoic acid species (L^-) in the bulk of the solution, indicating that these species are adsorbed. In

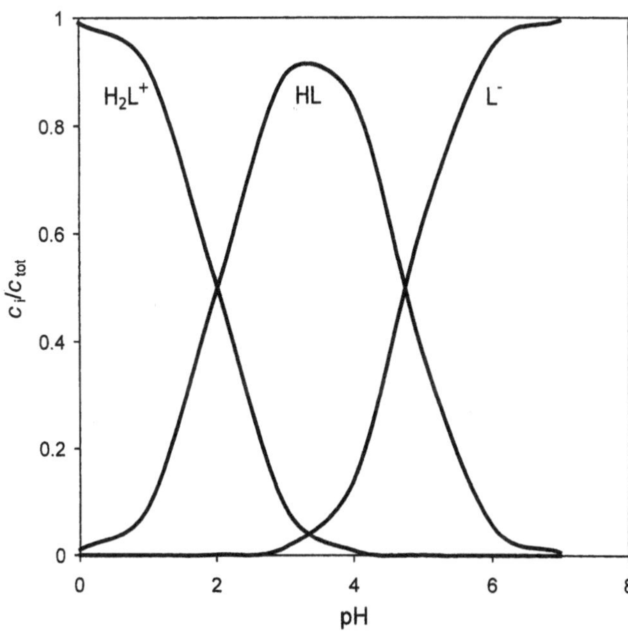

Fig. 3 "Speciation diagram" demonstrating the change in fractions of different ionic species of *o*-aminobenzoic acid with pH in the bulk of the solution

the pH region pH > 5 the adsorption of L^- species is suppressed owing to the reduction of the positive potential at the surface. Accordingly, the adsorption of amino acids may be described as

$$\equiv MOH_2^+ + L^- \rightarrow \equiv MOH_2^+ \cdot L^-$$

$$K_L = \exp(-F\phi_\beta/RT)\frac{\Gamma(MOH_2^+ \cdot L^-)}{\Gamma(MOH_2^+)a(L^-)} \quad , \tag{14}$$

where K_L is the thermodynamic equilibrium constant for specific adsorption of L^- species. Specifically adsorbed L^- species are assumed to be located in the β plane and are exposed to the ϕ_β potential. In the positive region the association of cations could be neglected. The net surface charge density is equal to

$$\sigma_s = \sigma_0 + \sigma_\beta = \sigma_0 - F[\Gamma(MOH_2^+ \cdot A^-) + \Gamma(MOH_2^+ \cdot L^-)] \quad . \tag{15}$$

In the case of higher ionic strength most of the $MOH_2{}^+$ groups are associated with counterions so the exchange of anions and L^- ions needs to be considered, so

$$\equiv MOH_2^+ \cdot A^- + L^- \rightarrow \equiv MOH_2^+ \cdot L^- + A^-$$

$$K_{A/L} = \frac{\Gamma(MOH_2^+ \cdot L^-)a(A^-)}{\Gamma(MOH_2^+ \cdot A^-)a(L^-)} \quad , \tag{16}$$

where $K_{A/L}$ is the corresponding thermodynamic equilibrium constant for the exchange reaction. In the model used in this study no potential drop between the β and d planes is assumed, $C_2 \rightarrow \infty$, so $\phi_0 \neq \phi_\beta = \phi_d \neq \zeta$.

The interpretation of the data, performed according to the SCM, was based on the following procedure:

$$\zeta \xrightarrow{s} \phi_d \xrightarrow{G.C.} \sigma_s(= -\sigma_d) \xrightarrow{-\Gamma(MOH_2^+ \cdot L)} \sigma_0 - F\Gamma(MOH_2^+ \cdot A^-), \tag{17}$$

$$pH \xrightarrow{\alpha} \phi_0 \xrightarrow{C_1} \sigma_0 \quad . \tag{18}$$

The first step Eq. (17) in the interpretation is the calculation of ϕ_d from the measured ζ-potential (Eq. 12) assuming different values of the electrokinetic slipping plane separation, s, from 5 to 20 Å. No significant effect of the choice of *s* value on the final results was found. The value of $s = 15$ Å was used in further calculations because it was found to be representative for the metal oxide aqueous interface [10–13]. The surface charge density, σ_s ($= -\sigma_d$), was calculated from the ϕ_d potential via Eq. (11). In the next step of the interpretation one calculates $\sigma_0 - F\Gamma(MOH_2{}^+ \cdot A^-)$ from σ_s and $\Gamma(MOH_2^+ \cdot L^-)$ using Eq. (15).

In a parallel procedure (Eq. 18) the value of σ_0 was calculated as follows. The surface potential, ϕ_0, was obtained on the basis of the Nernstian approximation as

$$\phi_0 \approx \alpha \frac{RT \ln 10}{F}(pH_{PZC} - pH) \quad . \tag{19}$$

The correction factor, α, was taken to be 0.8, which is the representative value for the metal oxide aqueous interface [7, 14] and the PZC is at pH 8.5. From the calculated ϕ_0 and ϕ_β ($= \phi_d$) (Eq. 12) and the assumed capacitance C_1 the surface charge density in the 0 plane, σ_0, can be obtained via Eq. (9). For the capacitance C_1, different values between 0.5 and 1 F m^{-2} were assumed [15–17]. The proper value of C_1 corresponds to the minimum in the standard deviation of the resulting $\log K_{A/L}$ value. These two procedures (Eqs. 17, 18) lead to the determination of $\Gamma(MOH_2{}^+ \cdot A^-)$ (Eq. 15). Consequently, the thermodynamic equilibrium constant for exchange, $K_{A/L}$, was calculated by Eq. (16). The constants are determined as $\log K_{A/L} = 2.62 \pm 0.13$ for *o*-aminobenzoic acid, $\log K_{A/L} = 2.37 \pm 0.11$ for *m*-aminobenzoic acid and $\log K_{A/L} = 2.85 \pm 0.14$ for *p*-aminobenzoic acid for capacitance C_1 of 0.65, 0.55 and 0.95 F m^{-2}, respectively. Among the acids examined, *p*-aminobenzoic acid has the largest exchange constant and, therefore, the highest adsorption affinity.

References

1. Yates DE, Levine S, Healy TW (1974) J Chem Soc Faraday Trans I 70:1807
2. Kovacevic D, Pohlmeier A, Özbas G, Narres H-D, Schwuger MJ, Kallay N (2000) Colloids Surf 166:225
3. Kovačević D, Pohlmeier A, Özbas G, Narres H-D, Kallay N (1999) Prog Colloid Polym Sci 112:183
4. Kovačević D, Kallay N, Antol I, Pohlmeier A, Lewandovski H, Narres H-D (1998) Colloids Surf 140:261
5. Kovačević D, Kobal I, Kallay N (1998) Croat Chem Acta 71:1139
6. Kallay N, Čop A, Kovačević D, Pohlmeier A (1998) Prog Colloid Polym Sci 109:221
7. Lyklema J (1995) Fundamentals of interface and colloid science. Academic, London
8. Kallay N, Kovačević D, Čop A (1999) In: Kallay N (ed) Interfacial dynamics. Dekker, New York, pp 249–271
9. Cornell RM, Schwertmann U (1996) The iron oxides. VCH, Weinheim
10. Healy TW, White LR (1978) Adv Colloid Interface Sci 9:303
11. Sprycha R, Matijević E (1989) Langmuir 5:479
12. Harding H, Healy TW (1985) J Colloid Interface Sci 107:382
13. Chow RS, Takamura K (1988) J Colloid Interface Sci 125:226
14. Kallay N, Sprycha R, Tomić M, Žalac S, Torbić Z (1990) Croat Chem Acta 63:467
15. Blesa MA, Kallay N (1988) Adv Colloid Interface Sci 28:111
16. Lumsdon DG, Evans LJ (1994) J Colloid Interface Sci 164:119
17. Nilsson N, Persson P, Lövgren L, Sjöberg S (1996) Geochem Cosmochem Acta 60:4385

Progr Colloid Polym Sci (2001) 117:37–41

B. Ruffmann
R. Zimehl

Liquid sorption and stability of polystyrene latices

R. Zimehl
Institut für anorganische Chemie der Universität Kiel, 24098 Kiel, Germany

B. Ruffmann (✉)
Institut für Chemie, GKSS Forschungszentrum Geesthacht, 21502 Geesthacht, Germany

Abstract The surface structure of the colloidal polymer particles and the liquid structure in the dispersion medium strongly affect the stability of a latex dispersion. Two kinds of polystyrene latices, stabilised by sulfonate groups, were synthesised by emulsion polymerisation. The reactions were started by a redox starter of potassium peroxodisulfate and sodium bisulfite. Type 1 latices have a smooth surface and were prepared from styrene. The type 3 latices were prepared by a two-step mechanism. In the first step, seed particles were prepared. In the second step, the seed particles were grown by further polymerisation of styrene and sodium styrene sulfonate. By this method oligomers of styrene and sodium styrene sulfonate accumulate in the outer spheres of the particles; thus a hairy surface is formed. The stability of the dispersions and the selective liquid sorption were investigated for both kinds of latices: selective liquid sorption from 1-propanol(1)/water(2) mixtures on the colloidal polymer particles with different surface structure was obtained by the reduced surface excess. The isotherms were analysed in order to determine the mole fraction of the adsorbed layer. The different surface structure of the particles leads to different stability behaviour of the dispersions. The critical coagulation concentrations (c_k) for barium perchlorate in 1-propanol/water mixtures were obtained by test-tube experiments. At low 1-propanol concentrations the stability of the type 3 latices increases, exhibits a maximum stability at about $x_1 \approx 0.1$ and decreases upon further addition of 1-propanol. The decreased c_k values for type 3 latices in water-rich mixtures are explained by swelling of the hairy surface. Imbedding 1-propanol/water clusters between the oligomers causes swelling of the outer sphere of the particles. Formerly not dissociated surface groups in the oligomer region are able to dissociate. This leads to a vaster double layer and thus higher c_k values. By further addition of 1-propanol, counterions are pushed back to the surface and the stability decreases.

Key words Adsorption · Coagulation · Liquid structure · Polystyrene latices · Surface structure

Introduction

The preparation of stable colloidal systems requires a subtle balance of surface forces of different nature. Among others, the stabilisation of dispersions by a corona of macromolecules (i.e. steric stabilisation) or by charges linked to the particle surface (i.e. electrostatic stabilisation) is most prominent. Electrostatic stabilisation of colloidal particles is very well described by the Derjaguin–Landau–Verwey–Overbeek (DLVO) theory.

The model is based on the formation of electrical double layers at the particle surfaces. Repulsion between approaching particles is due to the high ion density in the contact zone of the fully relaxed double layers. Electrolyte addition leads to the compression of the double layer and usually this decreases the stability of the dispersion.

If the electrical double layer is not in equilibrium with either the particle surface or the bulk electrolyte solution or both, the distribution of ions can no longer be described by the dogmatic equations of the DLVO theory. For example, alcohols as cosolvents of water decrease the solubility of electrolytes; counterions are pushed back to the surface and the stability decreases [1, 2]. The stability of different kinds of dispersions composed from mixtures of water and organic liquids has been investigated by several authors. None of them presented coherent formalisms in predicting the properties of colloidal particles in mixed aqueous media of different composition.

The situation is even more disappointing for colloidal systems of technological importance. Especially for industrial use, dispersions are made up from really complex systems. Many modifiers, regulators and cosolvents are added to optimise the properties of the final dispersion. This leads to deviations from the ideal behaviour described by DLVO theory for the stabilisation of dispersions. Even the addition of small amounts of cosolvents (e.g. lower aliphatic alcohol) to an aqueous dispersion provokes unusual stability behaviour of latices [3, 4].

In former studies the sorption of 1-propanol from water on polymer networks with different surface structure was investigated [5, 6]. By a different polarity of the polymer frame an additional swelling ability was introduced to the sorbents. Selective liquid sorption for the different polymer networks depended on the polarity as well; however, the influences of the surface structure and the polarity of the colloidal polymer particles on the swelling behaviour in the presence of 1-propanol has not been reported yet. Up to now, the adsorption behaviour of polystyrene resins and colloidal particles was investigated by microcalorimetric measurements and conformational changes in the adsorption layer of type 3 latices were noted [7]. Type 1 latices have a smooth surface (Fig. 1a), whereas type 3 latices are supposed to have a hairy surface (Fig. 1b). The different surface structure leads to different stability behaviour. As the influence of electrolytes on the stability of colloidal particles strongly depends on the change in the water structure by the ions, structural effects of the surface of the colloidal particles have an influence as well [8].

In this study the selective liquid adsorption on colloidal polymer particles with different surface structure from 1-propanol(1)/water(2) mixtures is investigated by the reduced surface excess. The excess isotherms are analysed to determine the mole fraction of the adsorbed layer, x_2^s. The critical coagulation concentration, c_k, for barium perchlorate in 1-propanol/water mixtures is reported. The aim of this study is to show the influence of surface structure on the adsorption behaviour and thus the stability of polystyrene latices.

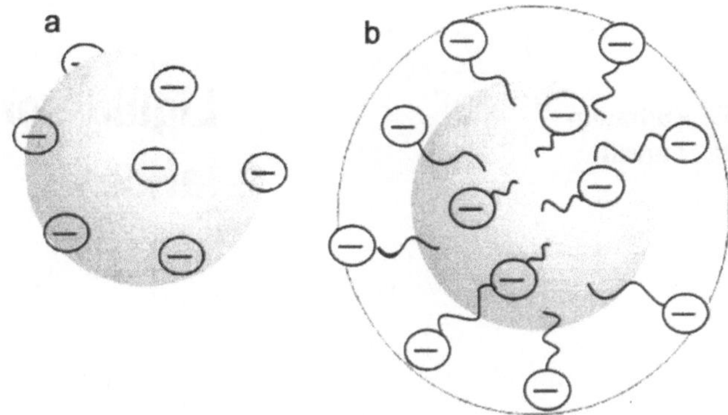

Fig. 1 Particles of different types of latices: **A** type 1 latex (smooth surface); **B** type 3 latex (hairy surface)

Experimental

The adsorption of 1-propanol (index 1) and water (index 2) was studied with different polymer particles.

Preparation of colloidal particles

Several latices were prepared by emulsion polymerisation in a three-necked vessel under a N_2 atmosphere without emulsifier.

Type 1 latex

Styrene (35 ml) was suspended in 500 ml water at 40 °C. The reaction was initiated by a redox starter of potassium peroxodisulfate and sodium bisulfite.

Type 3 latices

In a first step, seed particles were prepared by emulsion polymerisation of styrene using sodium styrene sulfonate as comonomer. The reaction was started by potassium peroxodisulfate and sodium bisulfite. In a second step, the seed particles were grown by further polymerisation of styrene and sodium styrene sulfonate initiated by the same starter. Different styrene/sodium styrene sulfonate ratios and temperatures were used [9, 10, 11]. This preparation leads to particles with oligomers consisting of styrene and sodium styrene sulfonate in the outer sphere of the particles, which resulted in a hairy surface [5].

Colloidal resin

A dispersion of a ground ion-exchange resin (PS/KM/1) was prepared from Amberlyst 15 (Fluka). The macroreticular cation-exchange resin has a polystyrene network cross-linked by 30% divinylbenzene and bears sulfonate groups in the matrix. It was ground in a ball mill to form a dispersion with particles of 180–340-nm diameter. The dispersed particles are assumed to have no oligomers on the surface.

All the dispersions were cleaned by centrifugation and redispersion, dialysis for 4 weeks and were transformed to the proton form by ion exchange.

Characterisation of colloidal particles

The mean diameters of the colloidal particles were determined by sedimentation in the centrifugal field (Bi-DCP particle sizer, Brookhaven Instrument Corporation). The surface area was calculated from the diameter. The surface charge of the particles was determined in a particle charge detector (PCD 02, MÜTEK) by titration with 0.001 N poly(diallyldimethyl ammonium chloride). The particles and their surface charges are listed in Table 1.

Surface-excess isotherms

Adsorption measurements were performed in well-sealed test tubes at room temperature. The binary liquid mixture (3 cm^3) was added to sediment containing 0.15 g polymer samples. The colloidal particles were redispersed by ultrasound. The change in the water concentration in the bulk liquid (x_2) was measured by the differential refractive index in a differential refractometer of a high-performance liquid chromatography system (Waters) after equilibration periods of 48 h and centrifugation.

Critical coagulation concentration

The c_k values were obtained by test-tube experiments [1]. A series of dispersions containing increasing amounts of barium perchlorate were prepared by adding 1 ml salt solution to 1 ml dispersion (final polymer volume fraction: 2×10^{-4}). After shaking, the dispersions were left for 24 h. It was then determined which salt concentration causes coagulation. When required, additional series were prepared by increasing the salt concentration in smaller steps.

Results

Composition and volume of the adsorbed phase

The selective liquid sorption on the different polymer particles was studied by specific-excess quantities which can be directly determined by experiment [12–14]. The experimental data are expressed as composite isotherms $n_i\sigma^{(n)} = f(x_i)$, in which the excess amount, $n_i\sigma^{(n)}$, adsorbed of one component is plotted as a function of its equilibrium concentration, x_i [15]:

$$n_i^{\sigma(n)} = \frac{n^0}{m}(x_i^0 - x_i) = f(x_i) \quad (\mathrm{mol/g}) \ , \tag{1}$$

Table 1 Specific surface charge, Σ_0, of the particles and composition of the adsorption layer in the linear part of the adsorption-excess isotherm

Dispersion	Σ_0 (Cg^{-1})	$\lvert n_1^s \rvert$ ($mmolg^{-1}$)	$\lvert n_2^s \rvert$ ($mmolg^{-1}$)	$\lvert n_1^s \rvert / \lvert n_2^s \rvert$
P(St/NaSS$_0$)R	2.87	0.7	2.8	0.25
P(St/NaSS$_5$)1	11.77	1.4	6.2	0.23
P(St/NaSS$_3$)4	6.43	1.0	3.35	0.29
P(St/NaSS$_4$)1	9.45	1.45	6.5	0.22
NSS/2/2.1	12.76	1.3	5.4	0.24
NSS/3/1.1	315.3	4.3	29	0.15
PS/KM/1	75.73	1.5	18.8	0.08

where n^0 is the initial number of moles of the binary liquid, m the amount of solid, x_i^0 the mole fraction before adsorption and x_i the mole fraction after adsorption.

The excess isotherms were analysed according to the Schay–Nagy extrapolation method [13, 14, 16]. The extrapolated values $|n_1^s|$ and $|n_2^s|$ give the composition of the adsorbed phase within the linear range of x_2. The volume occupied by the sorbed liquid is

$$V^s = |n_1^s|V_{m,1} + |n_2^s|V_{m,2} \ . \tag{2}$$

The adsorption capacity of pure component 2 is then

$$n_{2,0}^s = \frac{V^s}{V_{m,2}} \ . \tag{3}$$

The mole fraction of component 2 in the adsorption layer x_2^s is given by

$$x_2^s = \frac{n_2^s}{n_1^s + n_2^s} \ , \tag{4}$$

which is suitable for determining $x_2^s = f(x_2)$, if n_1^s and n_2^s have been derived from the adsorption-excess isotherms by

$$n_1^s = \frac{r' n_{2,0}^s x_1 - r' n_2^{\sigma(n)}}{x_1 + r' x_2} \quad \text{and} \quad n_2^s = \frac{n_2^{\sigma(n)} + r' n_{2,0}^s x_2}{x_1 + r' x_2} \ , \tag{5}$$

where $r' = M_2/M_1$ is the ratio of the molar masses of the two components.

The mole fraction of the adsorbed layer, x_2^s, was calculated from the excess isotherms according to Eq. (4). The equilibrium diagram (Fig. 2) shows the

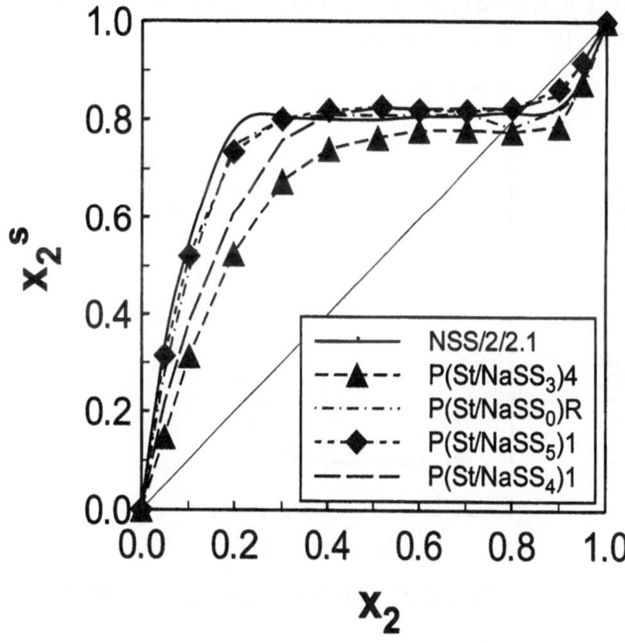

Fig. 2 Mole fraction of the adsorption layer, x_2^s, of the latices as a function of the bulk composition, x_2

composition of the adsorbed layer x_2^s versus the composition of the bulk phase x_2. Water is bound preferentially from the 1-propanol/water mixtures at low water mole fractions for all the particles. At higher water content, 1-propanol is partitioned preferably towards all the particles. The curves indicate that the uptake of water in the plateau region does not depend on the nature of the particles. In a wide range the composition of the adsorption layer does not depend on the surface charge of the particles (Table 1).

The adsorption capacity $n_{2,0}^s$ was calculated according to Eq. (3). The data were related to the surface area for comparison of particles with different diameters. For the type 3 latices the adsorption capacity increases with the surface charge (Fig. 3). This results in an increase in the volume of the adsorption layer, V^s (Table 2, Fig. 4).

Critical coagulation concentration

The critical coagulation concentration of the colloidal particles for barium perchlorate in 1-propanol/water mixtures was studied. In Fig. 5 the dispersions investigated are arranged according to the specific surface charge of the particles. The c_k values are shown as a function of the bulk concentration x_2.

For the dispersions of particles with a smooth surface (the type 1 latex P(St/NaSS$_0$)R and the milled resin PS/KM/1) the c_k values decrease with increasing mole fraction of 1-propanol in the liquid mixture ($x_2 = 1$–0.8). This effect is related to the decrease of the dielectric constant in the bulk mixture. Thus, the solubility of the electrolytes decreases and the counterions are pushed back to the surface. This compression of the double layer leads to a decrease in stability [1, 2].

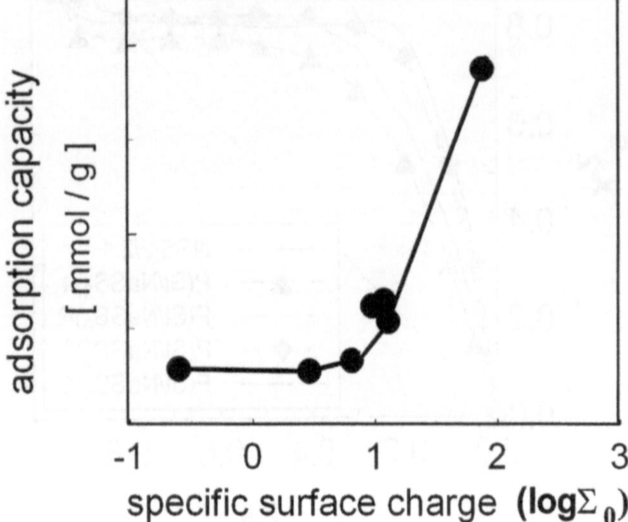

Fig. 3 Adsorption capacity, $n_{2,0}^s$, as a function of surface charge of the colloidal particles

The c_k values of dispersions of particles with a hairy surface (type 3 latices) do not decrease continuously with increasing 1-propanol concentration. Steplike discontinuities are observed in the curves of type 3 latices with low specific surface charge at $x_2 = 0.9$. For the type 3 latex with the highest surface charge a maximum of c_k values at $x_2 = 0.9$ is found. Here, the compression of the

Table 2 Surface charge density, σ_0, adsorption capacity, $n^s_{2,0}$, and volume of the adsorption layer, V^s, for the colloidal particles investigated

	σ_0 (μCcm^{-2})	$n_{2,0}^s$ (mmolm^{-2})	V^s (μlm^{-2})
Polymer material latice			
P(St/NaSS$_0$)R	15.29	0.44	7.94
P(St/NaSS$_5$)1	15.29	0.14	2.61
P(St/NaSS$_3$)4	19.16	0.22	4.03
P(St/NaSS$_4$)1	19.74	0.26	4.72
NSS/2/2.1	33.87	0.29	5.18
NSS/3/1.1	304.14	0.45	8.15
Dispersion			
PS/KM/1	252.4	0.83	38.15

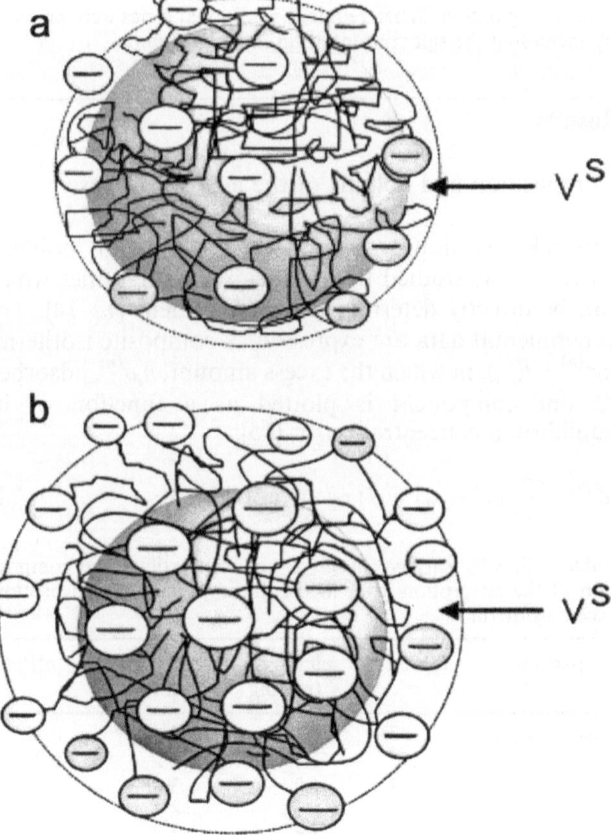

Fig. 4 Volume of the adsorbed phase, V^s, for low and high charged type 3 latex particles

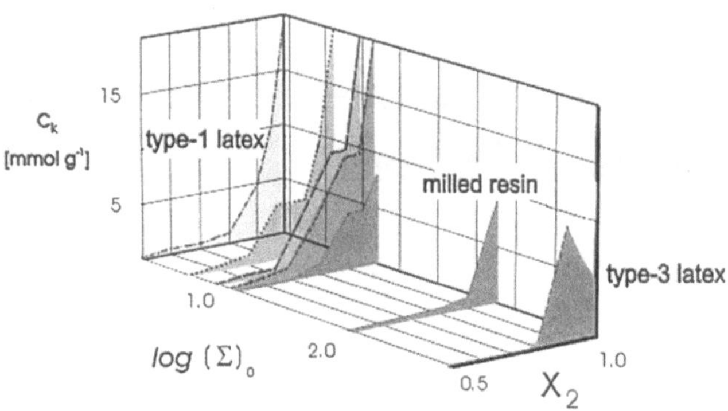

Fig. 5 c_k values of the dispersions for $BaClO_4$ in1-propanol (x_1)/ water(x_2) mixtures(*left* to *right*: P(St/NaSS$_0$)R, P(St/NaSS$_3$)4, P(St/ NaSS$_4$)1, P(St/NaSS$_5$)1, NSS/2/2.1, PS/KM/1, NSS/3/1.1

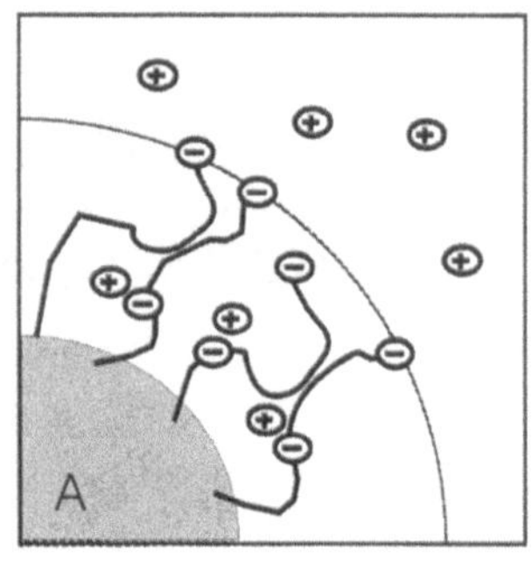

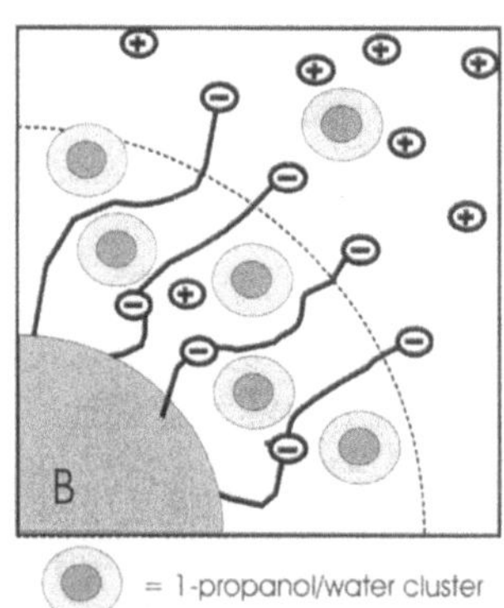

Fig. 6 Adsorption of 1-propanol/water clusters and swelling of the outer sphere of type 3 latices, oligomer region: **A** in water; **B** in 1-propanol/water at $x_2 \approx 0.9$

double layer due to the presence of 1-propanol is superimposed by a stabilising effect.

Conclusions

The aim of this study was to demonstrate the influence of surface structure on the stability of dispersions. The adsorption from 1-propanol/water mixtures onto colloidal sulfonated polystyrene particles and the stability of their dispersions were examined.

Two main structural effects have to be accounted for, the first being the structure of the dispersion medium. The composition of the adsorbed phase of all the latices presented here is similar to that of 1-propanol/water clusters formed in a wide range of bulk mixture composition ($x_2 = 0.6$–0.9) [17, 18].

The second effect considers the surface structure of the particles. A smooth surface (type 1 latex, milled resin) is assumed to have no swelling ability. By introducing oligomers to the particles surface it obtains a gel-like character and the ability to swell [7]. The swelling ability increases with surface charge density owing to the increased adsorption capacity (Table 2).

The different surface structures of the particles lead to different coagulation behaviour of the dispersions. The dispersions of particles with a smooth surface (type 1 latex, milled resin) show a decrease in stability with increasing content of 1-propanol owing to the double-layer compression.

For the type 3 latices, swelling of the hairy surface enables the dissociation of charges trapped in the oligomer region (Fig. 6). The additional counterions enhance the stability of the dispersions with low 1-propanol content ($x_2 = 0.9$). As the swelling ability increases with the adsorption capacity, this stabilising effect is more pronounced for particles with higher surface charge.

Usually the stability of latex dispersions is decreased by the addition of hydrophobic cosolvents; however, structural properties of the surface of the particles can cause stabilising effects. The analysis of the adsorption-excess isotherms contains important new information about the properties of colloidal particles, such as the composition and the volume of the adsorbed layer. This information is of crucial importance for understanding the behaviour of the dispersions.

References

1. de Rooy N, de Bruyn PL, Overbeek JTG (1980) J Colloid Interface Sci 75:542
2. Matijevic E, Ronayne ME, Kratohvil JP (1966) J Phys Chem 70:3830
3. Seebergh JE, Berg CB (1997) Colloids Surf A 121:89
4. Galembeck F, Braga M, da Silva MCVM, Cardoso AH (2000) Colloids Surf A 164:217
5. Zimehl R, Vózár A, Dékány I (1996) Thermochim Acta 271:59
6. Zimehl R (1998) Thermochim Acta 310:207
7. Ruffmann B, Ammann L, Zimehl R, Dékány I (1999) Thermochim Acta 337:55
8. Zimehl R, Lagaly G (1986) Prog Colloid Polym Sci 72:28
9. Okubo M, Nakagawa T (1992) Colloid Polym Sci 270:853
10. Kim JH, Chainey M, El-Asser MS, Vanderhoff JW (1989) J Polym Sci Part A Polym Chem 27:3187
11. Ruffmann B (1999) Dissertation. Kiel
12. Everett DH (1964) Trans Faraday Soc 60:1803
13. Everett DH (1965) Trans Faraday Soc 61:2478
14. Schay G (1976) Pure Appl Chem 48:373
15. Everett DH (1983) In: Ottewill RH, Rochester CH, Smith AL (eds) Adsorption from solution. Academic, London, pp 302
16. Schay G, Nagy LG (1972) J Colloid Interface Sci 38:302
17. Ebert G, Wendorff J (1970) Ber Bunsenges Phys Chem 74:1071
18. Großmann GH, Ebert KH (1981) Ber Bunsenges Phys Chem 85:1026

Progr Colloid Polym Sci (2001) 117:42–46

R. Zimehl
M. Hannig

Adsorption onto tooth enamel – the biological interface and its modification

Abstract The purpose of this study was to provide further information on several aspects of bioadhesion at the enamel surface. In the first part of the article the artificial modification of dental enamel is illuminated. The interaction of hydroxyapatite, which is the main constituent of tooth enamel, and tooth enamel with different kinds of carboxylic acids was investigated by Fourier transform IR spectrometry, scanning electron microscopy and X-ray photoelectron spectroscopy (XPS). The XPS spectra clearly reveal that the surface treatment substantially increases the relative intensities of Ca, P and O with respect to the surface of the polished enamel reference. In the second part of the article the examination of the natural modification of dental enamel by atomic force microscopy (AFM) is reported. This study was performed to evaluate the use of the aforementioned techniques in examining the surface of the native dental enamel and the protein layer (salivary pellicle) adsorbed in vivo on dental enamel as well. Enamel test pieces were attached to the buccal surfaces of the upper first molar teeth in two adults using removable intraoral splints. The splints were carried intraorally over periods ranging from 1 min to1 h. Pellicle structures could be identified by AFM on intraorally exposed specimens as compared to nonexposed enamel surfaces. The surface of the adsorbed salivary pellicle was characterised by a dense globular appearance. The diameter of the globulelike protein aggregates varied between 100 and 250 nm. This study indicates that the combination of microscopic and spectroscopic techniques is a powerful tool for high-resolution examination of enamel and the salivary pellicle surface structure.

Key words Atomic force microscopy · Salivary pellicle · Protein adsorption · Enamel

R. Zimehl (✉)
Institute for Inorganic Chemistry,
Christian Albrechts University of Kiel,
Olshausenstrasse 40, 24098 Kiel,
Germany

M. Hannig
Clinic for Operative Dentistry
and Periodontology,
Christian Albrechts University of Kiel,
Arnold-Heller-Strasse 16, 24105 Kiel,
Germany

Introduction

Dental diseases such as caries and periodontitis are serious problems for almost any human being all over the world and cause enormous costs for the public health care organizations. One reason for the development of dental diseases can be found from the structure of the tooth hard substances, namely enamel, and its interaction with the oral environment, saliva and different microorganisms. The surface of the teeth is characterised by a crystallised arrangement of mineral components with incorporated and chemisorbed biopolymers. The enamel is highly mineralised, about 96% by weight consists of tightly packed hydroxyapatite crystallites with only 1% of its weight being organic molecules and 3% being water; therefore, it is reasonable to consider dental enamel as a microporous solid with a superior specific surface. Thus, the tooth enamel is prone to microbial attack and can be dissolved and solubilised by acidic agents. Formation of a microbial biofilm (so-called plaque) causes the main dental diseases caries and periodontitis. The natural mechanism to protect the metastable biological structure of the tooth hard substance is the formation of the so-called salivary pellicle owing to adsorption of salivary proteins onto the enamel surface. This protective layer of different salivary biopolymers is formed in vivo on any solid surface exposed to the oral environment; however, even enamel covered by the natural protective layer is sensitive to acid attack. A preventive approach could be to cover the dental enamel by suitable materials serving as an artificial protective layer. To analyse the mechanism of bonding between the dental enamel and the protective layer we performed experiments with different chemical conditioning agents. The chemical adhesion of protective and restorative materials to tooth surfaces has been studied with different kinds of carboxylic and polyacrylic acid using hydroxyapatite and polished or ground enamel as a model system [1–4] and evidence of chemical binding has been reported recently [5]. Binding is achieved by carboxylate groups penetrating the apatite matrix and displacing calcium and phosphate ions. In recent work we have reported the topography of conditioned and untreated enamel surfaces by scanning electron microscopy (SEM) and X-ray photoelectron spectroscopy (XPS) [3]. We have demonstrated that the mechanical strength of the interface between restorative or protective materials and the tooth hard substance dental enamel could be improved by cleaning organic debris off with different carboxylic acids. Nevertheless the mechanism of adhesion of restorative and sealing material to the tooth surface is still under discussion. This study was performed to evaluate future developments and possible ways out of plaque-related dental diseases by conservative and restoring strategies. The use of SEM, atomic force microscopy (AFM) and Fourier transform IR (FTIR) spectroscopy in examining the surface of native and modified dental enamel is discussed.

Materials and methods

The interaction of hydroxyapatite with benzoic acid in acetone and poly(acrylic acid) in water was investigated by extraction techniques and FTIR spectroscopy. Porous hydroxyapatite (Fluka "high-resolution", analytical-reagent grade material) was used in form of a fine powder. The carboxylic acids were used as a 3% solution in either organic or aqueous media. The mineral powder (50 mg) and the solution (2 ml) were ground together with an agate mortar and pestle. The slurry was dried and then washed with a large excess of organic solvent or water. Finally the conditioned hydroxyapatite was recovered from the slurry by centrifugation. A small amount of the dried powder was mixed with potassium bromide and pressurised to form a suitable tablet. IR spectra of the original and conditioned hydroxyapatite were recorded using a FTIR absorption spectrometer (GENESIS, ATI Mattson Unicom).

Topographical scanning electron microscopic analyses of the enamel surfaces were carried out using an analytical scanning electron microscope (XL-30, Philips). Cleaned and dried enamel substrates were exposed for 2 min to a solution of benzoic acid (3% weight in acetone). After exposure, the enamel was washed with distilled water and subsequently dried with compressed air. The enamel specimen was sputter-coated with gold in an EMITECH sputter coater prior to SEM analysis.

For XPS investigations a native (nonpolished, nonconditioned) enamel surface, an enamel specimen cleaned with acetone and a conditioned enamel sample (treated with benzoic acid in acetone) were conducted using an OMICRON system under an ultrahigh vacuum of 10^{-9} mbar, using a MgKα soft X-ray source. Polished, nonconditioned enamel was used as a reference surface [3]. The peaks of the following elements were analysed: phosphorus, calcium, carbon, and oxygen. Owing to charging effects the peak positions were corrected with the energy of the 1*s* carbon peak at 284.5 eV as the reference. Background subtraction and area calculations of the different peaks were carried out with the software supplied.

For AFM investigation several enamel specimens with a surface area of about 2×2 mm^2 were cut from labial surfaces of freshly extracted bovine incisors. The enamel specimens were mounted on the buccal surfaces of the maxillary first molars in two subjects using intraoral acrylic appliances (minisplints [1, 6]) and were exposed to the oral environment by the subjects over periods of 1, 10, 30, and 60 min. After intraoral exposure the specimens were removed from the oral cavity, rinsed in distilled water and immediately mounted and analysed by AFM (Autoprobe CP operated in contact mode [7]). For reference purposes, enamel specimens not exposed in the oral cavity were investigated by AFM.

Results and discussion

The interaction of hydroxyapatite, which is the main constituent of tooth enamel, with different kinds of carboxylic acid was investigated by extraction techniques and FTIR spectroscopy. After treatment with different kinds of carboxylic acid the spectra of the hydroxyapatite powder provided evidence of chemical interaction. The appearance of IR peaks at about 1,560 cm^{-1} indicates that at least a part of the carboxylic acid is adsorbed at the mineral powder via ionised carboxylic groups. The large number of washing cycles needed to reduce the characteristic IR peaks was taken as an indicator for strong binding efficiency of the carboxylic acid (Table 1). From SEM we could demonstrate that the interaction of

Table 1 Number of washing cycles needed to reduce the characteristic IR peaks at about 1,560 cm^{-1} to a constant amplitude. The large number of washing cycles for the hydroxyapatite samples is indicative of the strong binding efficiency of the carboxylic acids

Hydroxyapatite sample with	Washing liquid	Number of washing cycles to reach constant value of IR peak
Poly(acrylic acid) in water	Water	10
Benzoic acid in acetone	Acetone	5

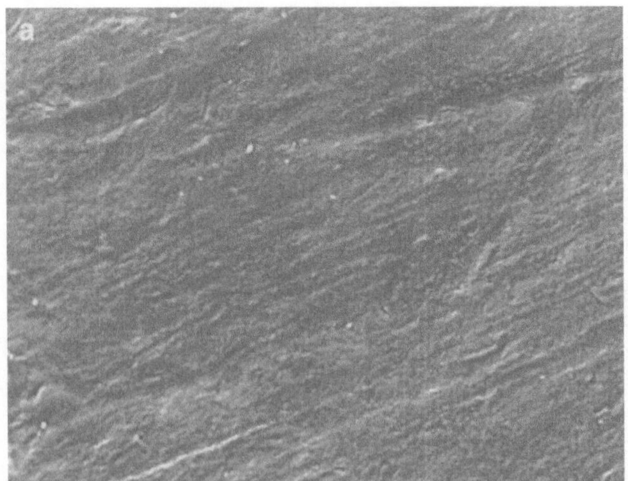

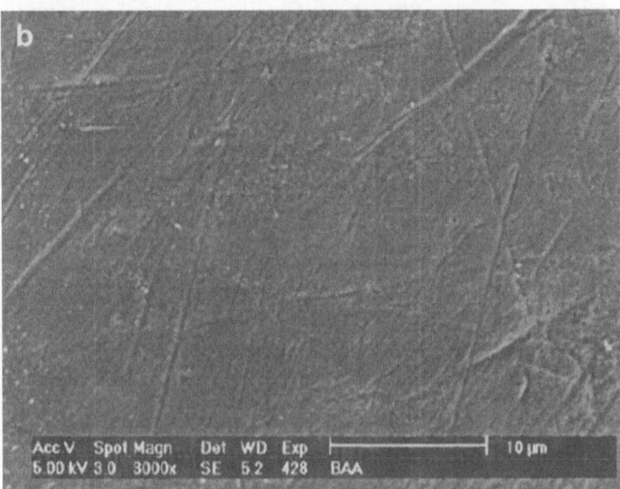

Fig. 1 a Scanning electron microscope picture of the characteristic topography of the polished part of the enamel surface. **b** Conditioning with acetone or benzoic acid in acetone did not lead to visible etching, and the surface topography appeared unchanged. Magnification × 2,500. For details of the preparation see Ref. [3]

enamel with several carboxylic acids in acetone (e.g. benzoic acid in acetone, BAA) does not produce etching patterns, whereas poly(acrylic acid) in water does [8]. The topography of the polished part of the enamel surface is shown in Fig. 1a. Conditioning with acetone or BAA did not lead to detectable etching and the surface topography remained unchanged (Fig. 1b), but from XPS investigations we obtained amplified peaks for Ca, P, and O with respect to the untreated reference surface [3, 8]. A closer look to the XPS spectra clearly reveals that the surface treatment substantially increases the relative intensities of Ca, P, and O with respect to the polished surface (Table 2). The carbon peak area of the acetone-and BAA-treated surfaces decreases. The area of the carbon peak is highest on the native surface. Acetone increases the area of the Ca peak by 20%, whereas an enrichment of about 60% is achieved by BAA conditioning. In the

Table 2 X-ray photoelectron spectroscopy results for the interaction of dental enamel with benzoic acid in acetone. The area under the peaks for Ca, P, C, and O is given with respect to the polished, nonconditioned enamel used as the reference surface

Sample	Relative peak area (%) for			
	Ca	P	C	O
Native dental enamel covered with pellicle	−25	−45	80	−4
Enamel conditioned with acetone	20	−2	9	7
Enamel conditioned with benzoic acid	63	23	−5	40

native state the Ca peak intensity is weaker than that of the polished surface because of the attenuation effects due to the outermost organic layer. In comparison to the polished surface, the relative intensity of the phosphorus peak increases on the BAA-treated surface by about 20%.

To elucidate further the interaction of tooth enamel with carboxylic and polycarboxylic acids and related polyelectrolytes (biopolymers) we investigated the topography of the tooth enamel surface by AFM. We used the contact mode of AFM to follow up the formation of salivary protein layers on tooth enamel in vivo. The characteristic surface pattern of a polished enamel specimen not exposed to the oral environment is shown in Fig. 2a. Individual enamel crystallites measuring 30–90 nm in width could be observed by AFM. The enamel crystallites appeared rounded and caused a spotted enamel surface pattern. The appearance of the enamel specimens after intraoral exposure for 60 min towards human salivary changed distinctly (Fig. 2b). The finely patterned surface of the enamel was masked owing to salivary protein adsorption and pellicle formation. The pellicle manifested itself in the AFM pictures as a tightly packed globular surface layer. The finely patterned surface of the enamel is masked owing to salivary protein adsorption and pellicle formation. The adsorbed protein layers reveals a globular-shaped surface structure that is also depicted in the corresponding transmission electron microscopy (TEM) micrograph of the ultrathin section depicted in Fig. 3a. From the AFM picture taken so far we could make a rough estimate of the coil dimensions of the protein molecules in the hydrated surface layer. The average diameter of the globular-like structures varied between 100 and 250 nm. The AFM results confirm previously published SEM and TEM studies showing that the salivary pellicle layer has a globular-like surface texture [2, 6, 9, 10]. The present AFM investigation provides evidence that the native salivary pellicle is mainly composed of globular-shaped protein agglomerates which are not artefacts resulting from

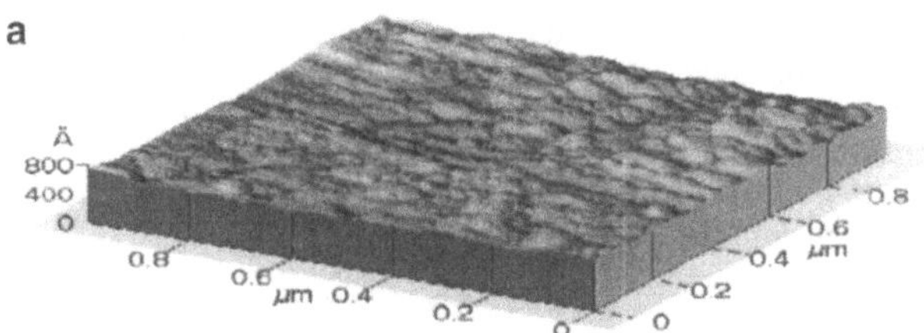

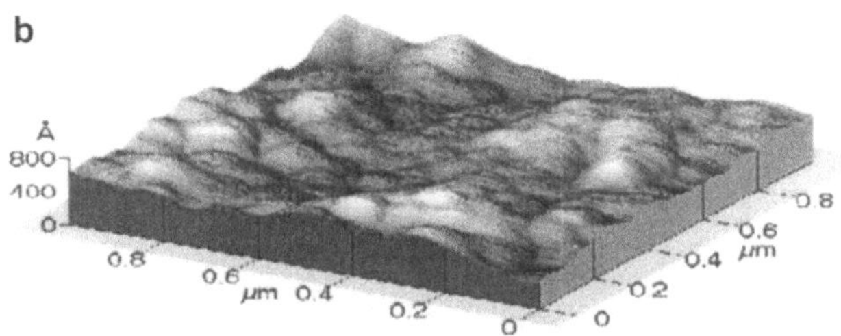

Fig. 2a, b The characteristic surface pattern of a polished enamel specimen as observed by atomic force microscopy (*AFM*) and transmission electron microscopy (*TEM*). **a** AFM picture of reference surface, not exposed to the oral environment. The individual enamel crystallites measure approximately 30–90 nm. **b** AFM picture of enamel sample exposed for 60 min to the oral environment. The finely patterned surface of the enamel is masked owing to salivary protein adsorption and pellicle formation. The adsorbed protein layers reveals a globular-shaped surface structure that is also depicted in the corresponding TEM micrograph (cf. Fig. 3a)

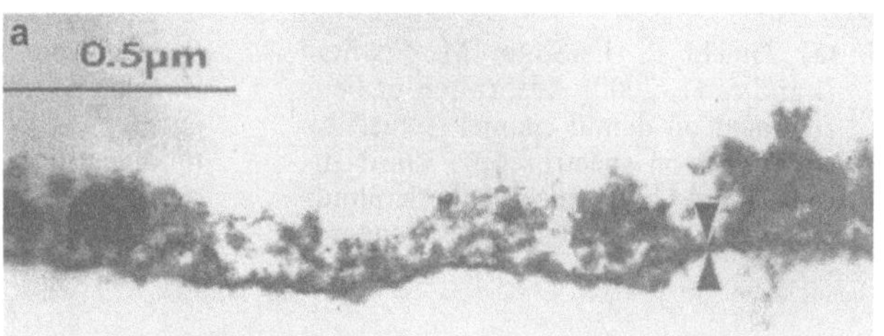

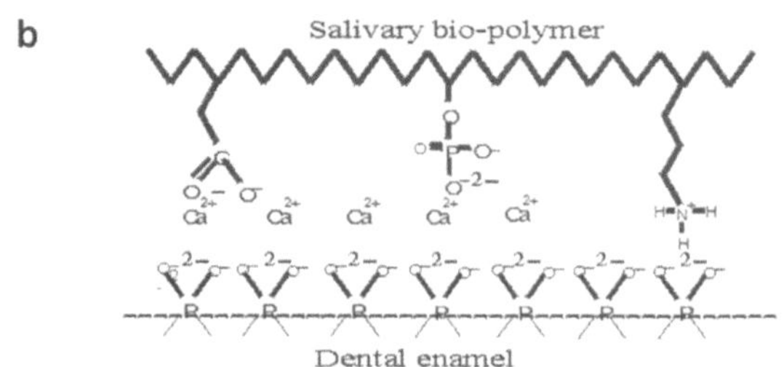

Fig. 3 a TEM picture of an ultrathin section of dental enamel exposed for 60 min to the oral environment. Note the electron dense pellicle basal layer adsorbed directly upon the enamel surface. The enamel was dissolved owing to ethylenediaminetetraacetate treatment prior to ultrathin sectioning. Magnification ×30,000. For details of preparation see Refs. [1, 6]. **b** Model for the orientation of biopolymer molecules on the surface of enamel crystallites

dehydration and agglomeration of condensed salivary proteins.

Summary and conclusion

On the basis of the results of the foregoing section we may speculate about the interaction of polycarboxylic acids and related biopolymers with the tooth surface. After treatment of hydroxyapatite powder with different carboxylic acids in acetone and repeated washing cycles the FTIR spectra of the hydroxyapatite still provided visual means of evidence for adsorbed carboxylic acid molecules. Perhaps the carboxylic acid coordinates with calcium ions located at the surface of the hydroxyapatite particles and adhesion results from the displacement of anions by the carboxylate groups (Fig. 3b). If this chemisorption model for the interaction between hydroxyapatite and carboxylate groups is correct, the use of carboxylic acids with different molecular geometry and electron density distribution enables us to tune the bonding strength between the mineral constituent of tooth enamel. A comparable mechanism is presumably operative for the adhesion and deposition of protein molecules onto the tooth hard substance (cf. Fig. 3a). The so-called pellicle formation is the result of salivary protein adsorption at the tooth–saliva interface [9, 11]. The composition of the surface layer is largely influenced by the structure and polarity of solid substrate (the enamel) and the hydrated biopolymers involved. Traditional electron microscopic techniques do not enable analysis of protein layers and adsorbed polyelectrolyte in their native state. Artefacts resulting from sample preparation as well as dehydration of the specimens in the high vacuum of the electron microscope chamber strongly limit the validity of electron microscopy in studying the initial stages of biofilm formation on dental tissues. AFM allows surface analysis of wet and fully hydrated specimens without prior evaporation or special fixation procedures and a lateral resolution in the lower-nanometre scale. This overview illuminates the usefulness of the combination of different techniques such as SEM, TEM, AFM and FTIR spectrometry in examining interactions at the enamel surface in contact with saliva or polyelectrolyte.

References

1. Wilson AD, Prosser HJ, Powis DM (1983) J Dent Res 62:590
2. Hannig M (1997) Eur J Oral Sci 105:422
3. Es-Souni M, Zimehl R, Fischer-Brandies H (1999) Colloid Polym Sci 277:382
4. Fischer-Brandies H, Scherer R, Theusner J, Häusler K (1992) Fortschr Kieferorthop 53:131
5. Yoshida Y, Van Meerbeek B, Nakayama Y, Snauwaert J, Hellemans L, Lambrechts P, Vanherle G, Wakasa K (2000) J Dent Res 79:709
6. Hannig M (1999) Clin Oral Invest 3:88
7. Hannig M, Herzog S, Willigeroth SF, Zimehl R (2001) Colloid Polym Sci 279:479

8. (a) Zimehl R, Es-Souni M, Fischer-Brandies H (2000) Adsorption of benzoic acid on dental enamel studied by photoelectron spectroscopy, Third International Conference of the Kolloid-Gesellschaft e.V., Budapest, Hungary; (b) Es-Souni M, Fischer-Brandies H, Zimehl R (2000) The role of tooth enamel as a biological interface and its modification: 2. Electron microscopy on dental enamel modified by benzoic acid. The 10th International Conference on Colloid and Interface Science, Bristol, UK
9. Lie T (1975) Arch Oral Biol 20:739
10. Lie T (1977) Scand J Dent Res 85:217
11. Sönju T, Rölla G (1973) Caries Res 7:30

Progr Colloid Polym Sci (2001) 117: 47–50

Zs. Lengyel
R. Földényi

Adsorption of chloroacetanilide herbicides on Hungarian soils

Zs. Lengyel · R. Földényi (✉)
University of Veszprém,
Department of Environmental Engineering and Chemical Technology, 8201 Veszprém,
P.O. Box 158, Hungary
e-mail: foldenyi@almos.vein.hu
Tel.: +36-88-422022
Fax: +36-88-425049

Abstract Acetochlor and propisochlor are chloroacetanilide herbicides that are produced in Hungary. Their prolonged use in agriculture can result in soil contamination. The study of adsorption on different soil types helps predict the fate of these compounds in the environment. Static equilibrium experiments were carried out with Hungarian soils having different pH and humus content (chernozem, brown forest and sandy soil) in 0.1 mol/l phosphate buffer (pH 7). The concentration of chloroacetanilide herbicides was measured by high-performance liquid chromatography in the aqueous phase. The amount of the compound adsorbed on a unit mass of the soils is higher in the case of propisochlor than in the case of acetochlor. The difference between the amounts bonded to chernozem and sandy soil is small. As propisochlor is more hydrophobic than acetochlor, and chernozem has the highest organic content, these results support the idea of hydrophobic interaction between the compounds studied and the soil constituents but it can be explained by the formation of hydrogen bonding as well. The resulting adsorption isotherms consist of two parts that can be assigned to two different Langmuir-type isotherms. This phenomenon can be explained by two causes: either by the two phases of the soils (organic and inorganic phases) or by the interaction of the dissolved material with the first monomolecular layer.

Key words Adsorption · Chloroacetanilide · Herbicide · High-performance liquid chromatography · Soil

Introduction

The solid phase of soil consists of two main fractions having high specific surface area: clay minerals and organic matter. The latter is supposed to coat the soil particles as an organic film, leading to a hydrophobic surface [1, 2].

The soil as an adsorbent can be characterized by means of a model system, by the so-called organoclay. Although the surface of the natural clays is hydrophilic, the metal ion content can be exchanged with large surfactant cations, resulting in a hydrophobic surface [3]. Novel investigations were carried out by Patzkó and Dékány [4] concerning the adsorption of hexadecylpyrimidinium chloride from aqueous solution onto various clay minerals. The structure and the sorption properties of modified silicates were studied in aqueous solutions of primer alcohols [5–7]. The adsorption on *n*-alkylammonium vermiculites and graphitized silica gel was investigated by microcalorimetry and X-ray diffraction. It was pointed out that from propanol–water mixtures the formation of alcohol–water clusters on the surfaces plays an important role [6]. In the case of phenol, cyclohexanol and *n*-hexanol (C_6-ols) a multilayer was formed from their diluted binary aqueous solutions on graphitized carbon black (GCB) [8].

Small adsorbed amounts of aniline and phenol were determined on untreated montmorillonite, but the adsorption of both organics was significantly enhanced after the modification of the clay mineral by cationic surfactant [9]. Mortland [3] pointed out that organoclays are very effective in sorbing nonionic organic pollutants. Many pesticides are nonionic organic compounds that may adsorb on modified clay minerals as well as on the soils owing to their organic matter content leading to hydrophobic surfaces.

The present work studies the adsorption behavior of two chloroacetanilide type herbicides (Fig. 1) on soils. Acetochlor and propisochlor are produced in Hungary and are used in large amounts in agriculture, leading to soil and water pollution. The investigations detailed here are of special importance owing to the frequent occurrence of these nonionic compounds in the environment [10, 11].

Experimental

Materials

Acetochlor (99.3%) and propisochlor (98.2%) standards were provided by Nitrokémia 2000. Potassium dihydrogen phosphate and disodium hydrogen phosphate were obtained from the Reanal Chemical Co. Tetrahydrofuran was purchased from Merck and acetonitrile was from Spektrum 3D. Some properties of the soil samples used in these studies are listed in Table 1.

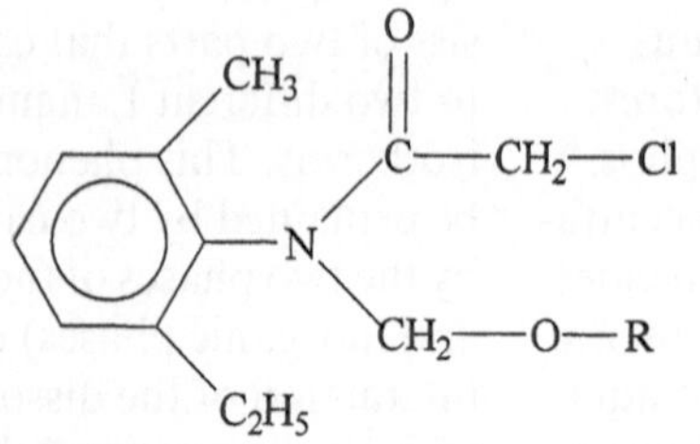

Fig. 1 The structure of the chloroacetanilide type herbicides produced in Hungary: **a** acetochlor, **b** propisochlor

Methods

Static equilibrium experiments were carried out in 0.1 mol/l phosphate buffer (pH 7). Soil samples (7 g) were left to swell in 5 ml distilled water for 24 h at room temperature, then 65 ml herbicide in an appropriate buffer solution (300, 240, 180, 150, 120, 96, 60, 30 μmol/l) was added. The suspension was shaken for 1 h, then separated by centrifuging. The aqueous phase was filtered, then analysed by high-performance liquid chromatography (HPLC). The standard solutions for calibration were the same as were used in the static equilibrium experiments.

HPLC conditions

Chromatographic separations were made by using a PerkinElmer Series 200 HPLC chromatograph equipped with a LiChrospher 100 column filled with 5 μm RP-18 packing material (125 mm × 4 mm) and with a diode array detector, set at 218 nm. Samples, injected using a 20-μl Rheodyne loop, were eluted by a mixture of 50% acetonitrile, 30% water and 20% tetrahydrofuran at a flow rate of 0.7 ml/min.

Results and discussion

The adsorption of acetochlor and propisochlor was studied on three different Hungarian soils. Although gas chromatography is mostly used for the determination of chloroacetanilides [12, 13], these analyses were carried out directly in the aqueous phase by HPLC with excellent reproducibility.

Concerning the acetochlor, the adsorption isotherms on the brown forest and sandy soil are very similar. Under 150 μmol/l equilibrium concentration chernozem bonded higher amounts of acetochlor than the other two soils but over this value the isotherm has almost the same shape (Fig. 2).

Table 1 Mineral composition and structural properties of the soil samples studied

Characteristics	Chernozem (Balatonfőkajár, Hungary)	Brown forest soil (Tés, Hungary)	Sandy soil (Dabrony, Hungary)
pH	7.73	6.35	6.27
Percentage of salt	0.08	0	0
Characteristic number defined by Arany	49	42	24
Percentage of organic matter	4.10	2.20	1.32
Brunauer–Emmett–Teller surface (m^2/g)	21.58	5.89	2.55
Main minerals			
Albite (%)	9.8	12.6	15.1
Chlorite (%)	5.3	6.6	2
Mica (%)	4.2	7.8	2.4
Quartz (%)	30.6	48.4	63.9
Smectite (%)	32.4	14.3	8.3

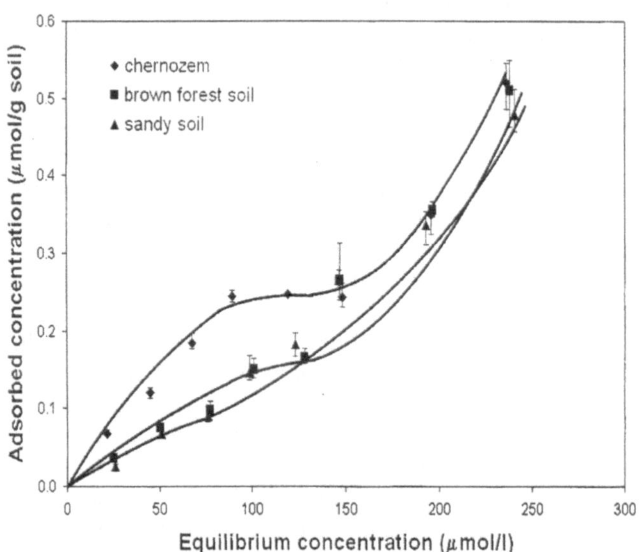

Fig. 2 Adsorption isotherms of acetochlor on three different soils (chernozem ◆, brown forest soil ■, sandy soil ▲) from 0.1 mol/l phophate buffer solution (pH 7.0)

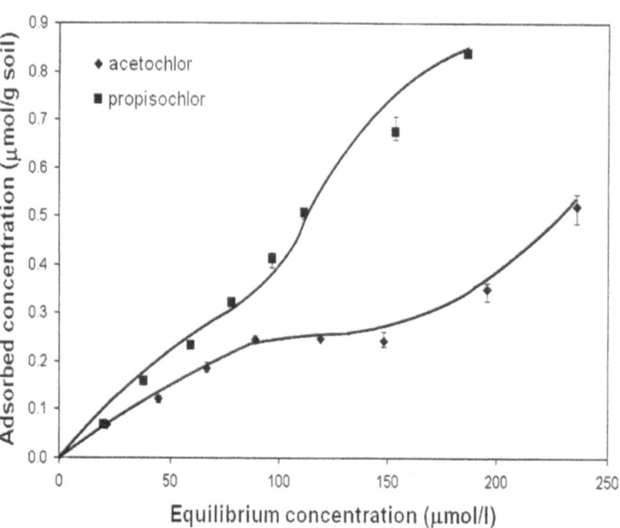

Fig. 4 Comparison of adsorption isotherms of acetochlor (◆) and propisochlor (■) from 0.1 mol/l phosphate buffer solution (pH 7.0) on chernozem

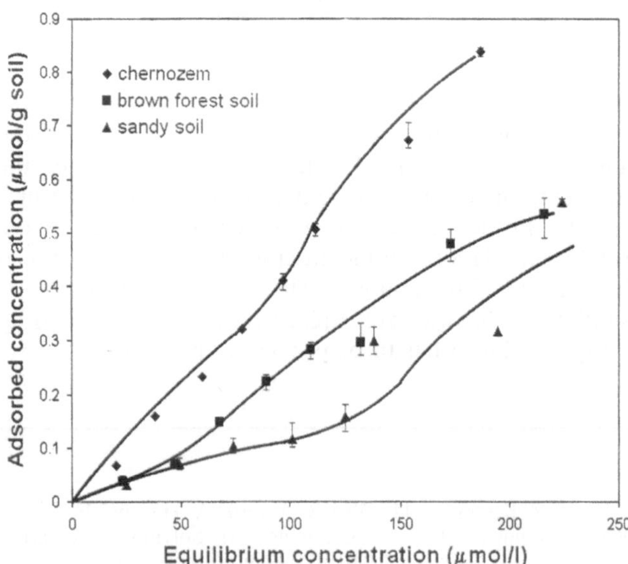

Fig. 3 Adsorption isotherms of propisochlor on three different soils (chernozem ◆, brown forest soil ■, sandy soil ▲) from 0.1 mol/l phophate buffer solution (pH 7.0)

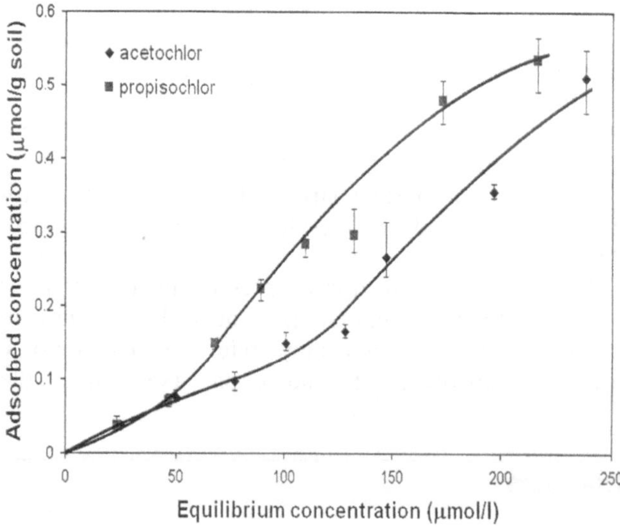

Fig. 5 Comparison of adsorption isotherms of acetochlor (◆) and propisochlor (■) from 0.1 mol/l phosphate buffer solution (pH 7.0) on brown forest soil

In the case of propisochlor, a larger difference can be seen (Fig. 3). Adsorbed amounts per unit mass of the soil decrease in the order chernozem, brown forest and sandy soil, respectively, but the shape of the isotherms is similar having two steps.

A comparison is made between the two chloroacetanilide type herbicides concerning their behaviour on the same soil in Figs. 4, 5 and 6. It can be emphasized that higher amounts of propisochlor than acetochlor are always adsorbed by the soil. Considering the shape of the isotherms, the results will be almost the same as those mentioned for propisochlor alone: the difference between the two isotherms decreases in the order chernozem, brown forest and sandy soil, respectively. In both cases two steps, which refer to two different Langmuir-type isotherms, are present.

The comparison between the various soils shows that both pesticides investigated are adsorbed by a unit mass

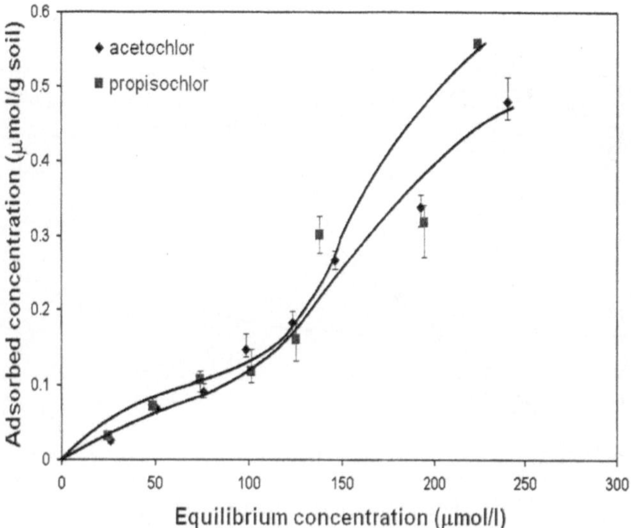

Fig. 6 Comparison of adsorption isotherms of acetochlor (♦) and propisochlor (■) from 0.1 mol/l phosphate buffer solution (pH 7.0) on sandy soil

of the soil in the order: sandy soil < brown forest soil < chernozem.

As chernozem has the highest organic material content it has the largest specific surface, and sandy soil has the smallest. From this point of view the unit surface of sandy soil adsorbs the highest amounts of these chloroacetanilides while chernozem adsorbs the least.

Propisochlor contains larger alkyl group than acetochlor (Fig. 1) and this results in more hydrophobic character.

If the surface of quartz present in the soils is supposed to be nonpolar (see neutral siloxanes in Ref. [14]), these results may refer to hydrophobic interaction between the adsorbent (soil) and the adsorbate (chloroacetanilide) concerning the adsorbed amount by unit mass (chernozem) as well as on the unit surface (sandy soil). Another possible explanation could be the hydrogen bonding between the molecules and the active sites.

The shape of the isotherms can be explained by the previously mentioned ideas. As the soil consist of two fractions (inorganic and organic), it has two main different types of active sites where hydrogen bonding as well as hydrophobic interaction can be responsible for the two steps.

If homogenous active sites are supposed to be on the surface, the hypothesis is in harmony with the working mechanism of the accepted model for the adsorbent coated with an organic film [1–3]. In the case of the adsorption of C_6-ols on GCB the formation of multilayer regimes was proved [8]. The isotherm of *n*-hexanol on GCB had a similar shape to those presented earlier for chloroacetanilides on soils. This agrees with the supposition that in the case of the herbicides investigated a monomolecular layer forms (first step) and subsequently the interaction between this layer and the dissolved molecules results in the second step.

Conclusion

Propisochlor adsorbs in higher amounts than acetochlor on the soils studied. Concerning the unit mass of the soils, chernozem adsorbs the highest amounts of the herbicides investigated – contrarily, concerning the unit specific surface, it can be observed in the case of the sandy soil. Hydrophobic interaction and/or hydrogen bonding could be supposed. The shape of the isotherms is similar and has two steps referring to that described by Langmuir but it has to be proved in the future.

References

1. Chiou CT, Peters LJ, Freed VH (1979) Science 206:831
2. Karichkoff SW, Brown BJ, Scott TA (1979) Water Res 13:241
3. Mortland MM (1970) Adv Agron 22:75
4. Patzkó Á, Dékány I (1993) Colloids Surf A 71:299
5. Regdon I, Király Z, Dékány I, Lagaly G (1994) Colloid Polym Sci 272:1129
6. Marosi T, Dékány I, Lagaly G (1994) Colloid Polym Sci 272:1136
7. Dékány I, Farkas A, Király Z, Klumpp E, Narres HD (1996) Colloids Surf 119:7
8. Király Z, Dékány I, Klumpp E, Lewandowski H, Narres HD, Schwuger MJ (1996) Langmuir 12:423
9. Zhang PC, Sparks DL (1993) Soil Sci Soc Am J 57:340
10. Kolpin WD, Goolsby AD, Thurman ME (1996) Environ Sci Technol 30:1459
11. Visi É, Ambrus Á, Hargitai É, Károly G, Ferenczi M, Solymos E, Berczi B (1998) Monitoring the pesticide residues in surface water in Hungary (presentation), ETECI'98, Budapest
12. Balinova A (1988) J Chromatogr 391
13. Sebök D (1991) Magy-Kém Lapja (in Hungarian) 46:422
14. Johnston CT, Tombácz E (in press) In: Dixon J, Schulze D (eds) Environmental soil mineralogy, chapter 2

Progr Colloid Polym Sci (2001) 117: 51–55

I. Pászli
K. László

Stagnation phenomenon of solid/fluid interfaces

I. Pászli
Department of Colloid Chemistry, Loránt Eötvös University, P.O. Box 32, 1518 Budapest, Hungary

K. László (✉)
Department of Physical Chemistry, Budapest University of Technology and Economics, 1521 Budapest, Hungary
e-mail: klaszlo.fkt@chem.bme.hu
Tel.: 36-1-4631893; Fax: 36-1-4633767

Abstract In this article we focus on the adsorption of binary nonelectrolytes, where we try to find that specific feature of the layer structure which supports the interpretation of experimental data. Our hypothesis, consistent with thermostatical considerations, is related to the criteria of the incidence of the phases and provides a uniform description for both the adsorption from liquid mixtures and electrokinetic interactions.

Key words Adsorption · Excess isotherm · Layer structure · Captation

Introduction

The study of the adsorption from binary nonelectrolytes has led to the conclusion that the interfacial layer cannot be exchanged completely by varying the concentration of the fluid bulk phase.

The unexchangeable adsorbed quantities are more or less analogous with the Stern layer or the stagnation layer joining directly to the solid phase in the case of electrical double layers. This phenomenon significantly influences the properties of the double layer. Its characteristic feature is the assumed constant composition [1], which still requires correct interpretation. The existence of a linear section appearing occasionally in the adsorption isotherms of binary nonelectrolytes and the shift of the apparent adsorption azeotropic composition can be associated with a portion of the interfacial layer with a relatively constant composition. The appearance of the linear section depends on the adsorbent and provides the essential basis of the Schay–Nagy method developed to derive the specific surface area [2]. Therefore, there is an analogy between the adsorption interactions in the interfacial layer and the low-speed electrokinetic processes, etc. Similar behavior was experienced by Dékány and coworkers [3–6] in swelling clay systems. When the adsorption takes place in the interlamellar layer, 10–30% of the liquid sorbed forms a "captive" phase which does not belong to the bulk liquid. If the surface of the clay mineral is hydrophobic enough, the alkyl chains assemble the adsorbed molecules in liquid-crystalline form [7]. All the interactions are independent and the effect is obviously related only to the structural behavior of the equilibrium phase boundary layer.

In this article we focus on the adsorption of binary nonelectrolytes, where we try to find that specific feature of the layer structure which supports the interpretation of experimental data. Our hypothesis, consistent with thermostatical considerations [8], is related to the criteria of the incidence of the phases and provides a uniform description for both the adsorption from mixtures and electrokinetic interactions.

The adsorption experiment and the Ostwald–de Izaguirre equation

The equilibrium state of binary fluids and a solid phase can be established in the so-called adsorption experiment. When contacting the two phases of known quantity, the state parameters change: the homogeneous fluid phase separates into bulk and interfacial phases, with an equilibrium composition of x_1 and x_1^s, respectively, which are supposedly different from the starting composition of the fluid phase (x_1^0). It should be mentioned that x_1^s cannot be measured directly. The sorption accompanies the modification of the homogeneous distribution of the fluid components, which is governed by the fields of the solid and fluid phases interacting. If the variation of the "input" quantities results in a continuous variation of the "output" data, the sorption is, necessarily, a competitive process. The sorption equilibrium can be described by the adsorption isotherm derived, for example, from the mass balance. If the quantities of the solid/fluid system in the experiment are constant, the sorption equilibrium is distinguished only by the spatial distribution of the components of the fluid phase.

The total number of moles in the fluid phase (n^0) is the arithmetic sum of the quantities of the components 1 and 2: $n^0 = n_1^0 + n_2^0$. In the equilibrium state these quantities split into n_i and n_i^s, characteristic of the bulk phase and the adsorption layer (Guggenheim adsorption), respectively. $n = n_1 + n_2$ and $n^s = n_1^s + n_2^s$, where n^s is the so-called adsorption capacity. The ratio of the components can be expressed by the molar fraction as

$$x_i^0 = \frac{n_i^0}{n_1^0 + n_2^0} \quad (1)$$

in the initial state. In the equilibrium

$$x_i = \frac{n_i}{n_1 + n_2} \quad (2)$$

in the bulk phase, while the average composition of the interfacial layer is

$$x_i^s = \frac{n_i^s}{n_1^s + n_2^s} \; . \quad (3)$$

If the system is closed,

$$n^0 = (n_1 + n_2) + (n_1^s + n_2^s) \; . \quad (4)$$

Therefore,

$$n^0 x_1^0 = n_1^0 = \begin{cases} n_1^s + n_1 \\ (n_1^s + n_2^s)x_1^s + (n_1 + n_2)x_1 \end{cases} , \quad (5)$$

$$n^0(x_1^0 - x_1) = [n_1(1 - x_1) - n_2 x_1] + [n_1^s(1 - x_1) - n_2^s x_1] \; , \quad (6)$$

$$(1 - x_1)n_1 - x_1 n_2 = 0 \; , \quad (7)$$

$$(1 - x_1)n^0 x_1^0 - x_1 n^0(1 - x_1^0) \equiv n^0(x_1^s - x_1) \; . \quad (8)$$

The definition of the reduced (or Defay) excess can be given directly from the mass balance:

$$n_1^\sigma = \begin{cases} n^0(x_1^0 - x_1) \\ (1 - x_1)n_1^s - x_1 n_2^s \\ n^s(x_1^s - x_1) \end{cases} . \quad (9)$$

The first expression in Eq. (9) can be taken as an instruction of measurement; the other two relations are the alternative forms of the Ostwad–de Izaguirre equation [9].

The partial domains in the equilibrium fluid phase can be distinguished by the x_i^s and x_i quasilocal compositions; thus, $N_1^{(s)} = n^s x_1^s$ and $N_1^{(fl)}$ are the contributions of component 1 from n^s in the state of equilibrium. According to Eq. (9), the Defay adsorption can be given as

$$n_1^\sigma = N_1^{(s)} - N_1^{(fl)} \; . \quad (10)$$

This quantity (of positive or negative sign) expresses the deviation (in the case of positive sign the excess) of the quantities of the same component in the interfacial and bulk phases, respectively. Using the definition given in Eq. (9) the so-called relative or Gibbs adsorption excess [10] can be given as

$$_G n_1^\sigma(x_1) = \frac{1}{1 - x_1} n_1^\sigma(x_1) = n_1^s(x_1) - \frac{x_1}{1 - x_1} n_2^s(x_1)$$
$$= n_1^s - \frac{n_1}{n_2} n_2^s \; . \quad (11)$$

All the Guggenheim, Defay or Gibbs excesses characterize the same interface in the same state, only their scaling is different; however, only the Defay excess can be calculated from experimental data and that is the reason for its metrological application.

The mass balances and the experimental isotherms

The isotherms may be derived directly from the measured data according to the first relation in Eq. (9). Various isotherms published earlier are collected in Fig. 1 [11–13]. In these examples the fluid phase is always a benzene(1)–ethanol(2) mixture, only the solid phase is different.

The Ostwald–de Izaguirre equation is incapable of describing the isotherms of these systems. That is, when the composition of the interfacial layer formed in the experiment is identical to that of the bulk phase, the excess, in accordance with the instruction of the measurement (first relation in Eq. 9), disappears:

$$n_1^\sigma(x_1) = 0 \; . \quad (12)$$

In these cases the fluid phase is an azeotropic mixture of a quasisingle component. In the case of benzene(1)–ethanol(2) mixtures $x_{ethanol}^{aze} = 0.24$ at 25 °C [14], i.e. at this specific bulk composition the Defay adsorption excess has to disappear in the case of all the systems; however, this condition is not fulfilled in any of the examples, as none of the isotherms exhibit an intersection at this composition. Apparently, n_1^σ and the average layer composition (x_1^s), respectively, are not a single-valued function of the bulk composition. This leads to the

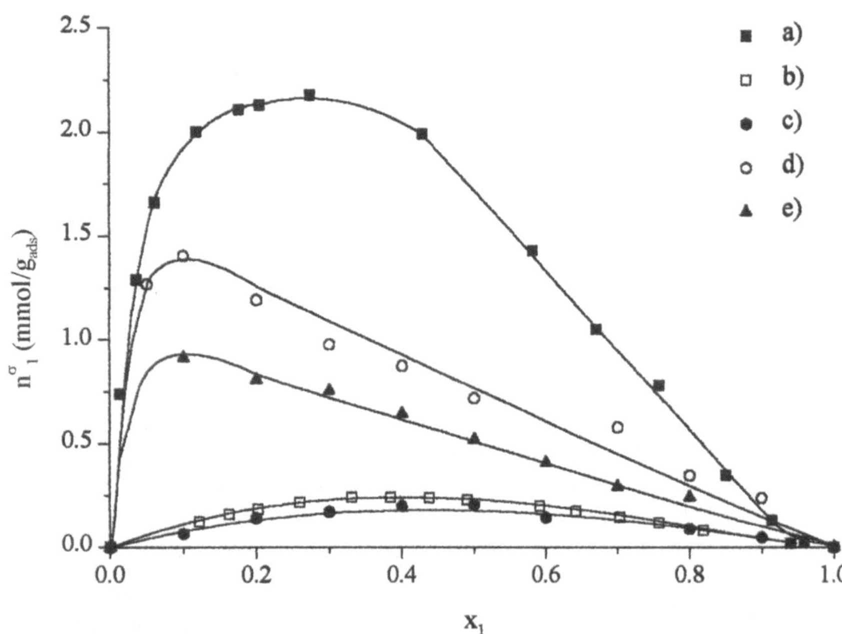

Fig. 1 Selected excess isotherms obtained with benzene(1)–ethanol(2) binary liquid mixtures at 303.15 K. The solid phase is (*a*) activated carbon, BPL (Pittsburgh Activated Carbon) [11], (*b*) graphitized carbon black, Graphon (Cabot) [12], (*c*) graphitized carbon black, Graphon (Cabot) [13], (*d*) silica gel, Cab-O-Sil (Cabot) [13], (*e*) silica gel, Hi-Sil 215 (PPG Industries) [13], respectively

contradiction that the Guggenheim adsorption is apparently not a single-valued function of x_1.

Interpretation of the captation

In order to solve this contradiction, it can be supposed that the molecules in the interfacial layer are adsorbed by two different mechanisms. A part $[N_i^s(x_1)]$ of the adsorbed amount obeys the sorption law, i.e. it varies with the x_1 composition of the equilibrium bulk phase (net adsorption), while the other part (N_i^*) is constant over the whole concentration range. This latter one, the so-called stagnation or captive layer does not take part in the mass transfer. It is possible only if the equilibrium interfacial layer consist of two isolated subsystems which are thermodynamically closed with respect to chemical interactions. Thus, in the fluid phase

$$n_i^S(x_1) = N_i^{(S)}(x_1) + N_i^* \tag{13}$$

and as

$$N_i^\sigma(x_1) = (1 - x_1)N_1^s(x_1) - x_1 N_2^s(x_1) \tag{14}$$

the Ostwald–de Izaguirre equation can be written as

$$n_i^\sigma(x_1) = N_1^\sigma(x_1) + \left[(1 - x_1)N_1^* - x_1 N_2^*\right] \ . \tag{15}$$

Therefore, the gross isotherm $[(n_1^\sigma(x_1)]$ can be given as the total of the net adsorption $[N_1^\sigma(x_1)]$ and the excess of the captivated amount. The latter can be given as

$$N^* = (1 - x_1)N_1^* - x_1 N_2^* = N_1^* - x_1\left(N_1^* + N_2^*\right) \tag{16}$$

where N^* is a linear function of x_1. The net adsorption is a more complex function of x_1, as all the Guggenheim quantities depend on x_1.

The captation of the particles takes place simultaneously with a phase separation in the fluid phase. Both processes imply energy exchange. The resulting state is defined by the relation of the two effects. If the process is influenced by the morphology of the solid phase, the captation may take place only on special structures and may correspond, for example, to a molecular sieve effect.

When the characteristic size or shape of the surface vacancies (holes, cracks, ends of pores, etc.) allows the access of only one of the adsorptive species, as the access of the other one is, for example, sterically hindered, the captivated amount is generally made up of the former component. The lowest energy state may be reached either by the captation of a single component or by the formation of a mixed layer. When mixed layers are formed, the larger particle fills in the larger sites (supposing that this is energetically favorable), while the other sites are covered by the smaller particles. If the energies involved in these interactions are comparable, in most of the cases the captation of the smaller particles is preferred.

The gross and net adsorptions are described by similar formulas; therefore, these quantities can be separated only if the captivated amount can be derived separately. The existence of an adsorption azeotropic point is useful for this purpose: the net adsorption disappears at this composition, and the captivated amount can be derived *post hoc*.

The gross isotherm degenerates at the azeotropic composition (x_1^{aze}), and according to Eq. (15), it can be expressed as

$$n_1^\sigma(x_1^{aze}) = N^*(x_1^{aze}) = \begin{cases} N_1^*(1-x_1^{aze}) - N_2^* x_1^{aze} \\ N_1^*(1-x_1^{aze}) \\ -N_2^* x_1^{aze} \end{cases} \tag{17}$$

The first relation in Eq. (17) applies in the case of the simultaneous captation of both components. If only one of the fluid components is captivated, the captivated amount can be derived from the intercepts of the $n_1^\sigma(x_1)$ versus x_1 captivated diagram.

The captivated excess calculated from Eq. (17) may be subtracted from the gross isotherm at each composition. The difference is characteristic only for the equilibrium fluid phase. The net isotherms derived from the gross isotherms of the different solids characterize only the fluid phase and thus the net isotherms related to the surface area result is a single curve (Fig. 2).

The similarity of the curves confirms the previous interpretation, as the net isotherms derived from the various gross functions coincide, except in the lowest composition range. An analogous behavior has been experienced in the case of other (e.g. methanol–benzene) mixtures.

The shape of the gross isotherm reflects a superimposed effect and can be applied to the empirical estimation of the existence of the constant captivated amounts.

Some consequences of the captivation

If the captivated amount within the interfacial layer – in a given concentration range – significantly exceeds the Guggenheim adsorption, the shape of the gross isotherm is necessarily linear. Alternatively, the condition of the linearity is that the net isotherm is also linear or negligible.

The gross isotherm is linear in the same composition range where the linearity conditions established by Everett are valid [15, 16]. $M_E(x_1)$ defined by Everett can be derived from (Eq. 15) as

$$M_E(x_1) = \frac{x_1 x_2}{n_1^\sigma} = \frac{1}{\frac{N_1^*+N_1^s}{x_1} - \frac{N_2^*+N_2^s}{x_2}} \ . \tag{18}$$

If N_1^s and N_2^s are significantly smaller than the real captivated amounts (i.e. the net sorption can be neglected within the gross sorption), but both components are captivated, then the Everett quantity is

$$M_E(x_1; N_1^*, N_2^*) = \frac{x_1 - x_1^2}{N_1^* - (N_1^* + N_2^*)x_1} \ , \tag{19}$$

i.e, a nonlinear function of x_1. If only one component is captivated, then, with the same conditions,

$$M_E(x_1; N_1^*) \cong \left(\frac{1}{N_1^*}\right) x_1 \ , \tag{20}$$

$$M_E(x_1; N_2^*) \cong \left(\frac{1}{N_2^*}\right) x_1 \ . \tag{21}$$

Both formulas give a linear plot (only the sign of their intercept is different), and the captivated amount can be estimated. Conversely, if the Everett representation is linear, practically only one of the components is captivated, and the consequences are related to the captivated amount and not to the net captivated of marginal importance.

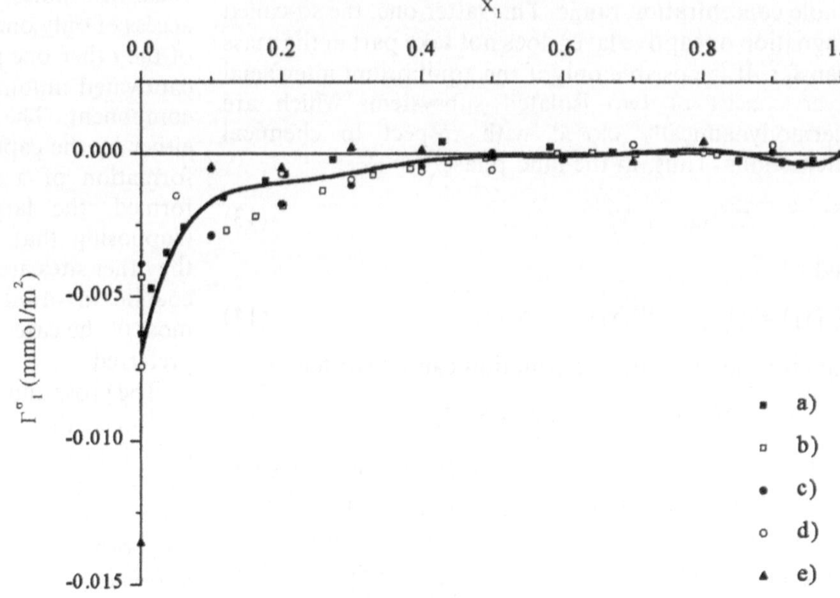

Fig. 2 Net isotherms calculated from the excess isotherms in Fig. 1. The *solid line* is the average derived from the individual plot

In the case of a given binary mixture, the gross adsorption $[n_1^\sigma(x_1)]$ disappears at different compositions, depending on N^*. Thus, the variability of the apparent azeotropic composition can be interpreted by the captivation.

The stagnation behavior of the interfacial layer is also manifested in the constant composition of the "inner plate" of the electrical double layer. The stagnation layer plays a fundamental role in the interpretation of either the zeta-potential and electrokinetic effects or the stability of sols.

Discussion

The aim of this article is to illustrate that some experimental results can be interpreted in an unusual manner as well. The hypothesis applied here can also be useful in establishing methods for surface area determination.

When the cross-sectional area (a_m) of the captivated compound is practically equal to that of the value used in vapor adsorption, the specific surface area of the solid surface can be calculated. The Schay–Nagy extrapolation method is based on this approximation [2]; however, paradoxically, this process is based not on the net but on the gross isotherm. The condition is that the captivated amount and the amount needed for the monomolecular coverage should be calculated from each other. In this way, a surface area equivalent to the one derived from the vapor adsorption by supposing a complete monolayer can be obtained. It should be noted that both methods are problematic for the uncertainty of the cross-sectional area values. The existence of a stagnation layer in the case of solid/vapor interaction needs further study.

The effects studied in this work – without giving a further discussion *hic et nunc* – can also be evaluated by the local thermostatics based on the phenomenological theory. The problem is related to the criteria of incidence of the physical quantities [8].

Acknowledgements This research was supported by the OTKA Fund (Hungary) no. T 025581. I.P. is thankful to the Szántó Ferenc Foundation for their financial support. The technical help of György Bosznai is gratefully acknowledged.

References

1. Lyklema J (2000) Electrokinetics: The dynamics of the stagnant layer (plenary lecture). In: Abstracts of the 10th International Conference on Colloid and Interface Science, 23–28 July 2000, Bristol, UK
2. Schay G, Nagy LG (1961) J Chim Phys 149
3. Dékány I, Weiss A (1990) Mag Kem Foly 96:206
4. Dékány I, Nagy LG (1990) Colloids Surf 58:252
5. Dékány I, Nagy LG (1991) J Colloid Interface Sci 147:119
6. Dékány I (1992) Pure Appl Chem 64:1499
7. Dékány I, Szántó F, Weiss A (1989) Colloids Surf 41:107
8. Pászli I (2000) In: Abstracts of adsorption and nanostructures – from theory to application, 25–28 September 2000, Budapest, Hungary, pp 149–151
9. Ostwald W, de Izaguirre R (1922) Kolloid-Z 279
10. Szekrényesy T, Fóti G (1981) Kem Kozl 55:397
11. Sircar S, Myers AL (1970) J Phys Chem 74:2828
12. Brown CE, Everett DH, Morgan CJ (1975) Trans Faraday Soc 71:883
13. Matayo DR, Wightman JP (1973) J Colloid Interface Sci 44:162
14. Ohe S (1989) Vapor–liquid equilibrium data. Elsevier, Amsterdam
15. Everett DH (1964) Trans Faraday Soc 60:1803
16. Klinkenberg A (1959) Rec Trav Chim 78:83

Progr Colloid Polym Sci (2001) 117: 56–62

M. Mielke
R. Zimehl

Measures to determine the hydrophobicity of colloidal polymers

M. Mielke (✉) · R. Zimehl
Institute of Inorganic Chemistry,
University of Kiel, Germany

Present address: M. Mielke
Unilever Research,
45 River Road,
Edgewater, NJ 07020
e-mail: mark.mielke@unilever.com
Tel.: 201-8402840
Fax: 201-8402180

Abstract The role of latex dispersions as well as water-soluble polymers in industrial and pharmaceutical applications (i.e. analytical and drug delivery systems) is growing fast. An important attribute of polymer particles is the high adsorption capacity for organic moieties (e.g. of molecules of biological origin). It is important that the biologically active molecules remain fully hydrated during the adsorption by the colloidal system; therefore, it is essential to have a measure for the hydration of both adsorbent particles and adsorptive molecules and to correlate its value with other properties of the polymeric material. We have estimated the hydrophobic/hydrophilic balance with adsorption isotherms of polymethine dyes (surface probes). The adsorption of the anionic dyes is 10 times stronger on positively charged polystyrene particles compared to the adsorption of the corresponding cationic dyes on negatively charged particles. We found a redshift in the absorption maximum of the anionic dyes after adsorption onto the positively charged particles. A blueshift was observed for cationic dyes bound to negatively charged particles. We deduce from our measurements that the nature of the surface charges present at the particle–liquid interphase alters the surface characteristics of the particles. We suppose that cationic charged polymer particles are more hydrophobic than particles from the same matrix with negative surface charges. In a qualitative model, we illustrate the governing role of hydration forces on particle aggregation.

Key words Microgel · Dye adsorption · Polymer colloids · Water structure · Hydrophobic hydration

Introduction

There are only a few ways to measure or compare the hydrophobicity of surfaces in disperse systems. Contact-angle measurements cannot be applied to the hydrated colloidal particle in its original dispersion medium. Polymer particles need to be dissolved in an organic liquid to cast the polymer film for the contact-angle measurements. The properties of the dry polymer film certainly differ from the properties of a strongly curved surface; therefore, it is insufficient to determine only the contact angle on macroscopic surfaces with conventional measurements, by either the sessile-drop or the inverted-plate method. For polymer colloids in nonmiscible fluids, depending on the hydrophobicity the particles concentrate in either phase. Over the last few decades, several improvements to the investigation techniques have been suggested, but there is virtually no easy method to provide the desired information.

In the last few years, we have focused our intention on the determination of the so-called surface hydrophobicity of colloidal particles and biopolymers [1]. We were searching for simple and effective methods to measure the extent of hydrophobicity of polymer latices and developed a preferential solvent adsorption method from binary liquid mixtures. In this technique, the colloidal particles can preferentially adsorb either the polar or nonpolar liquid from the liquid mixture. Careful analysis of the adsorption-excess isotherm gave us some useful information about the stagnant layer at the surface of the colloidal particles and the properties of the polymer network as well [14].

In a second set of experiments we investigated the adsorption of hydrophobic yet soluble organic dyes onto different polymer latices. The measurements were conducted using a UV–vis spectrophotometer. The advantage in measuring in the visible spectrum is that impurities (e.g. surfactants, monomers) released by the colloidal particles over time do not disturb the spectrum of the dyes. The shift in the absorption spectra of precipitated polymer dye conjugate and the shape of the adsorption isotherm of the dye provides information about the particles hydrophobicity.

Experimental

Preparation of particles

In general the polymerization was carried out in a 500-ml round-bottomed four-necked flask with stirrer, cooler and nitrogen purge.

Preparation of negatively and positively charged polystyrene

Water (200 ml, double deionized, Seradest) was stirred at 350 rpm and heated to 90 °C under a nitrogen blanket. After 10 min 25 ml, for negatively charged polystyrene, PS(−), or 12.5 ml, for positively charged PS, PS(+), styrene (stabilized with 12 ppm catechol, Hüls) was added. Potassium persulfate (0.25 g), for PS(−), or 0.5 g ADMBA, Wako), for PS(+)], in 25 ml water was added after 5 min. The stirrer was adjusted to 60 rpm. The warm dispersion was filtered through glass wool and after cooling was filtered through a nylon mesh.

Preparation of negatively charged poly(butyl methacrylate)

Water (480 ml, double deionized, Seradest) was stirred at 600 rpm and heated to 80 °C under nitrogen. Butyl methacrylate (BMA, 260 ml, Fluka), sodium dodecyl sulfate (2 g) and acrylic acid (2 ml) were added. After 30 min, 2 g potassium peroxodisulfate in 25 ml water was added. The stirrer was adjusted to 400 rpm. The reaction was stopped after 4.5 h and the latex was filtered.

Preparation of positively charged poly(butyl methacrylate)

Water (480 ml, double deionized, Seradest) was stirred at 600 rpm and heated to 65 °C under nitrogen. BMA (130 ml, Fluka), *N*-ethyl-*N*-hexadecyl-*N*,*N*-dimethyl ammonium bromide (1.3 g, Fluka), acrylamide (1 g) and, after 15 min, ADMBA (1.2 g) were added. The temperature was raised to 70 °C and the reaction started. Bluish opalescence was observed. The temperature was reduced to 60 °C. After 3 h the reaction was stopped and the latex was filtered.

Preparation of negatively charged poly(methyl methacrylate)

Sodium dodecyl sulfate (0.45 g) in water (500 ml, double deionized, Seradest) was stirred at 600 rpm and heated to 80 °C under nitrogen. A 1.3-ml aliquot of a 5.2% ammonium peroxodisulfate solution and 1.5 ml of a 7% potassium phosphate solution were added. One quarter of the monomer mixture as 124.8 ml methyl methacrylate (Fluka) and 4.7 ml ethylene glycol dimethacrylate (Fluka) was added. The rest of the monomer mixture was added over 90 min from an addition funnel. After the complete addition of the monomers over 2h the reaction was stopped and the latex filtered.

Preparation of poly(methyl methacrylate)

N-Ethyl-*N*-hexadecyl-*N*,*N*-dimethyl ammonium bromide (0.6 g) in water (500 ml, double deionized, Seradest) was stirred at 600 rpm and heated to 70 °C under nitrogen. ADMBA (0.13 g) and one quarter of the monomer mixture (125 ml methyl methacrylate and 5 ml ethylene glycol dimethacrylate) was added. The rest was added over 90 min. After 3 h the reaction was stopped and the latex filtered.

The characterization of these particles is consistent with the data published in Ref. [14] and the references therein.

Adsorption experiments

The Rose Bengal method

Rose Bengal (DAB 8, 46, 132–135) is a standard reagent to dye biological samples. The xanthene dye is adsorbed depending on how lipophilic the material is. The molecule is dipositively charged and soluble up to 1 mmol/l. Owing to its hydrophobic nature it is strongly adsorbed by polymer surfaces.

Standards of Rose Bengal (0.6–0.02 mmol/l) were used to measure a calibration curve. These standards (6 ml) were mixed with 1 ml PS latex (2.5 and 6% wt/wt), equilibrated for 3 h and centrifuged. The concentration of Rose Bengal in the supernatant (c_{equil}) was determined using a UV–vis spectrophotometer (Hitachi 2000) and the adsorbed amount, Γ (millimoles per gram), was calculated.

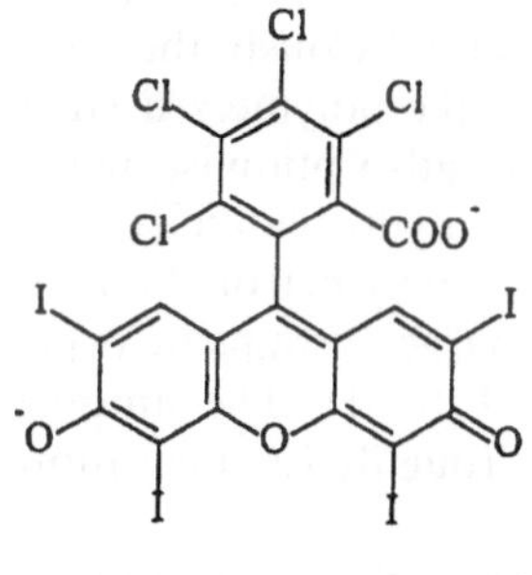

Cl Cl Cl Cl COO⁻ I I ⁻O O O I I

Rose Bengal, di-Sodium-tetrachlortetraiodfluorescein

Adsorption of polymethine dyes

The polymethine dyes were prepared by Dähne and coworkers [3–5]. They are available as both cations and anions. For the measurements, standards of the dyes were prepared and a calibration curve recorded. A 6-ml aliquot of the standards (4–0.0003 mmol/l) was mixed with 1 ml latex (solids: 2.5–6%), equilibrated for 3 h and centrifuged. The concentrations, c_{equil}, in the supernatant were measured with a UV–vis spectrophotometer (Hitachi 2000) and the adsorbed amount, Γ, was determined by subtraction. The short equilibration times were chosen because the lithium and potassium salts of C7(−) are sensitive towards hydrolysis. The measurement of the sample had to be performed quickly after the calibration in order to achieve reproducible data.

Li/K 1,7-Bis(dicyanomethylene) Heptamethin, Abr.: Li, or K C7

K 1, 5-Bis(dicyanomethylene) Pentamethin, Abr: K C5

1,7-Bis(dimethylamino) heptamethinchlorid, Abr: C7 Cl

1,5-Bis(dimethylamino) pentamethinchlorid, Abr: C5 Cl

Results and discussion

Adsorption isotherms of Rose Bengal for several polymer latices are presented in Fig. 1. PS and poly butyl methacrylate (PBMA) latices exhibit a high-affinity isotherm and for PS there is a very steep initial increase, whereas the initial increase for the poly methyl methacrylate (PMMA) latices is really low and the shape of the isotherm is not sophisticated at all. Obviously the Rose Bengal is adsorbed differently by the polymers used in this study. The plateau value of the adsorption isotherm, Γ_{max}, which determines the relative hydrophobicity scale strongly depends on the backbone polymer of the latex particles and on the sign of the charged groups incorporated in the polymer network (Table 1). The amount adsorbed by the PS(−) particles is roughly 1.6 times more than for the PS(+) particles.

To level out variations in the specific surface area of the different latex samples it is necessary to calculate Γ in millimoles per square centimeter. The recalculated values for cationic and anionic PS latices are presented in Fig. 2.

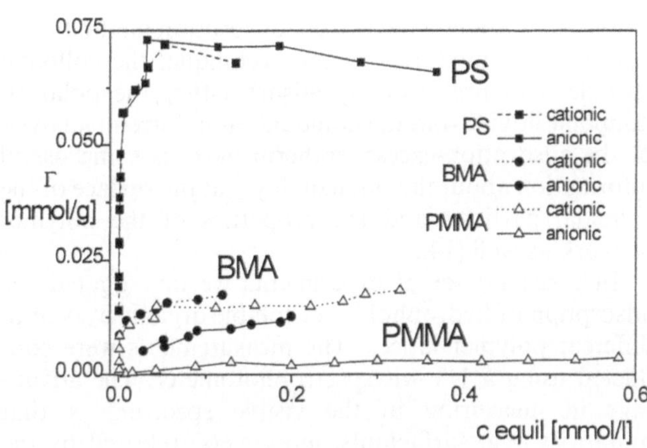

Fig. 1 Adsorption isotherms of Rose Bengal onto polystyrene (*PS*), poly(butyl methacrylate) (*PBMA*), and poly(methyl methacrylate) (*PMMA*) latices at 20 °C

It is evident from Fig. 2 that the plateau value for adsorption of Rose Bengal onto the positively charged polymer is close to the number of surface charges on the PS particles.

The difference in hydrophobicity of PS, BMA and PMMA becomes apparent while cleaning the dispersions extensively. BMA and PMMA latices form gels during cleanup of the colloids by dialysis and turn transparent, while PS dispersions form no gel under any conditions and after dialysis stay turbid. The influence of the cleaning of the Latex particles on their chemical behavior is discussed in reference [14]. This different character of the polymers is clearly revealed by the adsorption isotherms and show that Rose Bengal is suited to document big differences in hydrophobicity. It reconfirms that PS is the most hydrophobic of the systems analyzed. The decay of the adsorption isotherm of PS latices at higher equilibrium concentrations is due to aggregate formation of the dyes [2].

In contrast to the less specific Rose Bengal we wanted to introduce surface probes which are adsorbed highly specifically. For that reason we choose different polymethine dyes. They were prepared by Dähne and coworkers [3–5] and are available as both cationic and anionic forms. The main advantage lies in the fact that the dyes can be used as an adsorptive counterion for the corresponding latex particle. Thus, the effect of repulsive forces between negatively charged PS particles and the anionic Rose Bengal is leveled out. Despite the fact that Dähne and coworkers [3–5] also prepared C3 bodies the dyes used in our study had a C5- and

Table 1 Plateau value, Γ_{max}, of Rose Bengal with different latices

Latex	PS(+)	PS(−)	PBMA(+)	PBMA(−)	PMMA(+)	PMMA(−)
Γ_{max} (mmol/g)	0.072	0.072	0.018	0.012	0.017	0.0037

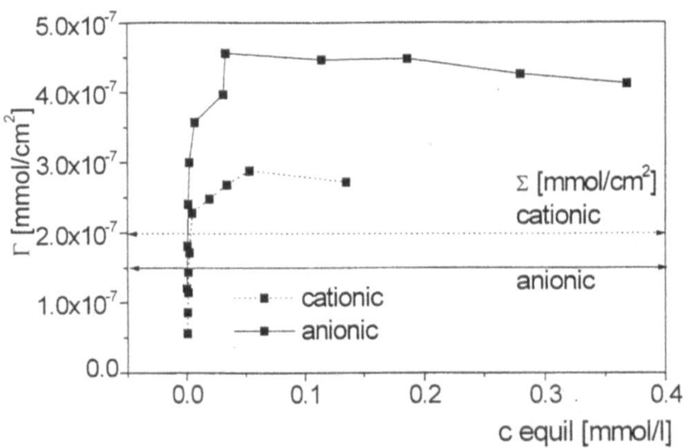

Fig. 2 Adsorption of Rose Bengal onto PS latices at 20 °C (Γ/mmol/cm²). The *double arrows* show the surface charge amount, Σ, of the PS particles

C7-carbon skeleton. The adsoprtion data obtained with the C7 and C5 dyes were highly reproducible. The isotherms obtained from PS latices as adsorbents for C7 and C5 dyes used as counterions are shown in Fig. 3. The plateau values, Γ_{max}, are very different. Now the difference between PS(+) and PS(−) with C7 and as well as with C5 becomes apparent. The difference in adsorption is almost 1 order of magnitude. The cationic PS(+) latex particles are the most hydrophobic and C5 is adsorbed less than C7. The adsorption capacity of PS(−) particles, i.e. the saturation concentration, Γ_{max}, at the plateau of the adsorption isotherm, is roughly 1.6 times bigger than the capacity of the PS(+) particles (Table 2).

After recalculation of Γ to millimoles per square centimeter, the differences between PS(−) and PS(+) are still prominent enough to eliminate the influence of particle size on Γ (Figs. 4. 5). The Γ value of C7(−) with PS(+) is higher by a factor of 16 and the one obtained with C5(+) is 10 times higher than the particle charge amount.

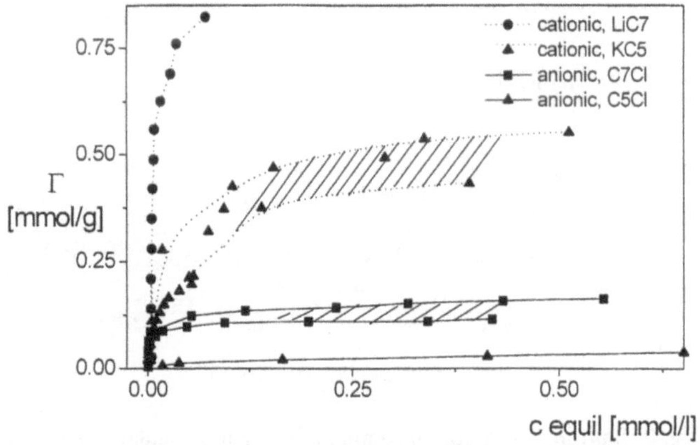

Fig. 3 Adsorption isotherms of polymethine (C7 and C5) onto PS latices at 20 °C. PS particles and dye molecules are oppositely charged

Table 2 Saturation concentration, Γ_{max}, of C7 and C5 polymethine dyes onto polystyrene latices

Latex	PS(+), C7(−)	PS(+), C5(−)	PS(−), C7(+)	PS(−), C5(+)
Γ_{max} (mmol/g)	0.83	0.48	0. 14	0.038

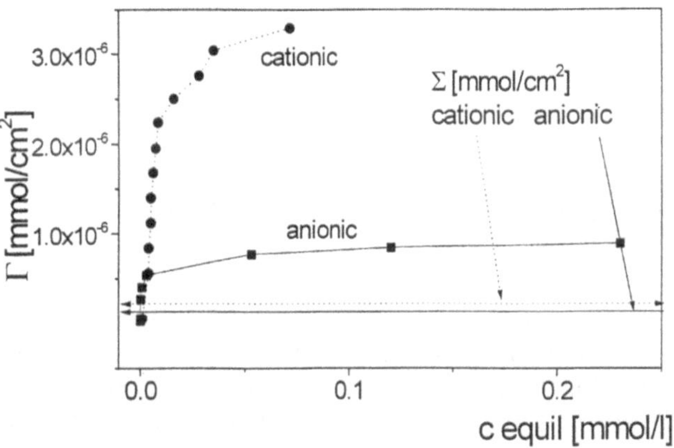

Fig. 4 Adsorption of C7 onto PS latices at 20 °C (Γ/mmol/cm²). The *double arrows* show Σ of the PS particles

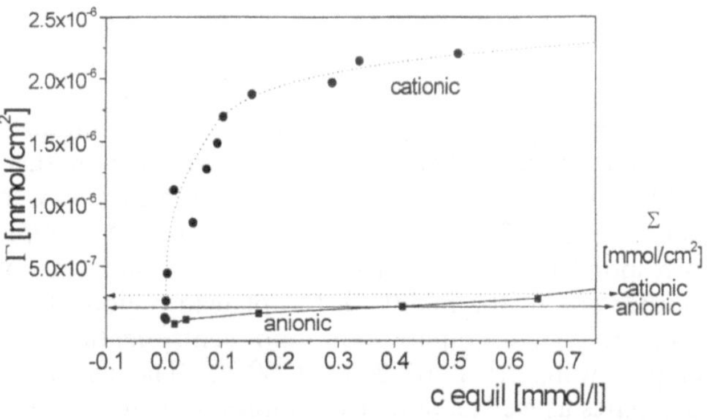

Fig. 5 Adsorption of C5 onto PS latices at 20 °C (Γ/mmol/cm²). The *double arrows* show Σ of the PS particles

The PS(−) particles adsorb 9 times more of the cationic dye C7(+) than it has charges. On the other hand, the plateau value of the adsorption isotherm of C5(+) with PS(−) coincides with the number of surface charges, Σ, of the latex particles. The polymethine dyes seem to be suitable as surface probes and may scan hydrophobic surfaces in the aqueous phase.

Isotherms obtained from microgels (BMA, PMMA) as adsorbents and C7 and C5 dyes used as counterions are shown in Fig. 6. The difference in Γ_{max} between positively and negatively charged latices is a factor of 2 (PMMA) and 4 (PBMA). The adsorbed amounts are low even for C7 owing to the hydrophilic character of the microgels. This is the reason why no data could be

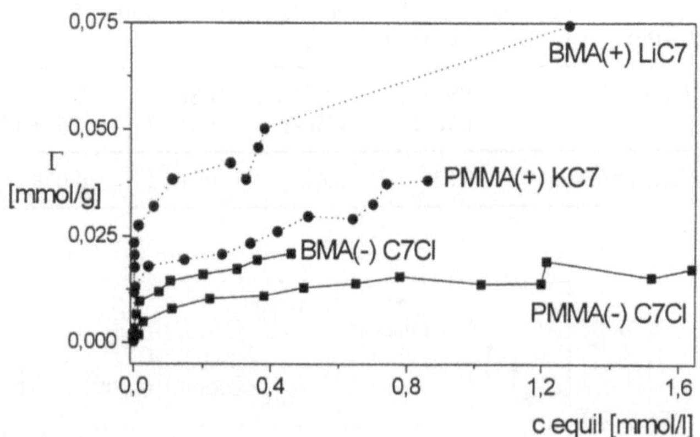

Fig. 6 Adsorption of polymethine (C7 and C5) onto PBMA and PMMA latices at 20 °C

achieved with C5. The curve of PMMA(−) versus C7(+) ($\Gamma_{max} = 0.018$) shows how sensitive the photometric method is (cf. Table 3). The sensitivity of the experimental setup is further supported by the high extinction coefficient of the polymethine dyes (Table 4).

Polymethine dyes as surface probes

To gain further insight into the adsorption mechanism and the molecular arrangement in the adsorbed dye layers, we prepared latex sediments from the polymer dispersions by centrifugation. The samples contained coagulated latex particles with the dyes adsorbed on their surface. The dye concentration was chosen in a way so that the adsorbed amount of dye of one sample corresponded to the beginning and the amount of another one to the middle of the plateau region of the isotherm. After equilibration and centrifugation the samples were dried for 24 h at 50 °C. The dye–latex coagulates as well as pure dye samples were ground and pressed in KBr. The adsorption maxima of the samples were measured using a UV–vis spectrophotometer (Cary 5, Varian).

Table 3 Saturation concentration, Γ_{max}, of C7 and C5 polymethine dyes onto microgels

Latex	PBMA(+), C7(−)	PBMA(−), C7(+)	PMMA(+), C7(−)	PMMA(−), C7(+)
Γ_{max} (mmol/g)	0.074	0.02	0.038	0.018

Table 4 Extinction coefficient, ε, of the polymethine dyes [3]

Polymethine	C7(+)	C7(−)	C5(+)	C5(−)
ε (l/mol × cm)	189,000	179,000	117,000	148,000

Some results of the spectrophotometric measurements are summarized in Fig. 7. The absorption maximum of the anionic polymethine dyes C7 and C5 adsorbed on cationic PS particles is shifted to the red region of the absorption spectrum. In contrast, the spectrum of the C7 cation adsorbed to anionic PS particles shows a blueshift. All the other polymers used exhibit a blueshift of the absorption maximum of the dyes. This phenomenon is well known from adsorption measurements of cationic dyes on clay minerals [6–8]. A blueshift is expected when dyes with delocalized electrons form dimers in solution. Charge separation elevates the energy level of the $\pi - \pi^*$ transition in the polymethine molecule [9]. A redshift can only be observed when the polymethine dyes are present in a nonpolar medium [5] or after adsorption on a nonpolar surface [3]. In a nonpolar environment both mesomeric structures of the polymethine dyes are energetically equal and the $\pi - \pi^*$ transition is lowered. Only the PS(+) particles are hydrophobic enough to lower the electron potential in the dye molecule.

Conclusions

To summarize our results with dye adsorption measurements we note that the more hydrophilic microgel latex particles adsorb less dye compared to their hydrophobic PS counterparts. However, all the positively charged samples (PS, PBMA, PMMA) showed a significantly higher absorbed amount of dye than the corresponding negatively charged polymer of the same matrix. The

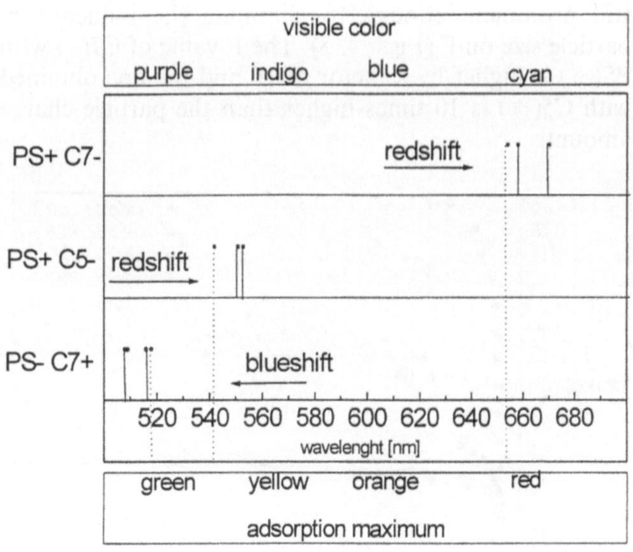

Fig. 7 Shift in absorption maximum of dye/latexcoagulates. Absorption maximum of pure dye in KBr (*dotted line*); absorption maximum coagulates at two different dye concentration in the plateau region (*solid lines*)

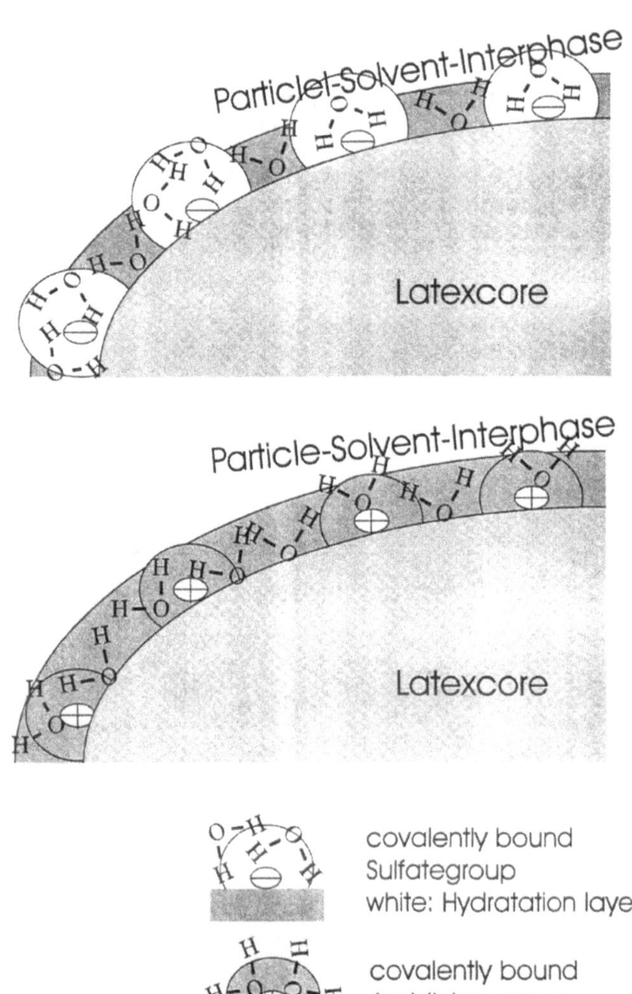

Fig. 8 Simple hydration model of a charged polymer surface. The covalently bound charges are surrounded by a water layer. *Grey*: water molecules point with hydrogen atoms towards the bulk phase (hydrophobic hydration); *white*: hydrogen atoms of water molecules point towards the negative charge

difference in Γ_{max} between positively and negatively charged particles is a factor of 6 for PS, 4 for PBMA and 2 for PMMA. Thus we assume that positively and negatively charged latex particles are surrounded by a different particle–solvent interphase (Fig. 8).

This simple model is further supported by earlier adsorption studies on the adsorption of tetraphenylphosphonium and tetraphenylborate ions onto PS latices and ion-exchange beads [2, 10]. When positively charged adsorbents adsorb negatively charged organic counterions, type 2 and type 3 isotherms (no saturation adsorption) are likely to occur. Negatively charged surfaces or particles with wel-hydrated poly(ethylene glycol) brushes and positively charged counterions exhibit type 1 isotherms with an adsorption maximum and a decay of the isotherm at higher ion concentrations [11, 12].

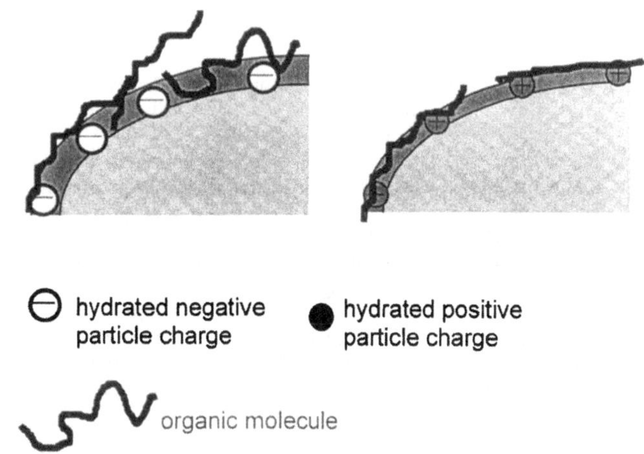

Fig. 9 Coagulates from PS particles and adsorbed polymethine dyes as counterions

We conclude that with hydrophobic latices, like PS, the hydrophobic/hydrophilic balance governs the particle's behavior mainly in the particle–solvent interphase. The water molecules around the positive surface charges are similar to the water structure adjacent to a hydrophobic surface. Positive charges fit well into the hydrophobic hydration water layer, whereas negative surface charges behave oppositely. The water molecules are forced by the negative charge to orient in a reverse manner, which leads to a more random water structure. This structure is comparable to the interphase of swollen, more hydrophilic polymers. When organic molecules are adsorbed onto positively charged polymer particles, the systems gain more entropy compared to a system with negatively charged polymers as adsorbents surrounded by an unorganized water structure.

Fritz et al. [13] studied a new class of drug carrier for oligodeoxyribonuclotides. The particles carried positive groups (amidinopropane) and showed high drug load capacity. It is interesting that the organic adsorptives were immobilized and were not attack by nuclease P1 enzyme in contrast to unadsorbed molecules. These results of Fritz et al. are in good agreement with our water structure model (Fig. 8).

Owing to the surface morphology of the positively charged particles, which are compact and contain less water, organic molecules can get closer to the particle surface and bind. A redshift is observed. Negatively charged particles carry a more polar and swollen interphase. Organic molecules get into poorer contact with the particle surface. Also less water is set free by hydrophobic interaction. The polarity of the particle–solvent interphase results in a blueshift of the adsorbed dye molecules.

Acknowledgements We are grateful to L. Dähne for preparation of the dyes and to G. Lagaly for helpful advice.

References

1. Mielke M, Zimehl R (1998) Ber Bunsenges Phys Chem 102:1698–1704
2. Priewe J (1994) Thesis. University of Kiel
3. Dähne L, Biller E (1998) Adv Mater 10:241–245
4. Dähne L, Horvath A, Weiser G, Reck G (1996) Adv Mater 8:486–490
5. Dähne L, Reck G (1995) Angew Chem 107:735–737
6. Grauer Z, Grauer GL, Avnir D, Yariv S (1987) J Chem Soc Faraday Trans 83:1201–1207
7. Yariv S, Ghosh DK, Hepler LG (1991) J Chem Soc Faraday Trans 87:1201–1207
8. Yariv S (1988) Int J Trop Agric IV:1–19
9. Klessinger M, Michl J (1989) Lichtabsorption und Photochemie organischer Molekuele. VCH, Weinheim
10. Mielke M, Lagaly G, Zimehl R (1997) Stability of colloidal polymers: aggregation of cationic and anionic latex particles with organic gegenions. Proceedings of the 7th Conference on Colloid Chemistry 200–203, Budapest
11. Hiemenz PC (1977) Principles of colloid and surface chemistry. Dekker, New York
12. Rouquerol F, Rouquerol J, Sing K (1999) Adsorption by powders and porous solids. Academic, London
13. Fritz H, Maier M, Bayer E (1997) J Colloid Interface Sci 195:272–288
14. Mielke M (1999) Thesis. University of Kiel

Progr Colloid Polym Sci (2001) 117: 63–69

A. Farkas
I. Dékány

Interlamellar adsorption of organic pollutants on hydrophobic vermiculite

A. Farkas · I. Dékány (✉)
Department of Colloid Chemistry and Nanostructured Materials Research Group of the Hungarian Academy of Sciences,
University of Szeged, Aradi v.t.1, 6720 Szeged, Hungary

Abstract The adsorption isotherms of nitrobenzene, 2-chlorophenol and 4-chlorophenol from water on hydrophobic vermiculite were studied. The excess isotherms were obtained by the immersion method. The basal spacings of the hydrophobic vermiculite organocomplexes in the adsorption equilibrium were determined by X-ray diffraction. By combining these two independent methods, the composition and the structure of the interlamellar space were evaluated. The free enthalpy of adsorption and the adsorption capacity were calculated by analyzing the adsorption isotherm on the basis of the Gibbs equation and by the Everett–Schay method.

Key words Adsorption · Hydrophobic clays · Organic pollutants · X-ray diffraction

Introduction

The adsorption of organic molecules on soil components – particularly on clay minerals – and their organophilized derivatives has attracted much interest lately [1–6]. Clay minerals are the most reactive soil components which, owing to their large specific surface area and ion-exchange capacity adsorb different organic pollutants [7–10]. In the last few years studies became increasingly focused on solutions of water with organic molecules which are very poorly miscible with water, not least because of their outstanding environmental significance [11–13]: these solutions are excellent models for water polluted with toxic aromatic compounds The exact description of their adsorption on clay minerals is therefore essential for environmental protection. The experimental results will open the way for the development of new procedures for the immobilization of pollutants in static and flowing systems.

The surface polarity of clay minerals is altered by hydrophobization via treatment with cationic surfactants. As a consequence, they become capable of adsorbing organic components very effectively [9, 10, 14]. The concentration of organic molecules in the interlamellar space enclosed by the silicate lamellae is calculated from the adsorption-excess isotherm. Knowledge of experimentally determined excess isotherms also allows the calculation of some of the thermodynamic properties of the adsorption layer at the solid/liquid interface. Using the Gibbs equation [15, 16] the free enthalpy of adsorption may be determined and combined with the excesses to calculate adsorption capacities [6–8], which, in the case of adsorption in dilute solutions, are often impossible to determine from the Langmuir adsorption isotherm. Changes in interlamellar distances owing to the interlamellar adsorption are monitored by X-ray diffraction (XRD) measurements. Values of basal spacing determined by XRD make possible the calculation of the volume of the interlamellar space and – knowing the amount and volume of alkyl chains and using the excess isotherms – also its composition over the entire concentration range, Thus, combining the results obtained by two independent methods yields detailed information on the adsorption equilibrium and the structure of the adsorption layer.

Determination of the adsorption capacity from adsorption isotherms of dilute solutions

Adsorption on a solid/liquid interface is a displacement process. Consequently the positive adsorption of one component (i.e. the increase in its concentration in the interfacial layer) is accompanied by the negative adsorption of the other. On the basis of the relationships postulated by the Gibbs model of solid/liquid interfaces, the reduced excess amount normalized to unit mass of adsorbent ($n_1^{\sigma(n)}$) and the Ostwald–de Izaguirre equation derived from the material balance of the components are formulated as [15, 16]

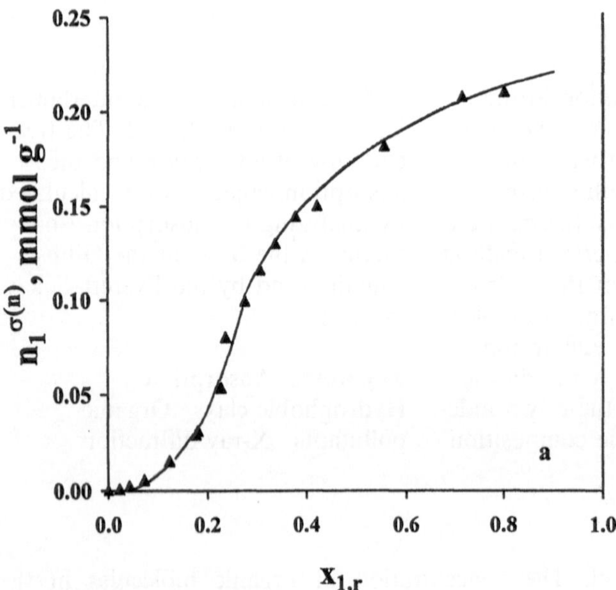

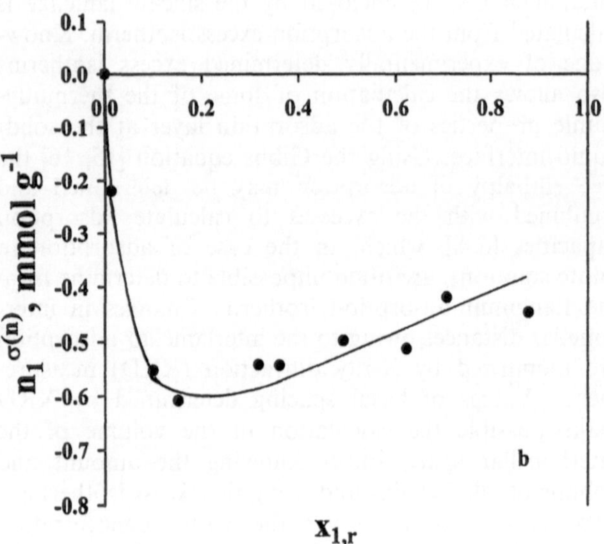

Fig. 1 **a** Adsorption-excess isotherms of nitrobenzene from water on *n*-octadecylammonium vermiculite (*C_{18}-vermiculite*) (◆). **b** Adsorption-excess isotherms of nitrobenzene from water on sodium montmorillonite (●)

$$n_1^{\sigma(n)} = n^0(x_1^0 - x_1)/m = n^0\Delta x_1/m \ , \quad (1)$$

$$n_1^{\sigma(n)} = n_1^s x_2 + n_2^s x_1 = n_1^s - n^s x_1 = n^s(x_1^s - x_1) \ , \quad (2)$$

where n^0 is the total material amount of the liquid, x_1^0 and x_1 are the initial and equilibrium molar fraction, respectively, of component 1 in the liquid phase, m is the mass of the adsorbent, $n^s = n_1^s + n_2^s$ is the material content of the adsorption layer and $x_1^s = (1 - x_2^s) = n_1^s/n^s$ is the equilibrium molar fraction of component 1 in the adsorption layer. If component 1 is preferentially adsorbed as compared to component 2, the excess, $n^{\sigma(n)}$, approximately equals the material amount, n_1^s[17]. In dilute solutions, in the case of preferential adsorption of the dissolved material the volume of the dissolved material in the adsorption layer is approximately $V_1^s = n_1^s V_{m,1} \approx n_1^{\sigma(n)} V_{m,1}$ [7, 8]. The adsorption capacity of pure component 1 is

$$n_{1,0}^s = n_1^s + rn_2^s = n^s x_1^s + rn^s x_2^s \ , \quad (3)$$

where $r = V_{m,2}/V_{m,1} = n_{1,0}^s/n_{2,0}^s$, i.e. the ratio of the molar volumes of the pure components. After the introduction of the separation factor, $S = x_1^s x_2/x_2^s x_1$, the equation $n^s = n_{1,0}^s(Sx_1 + x_2)/(Sx_1 + rx_2)$ is obtained for the material content of the adsorption layer. The value of $n_{1,0}^s$ may be determined from the slope and the intersection of the Everett–Schay linear representation [15, 16], i.e.

$$\frac{x_1 x_2}{n_1^{\sigma(n)}} = \frac{1}{n_{1,0}^s}\left[\frac{r}{S-1} + \frac{S-r}{S-1}x_1\right] \ . \quad (4)$$

Using the integrated form of the Gibbs equation describing solid/liquid interfaces [15–17], the free enth-

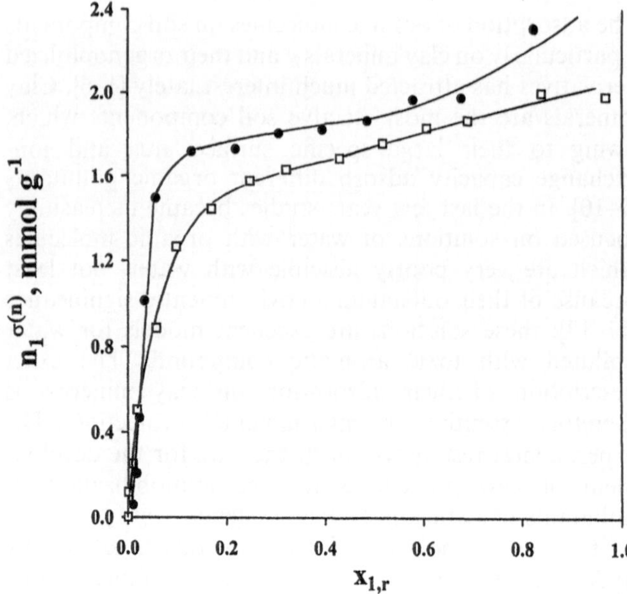

Fig. 2 Adsorption-excess isotherm of 2-chlorophenol (●) and 4-chlorophenol (□) from water on C_{18}-vermiculite

alpy of adsorption may be formulated in the following way:

$$\Delta_{21}G = -RT\int_0^{x_1}\frac{n_1^{\sigma(n)}}{(1-x_1)x_1}\left(1+\frac{\mathrm{d}\ln\gamma_1}{\mathrm{d}\ln x_1}\right)\mathrm{d}x_1 \ . \qquad (5)$$

The index 21 indicates that the molecules of component 2 are exchanged for those of component 1 and γ_1 is the activity coefficient of component 1 in the bulk phase. In ideal solutions or in solutions considered to be ideally dilute (like our systems), the value of the parenthetic term containing the derivative of the activity coefficient is 1. Using the molar free enthalpies of the components (g_i^s), the equation $\Delta_{21}G = n_1^s(g_1^s - g_2^s/r)$ holds for the entire concentration range. After combination with Eqs. (2) and (5) and appropriate rearrangements, the following relationship is obtained [7]:

$$\frac{\Delta_{21}G}{n_1^{\sigma(n)}} = \Delta_{21}g + n^s\Delta_{21}g\frac{x_1}{n_1^{\sigma(n)}} \ . \qquad (6)$$

The linear representation of the equation yields the molar free enthalpy of adsorption, $\Delta_{21}g$, and the material

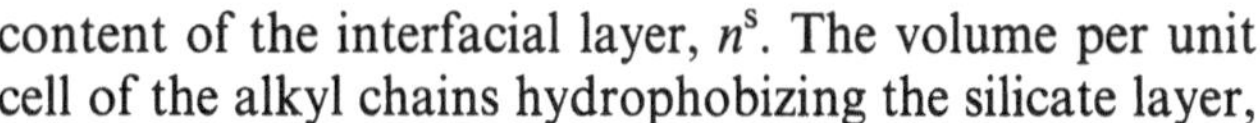

content of the interfacial layer, n^s. The volume per unit cell of the alkyl chains hydrophobizing the silicate layer,

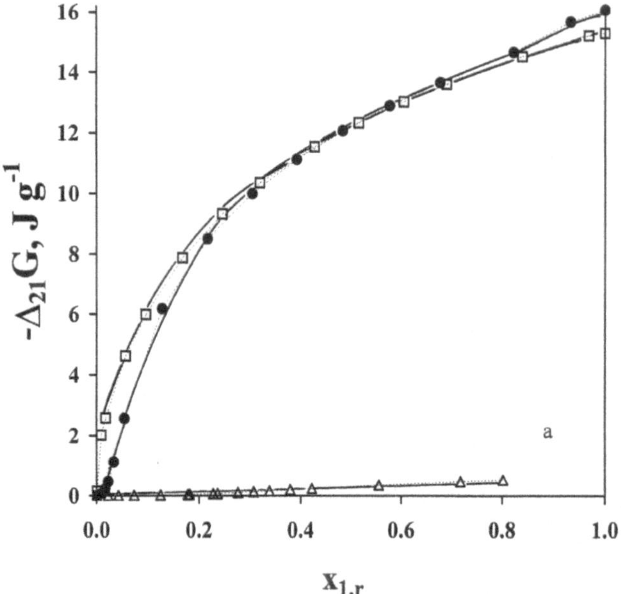

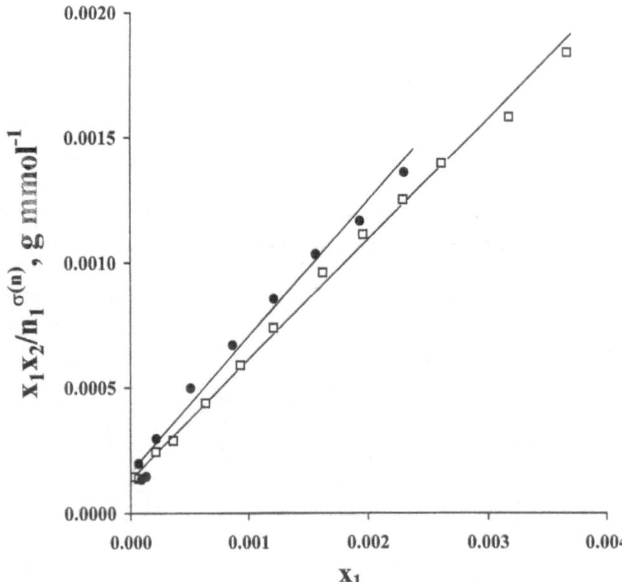

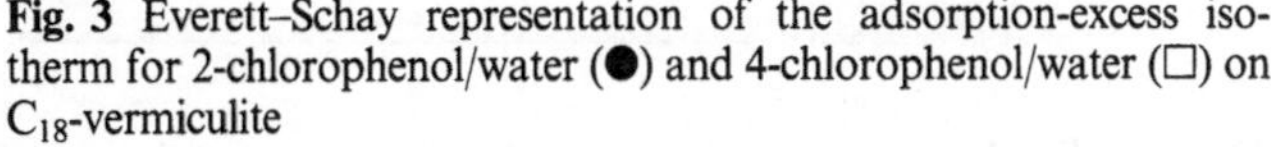

Fig. 3 Everett–Schay representation of the adsorption-excess isotherm for 2-chlorophenol/water (●) and 4-chlorophenol/water (□) on C_{18}-vermiculite

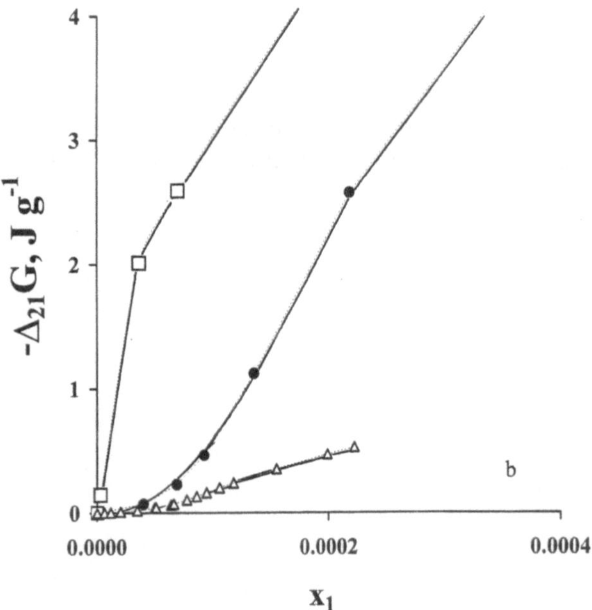

Fig. 4 **a** Free enthalpy of adsorption of 2-chlorophenol (●), 4-chlorophenol (□) and nitrobenzene (△) from aqueous solutions on C_{18}-vermiculite plotted against $x_{1,r}$. **b** as **a** for $x_{1,r}$ < 0.0004

Table 1 Adsorption capacity from the Everett–Schay representation, the results of combination of adsorption-excess and free-energy data and the distribution ratio, S_{sep}

Pollutant	$n_{1,0}^s$ (mmolg^{-1}) (Eq. 4)	n^s (mmolg^{-1}) (Eq. 6)	$-\Delta_{21}g$ (Jmmol^{-1}) (Eq. 6)	S_{sep} (x_1 = 0.0002)	S_{sep} (x_1 = 0.002)
Nitrobenzene	0.72	–	–	90	–
2-Chlorophenol	2.09	1.08	4.89	600	70
4-Chlorophenol	2.12	0.70	5.74	400	70

in units of cubic nanometers per $(Si,Al)_4O_{10}$, for *n*-alkylammonium vermiculite as adsorbent is

$$V_{alk} = \{0.205[0.127(n_{CC} + n_{CN}) + 0.28]\}\xi \ . \tag{7}$$

The term in parenthesis expresses the length (in nanometers) of the alkyl chains attached to the surface (n_{CC} and n_{CN} are the number of carbon–carbon and carbon–nitrogen bonds). The cross-sectional area of the all-trans alkyl chain is taken as 0.205 nm^2. The number of charges per formal unit (surface charge density, $\xi = 0.76$) was obtained from the amount of organic material determined by thermogravimetric analysis. The thickness of the silicate layer is 0.94 nm and the surface area of the formal unit is 0.2475 nm^2; thus, the interlamellar volume is

$$V_{int} = 0.2475(d_L - 0.94) \quad [nm^3/(Si, Al)_4O_{10}] \ . \tag{8}$$

The volume of the "free" interlamellar space is the difference $V_{int} - V_{alk}$.

$$V_{int} - V_{alk} = V_1^s + V_2^{s,b} \ , \tag{9}$$

$$1 = \Phi_{alk} + \Phi_1^s + \Phi_2^{s,b} \ , \tag{10}$$

where $V_2^{s,b} = V_2^s + V^b$ is the combined volume of the water molecules present in the adsorption layer and the so-called bulk liquid filling the adsorption space [5]. The latter does not participate in the formation of the adsorption layer; it is, however, controlled by the forces of adsorption, like all molecules in the interlamellar space. Owing to preferential adsorption of the dissolved material, V^b consists mostly of water, which is the reason why it may be combined with V_2^s. Division of Eq. (9) by V_{int} and subsequent rearrangement yield

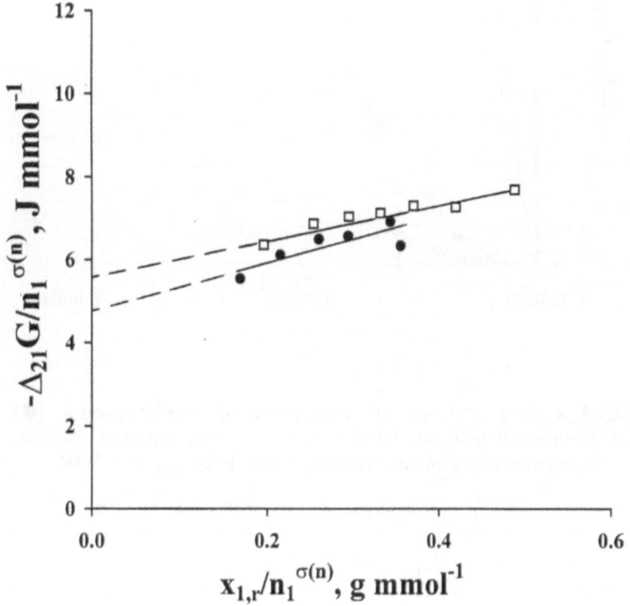

Fig. 5 Combination of adsorption-excess and free-enthalpy data in 2-chlorophenol/water (●) and 4-chlorophenol/water (□) on C_{18}-vermiculite

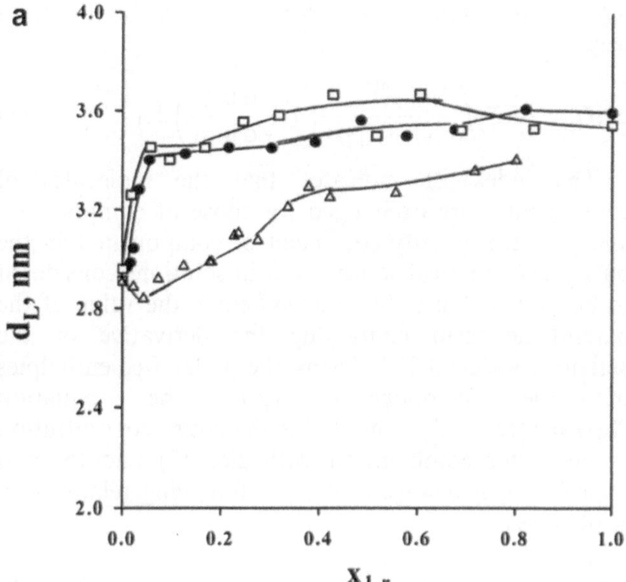

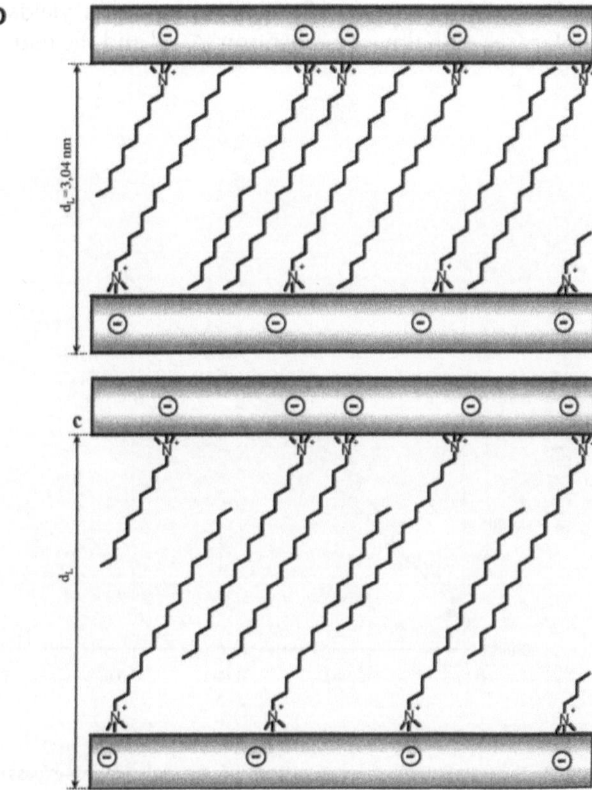

Fig. 6 **a** Basal spacing of C_{18}-vermiculite in aqueous solution of 2-chlorophenol/water (●), 4-chlorophenol/water (□) and nitrobenzene/water (△). **b** Structure of C_{18}-vermiculite with an alkylammonium monolayer with aromatic components. **c** Structure of C_{18}-vermiculite with an expanded monolayer with aromatic components

Table 2 Basal spacings of *n*-octadecylammonium vermiculite in water and in different solutions

	Experimental d_L (nm)
Dry sample	2.77
Suspended in water	2.91[a]
Nitrobenzene	3.41
2-Chlorophenol	3.60
In saturated solution of 4-chlorophenol	3.55

[a] Calculated for the bilayer of octadecylammonium ions with 55° chain orientation

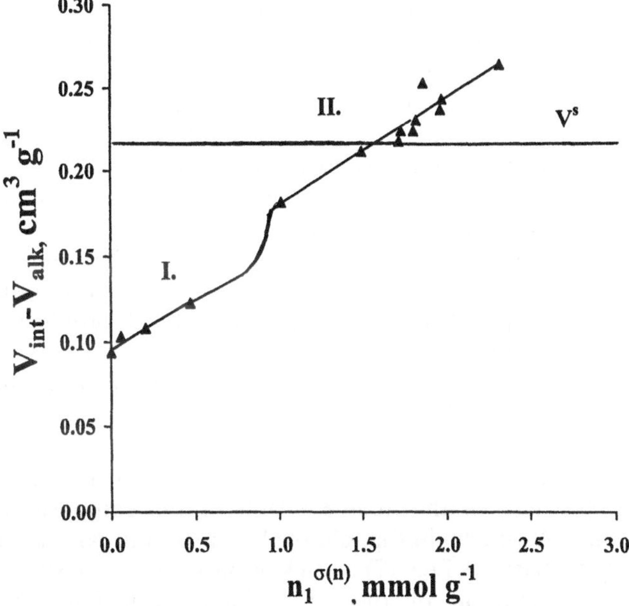

Fig. 7 Free interlamellar volume of C_{18}-vermiculite in 2-chlorophenol/water solution

Eq. (10) for the volume fractions of the alkyl chain, Φ_{alk} and the components, Φ_{s_i}. Representation of these data as a function of the entire range of relative concentrations or the real volume fraction of the bulk phase gives a demonstrative view of the interlamellar concentration of the various components of the system.

Experimental

Materials

The dilute solutions studied were saturated solutions of aromatic pollutants: nitrobenzene, 2-chlorophenol or 4-chlorophenol (component 1), distilled water (component 2). A series of dilutions were made for 10–12 points in the range $x_{1,\text{relative}} = x_1/x_{1,\text{saturation}} = 0$–$1$. Saturated solutions were made at 20 ± 0.5 °C by addition of the organic compound to distilled water under continuous mechanical shaking for 2 weeks. The upper, aqueous phase of lower density is the stock solution and was stored in a dark bottle for protection against photodecompostion.

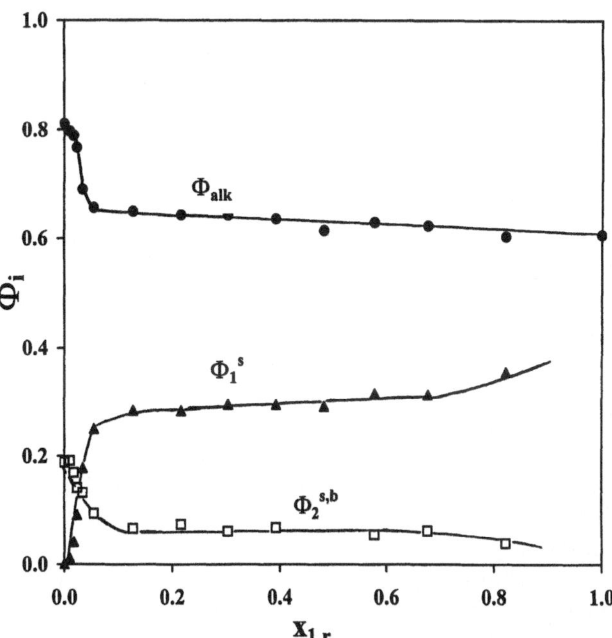

Fig. 8 Composition of the interlamellar space of the C_{18}-vermiculite in 2-chlorophenol/water solution

Adsorption of the pollutants was studied on *n*-octadecylammonium vermiculite (C_{18}-vermiculite). The original vermiculite samples (South Africa, cation-exchange capacity of 1.97 mmol/g) was organophilized by the addition of octadecylammonium chloride (in a 1.5-fold excess over the cation-exchange capacity) for an ion-exchange time of 1 month at 65 ± 0.5 °C. The organic derivative was washed, extracted, dried and cut into small pieces (about 1 × 1 mm).

Methods

Adsorption-excess isotherms were determined in the static system (immersion method) at 25 ± 0.1 °C, Known volumes (9 cm^3) of the solution were added to 0.03–0.3 g C_{18}-vermiculite. After equilibration the concentration of the supernatant and original solution was determined using a Zeiss liquid interferometer at 25 ± 0.1 °C.

X-ray diffraction measurements were carried out using a Phillips X-ray diffractometer (PW 1930 generator, PW 1820 goniometer) with Cu Kα radiation at $2\Theta = 1$–$10°$. In order to prevent the evaporation of the liquids, the adsorbent in contact with the equilibrium solution was coated by a thin Mylar foil. The basal spacing (d_L) was calculated from the (001) reflection using the Bragg equation.

Results and discussion

Adsorption of nitrobenzene from water solution on organophilized octadecylammonium–vermiculite is positive (preferential) with respect to the organic component over the entire composition range. In other words, the interfacial layer is enriched by the aromatic molecules. The isotherm (Fig. 1a) reaches saturation at an excess around $n_1^{\sigma(n)} = 0.2$ mmol/g. The course of adsorption is significantly affected by the surface polarity of the clay mineral, since on surfaces of polar character either the entire adsorption isotherm or its initial segment indicates preferential adsorption of water. The example shown in Fig. 1b is again the adsorption of nitrobenzene from an aqueous solution, but on sodium montmorillonite as adsorbent. The excess isotherm shows preferential adsorption of water (negative adsorption for nitrobenzene). The reason behind this phenomenon is the excellent hydratation of Na^+ ions and the hydrophilic character of the surface.

The 2- and 4-chlorophenols are preferentially adsorbed on the surface of hydrophobic vermiculite. Comparison of the adsorption isotherms of different chlorophenols, however, reveals that both 2-chlorophenol and 4-chlorophenol have saturation-type isotherms within the same concentration range (2.0–2.2 mmol/g) (Fig. 2).

Adsorption capacities calculated using the Everett–Schay linear representation (Eq. 4) are $n^s_{1,0} = 2.09$ mmol/g for 2-chlorophenol and 2.12 and 2.18 mmol/g for 4-chlorophenol (Fig. 3). The adsorption capacity calculated from the excess isotherm of nitrobenzene is only $n^s_{1,0} = 0.72$ mmol/g because of the very low solubility of nitrobenzene in water (Table 1).

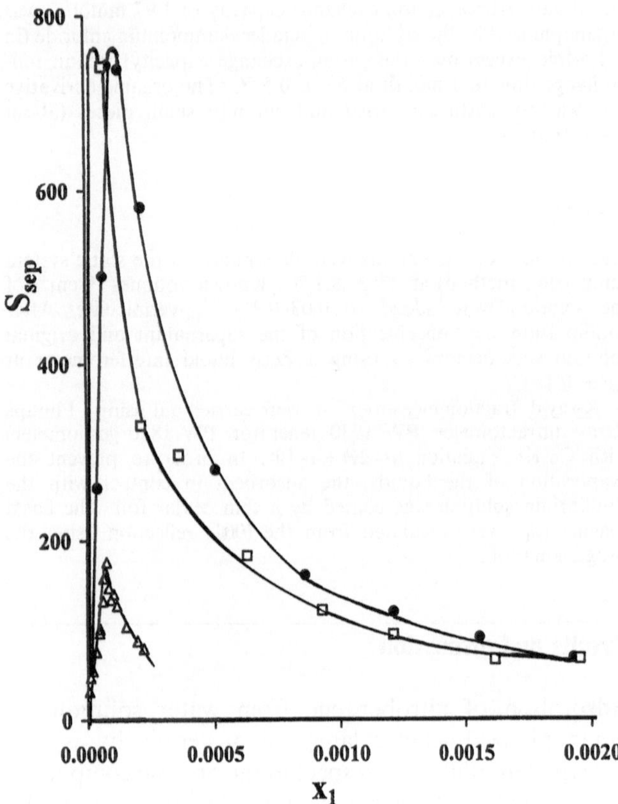

Fig. 9 Distribution ratio, S_{sep}, for 2-chlorophenol/water (●), 4-chlorophenol/water (□) and nitrobenzene/water solution on C_{18}-vermiculite

The free-enthalpy functions determined according to Gibbs' equation (Eq. 5) are plotted against relative molar fractions in Fig. 4. The curves for the two chlorophenols are quite similar at $x_{1,r} > 0.1$. Owing to the lower solubility and reduced adsorption, the free enthalpy is much less affected for the nitrobenzene solutions. The representation $\Delta_{21}G/n_1^{\sigma(n)}$ based on the free-enthalpy curve yields the values of molar free enthalpies and n^s (Fig. 5, Table 1). The basal spacing of C_{18}-vermiculite as a function of the relative molar fraction of the solutions is shown in Fig. 6a. The excess isotherms are in good correlation with the results of the X-ray diffraction measurements: a significant increase in the basal spacing is observed in each system and the breakpoints of the curve are in accordance with the increase in the adsorption excesses. The increase is caused by the intercalation of the aromatic molecules in the C_{18}-vermiculite. The lamellae, together with the alkyl chains bound to them, move apart while the concentration of the organic component is increasing in the adsorption layer (Fig. 6).

The basal spacing of C_{18}-vermiculite is 2.91 nm in water, 3.41 nm in nitrobenzene, close to saturation concentration, 3.60 nm in the presence of 2-chlorophenol and 3.55 nm in the presence of 4-chlorophenol (Table 2).

Owing to the ortho position of the chlorine atom in 2-chlorophenol, this molecule is smaller than the 4-chlorophenol molecule with the chlorine atom in the para position; therefore, the interlamellar space opens somewhat wider in the middle region of the composition range. As shown in Fig. 6b, the basal spacing calculated on the basis of Eqs. (7) and (8) is 3.04 nm, supposing a 55° orientation of the alkyl chains [18]. Figure 6c represents the state when the silicate lamellae have moved further apart and the alkyl chains overlap only partially the intercalation of additional aromatic molecules forming a multimolecular adsorption layer.

Representation of the so-called "free" interlamellar space ($V_{int}-V_{alk}$, Eq. 9) as a function of the material excess of the layer (Fig. 7) allows combination of the results of adsorption and X-ray diffraction measurements as well as comparison with the volumes $V^s = n_{1,0}V_{m,1}$ calculated from the Everett–Schay representation. The first molecules are adsorbed mostly on the external surfaces and in the monomolecular layer and the interlayer space is only slightly opened (stage I). When the concentration in the solution is increased, adsorption is enhanced, the monolayer structure is expanded and organic molecules therefore gain easy access to the V^s volume of the adsorption layer in the interlamellar spaces (stage II in Fig. 7).

The results of calculations regarding the structure of the interlamellar space (Eqs. 7–10) and the analysis of diagrams of interlamellar composition reveal the ratios of the various components (alkyl chains, organic component, water) in the accessible space (Fig. 8).

This conception of adsorption layer structure is also supported by the fact that the volume fraction relative to water, $F_2^{s,b}$ is larger in the case of 4-chlorophenol than of 2-chlorophenol; in other words, the extent of interlamellar swelling is larger when 4-chlorophenol is adsorbed than in the case of the adsorption of 2-chlorophenol.

The enrichment of the organic component may be characterized by the ratio $S_{sep}=[\phi_1^S/(1-\phi_{alk})]/\phi_1$. This value is a function of composition (Fig. 9). S_{sep} very descriptively gives information on the ratio of the volume of the bulk phase to that occupied by the liquid component adsorbed in the surface layer. This ratio reaches a maximum as high as 600–800 at low molar fractions of the aromatic component. Of the three aromatic compounds, nitrobenzene has the lowest values of S_{sep} (Table 1). The functions characterizing the two chlorophenol isomers have similar courses; the values for 2-chlorophenol are slightly higher as a consequence of higher adsorption.

Conclusion

Nitrobenzene and chlorophenols are effectively adsorbed from aqueous solutions on vermiculite hydrophobized with a cationic surfactant. The adsorption-excess isotherms (obtained by the immersion method) show the positive adsorption of the organic components over the whole range of composition up to $x_{1,relative}=1$. The results of X-ray diffraction measurements are in good agreement with excess isotherms: whenever a region of the isotherm indicates an increase in the organic component, an increase in basal spacing is also observed. The reason for the increase is the influx of aromatic molecules into the interlamellar space. The adsorption capacity and the free enthalpy of adsorption were calculated by analyzing the adsorption isotherm on the basis of the Gibbs equation and by the Everett–Schay method. The combination of the two independent measurement techniques yields further information on the adsorption equilibrium, the structure of the adsorption layer, the orientation of alkyl chains and the intercalation of organic contaminants. The results are readily utilizable for planning environmental protection systems and procedures as well as in the field of water purification and soil protection.

Acknowledgements The authors give their thanks for financial support to the National Scientific Research Foundation (OTKA) T034430 and to the Ministry of Education FKFP 0402/1999 projects.

References

1. Schieder D, Dobias B, Klumpp E, Schwuger MJ (1994) Colloids Surf 88:103
2. Klumpp E, Heitmann H, Lewandowski H, Schwuger MJ (1992) Prog Colloid Polym Sci 89:181
3. László K, Bóta A, Nagy LG (1997) Carbon 35:593
4. László K, Bóta A, Nagy LG, Subklew G, Schwuger MJ, (1998) Colloids Surf 138:29
5. Dékány I, Farkas A, Regdon I, Klumpp E, Narres HD, Schwuger MJ (1996) Colloid Polym. Sci 274:981
6. Dékány I, Farkas A, Király Z, Klumpp E, Narres HD (1996) Colloids Surf A 119:7
7. Regdon I, Dékány I, Lagaly G (1998) Colloid Polym Sci 276:511
8. Regdon I, Király Z, Dékány I, Lagaly G (1998) Prog Colloid Polym Sci 109:214
9. Dékány I, Szántó F, Weiss A, Lagaly G (1985) Ber Bunsenges Phys Chem 89:62
10. Dékány I, Szántó F, Weiss A, Lagaly G (1986) Ber Bunsenges Phys Chem 90:427
11. Bóta A, László K, Nagy LG, Copitzky T (1997) Langmuir 13:6502
12. László K, Bóta A, Nagy LG, Takács M (1997) Acta Chim Hung Models Chem 134:81
13. László K, Bóta A, Nagy LG (2000) Carbon 38:1965
14. Stul MS, Uytterhoeven JB, De Bock J, Huykens PL (1979) Clays Clay Miner 27:377
15. Schay, G (1976) Pure Appl Chem 48:373
16. Everett DH (1981) Pure Appl Chem 53:2181
17. Király Z, Dékány I (1988) Colloids Surf 34:1
18. Lagaly G, Witter R (1982) Ber Bunsenges Phys Chem 86:74

Progr Colloid Polym Sci (2001) 117: 70–75

A. Dąbrowski
M. Bülow
P. Podkościelny

Adsorption against pollution: current state and perspectives

A. Dąbrowski (✉) · P. Podkościelny
Faculty of Chemistry,
M. Curie-Sklodowska University,
20-031 Lublin, Poland
e-mail: dobrow@hermes.umcs.lublin.pl
Tel: +48-81-5375605
Fax: +48-81-5375685

M. Bülow
BOC Gases, 100 Mountain Ave.,
Murray Hill, NJ 07974, USA

Abstract An overview of modern trends in the application of adsorption science for protecting the human environment is presented. The essential subject comprises fundamental information on conventional pollutants, environmental challenges to adsorption, basic types of environmentally relevant adsorbents and information that refers to adsorbent properties with regard to practical applications. Current understanding and perspectives pertaining to environmental applications of adsorption phenomena are discussed as well.

Key words Adsorption · Environmental protection · Conventional pollutants · Environmental adsorbents · Trends and perspectives

Introduction

During the few last decades much effort has been devoted to the investigation of adsorption phenomena that occur at various fluid–solid interfaces. A variety of reasons have caused this strong interest both in fundamental research in the adsorption area and in practical application of its results (e.g., Refs. [1–3] and references therein). One of the most important utilitarian aspects of adsorption technologies emerges in local environmental control and global environmental conservation [4]. The term global environmental problem is used to characterize, in particular, emission of ozone-depleting gases, such as chlorofluorocarbons (CFCs), volatile organic (chlor-containing) compounds (VOCs) and the emission of "greenhouse" gases (CO_2, CH_4, N_2O, etc.). The term local environmental problem comprises first of all, the phenomena of flue gas recovery (SO_x and NO_x), solvent vapor fractionation and solvent vapor recovery, wastewater treatment and supply/production of drinking water. Other major environmental issues are related to industrial solid aerosols, which stem from incomplete combustion of a major number of fuels. These aerosols are harmful as precursors to (uncontrolled) synthesis of strong toxins, cancerogenes and mutagenes [5].

Nowadays, the pressure on industry is becoming increasingly strong in order to decrease emission of various pollutants into the environment, and a number of international agreements and protocols related to that issue are being enforced. A broad range of methods is available to control the removal from process streams and exhausts of both natural and anthropogenic (municipal, agricultural, commercial) and other pollutants [6]. With regard to their cost–performance relation, adsorption technologies represent the most important techniques to overcome a still ongoing degradation of environmental quality [7]. These techniques play a significant role both in environmental and in human health control and in minimizing and/or preventing emission of warming and ozone-layer depleting gases. Adsorption phenomena are essential for a large range of methods for environmental analysis, which is a discrete and sophisticated branch of modern analytical chemistry [8]. Adsorption can also be expected to play a significant role in dessicant dehumidification technologies [9] and life-support systems aboard spacecraft and planetary bases. In the latter cases, adsorbents may be used to process the habitat air and to recover useful substances such as drinking water, etc., from local environments [10].

Major applications of adsorption for protecting the environment are as follows [2, 3, 11]:

1. Solvent/chemical recovery.
2. Removal of VOCs, CFCs and other hazardous chemical compounds from air and industrial waste gases.
3. Gas separation and purification.
4. Surface water management and wastewater treatment.

Progress in research into the fundamentals of adsorption phenomena achieved during recent years has very much advanced our understanding of adsorption processes related to environmental protection [12]. A brief review of the state of this art is given in this article. In addition, challenges related to adsorption for the protection of the human environment are outlined.

Environmental pollutants

The term environmental pollutant refers to any cause of physical, chemical or biological disturbance of the ecological equilibrium in the environment. Environmental pollution results from random, accidental events, emission of certain pollutants owing to the activity of nature itself or owing to human activities. Natural pollutants stem from activities of nature itself and they contain certain sediments in air (dust) and surface waters, dissolved minerals (sometimes toxic metals) in ground and surface water and trace elements in air from volcanic activities.

Anthropogenic pollutants are emitted as a result of human activity. Their main types are as follows: industrial, municipal/domestic, and agricultural, institutional and commercial activities. On the other hand, the anthropogenic pollutants can be divided into point and nonpoint sources. Nonpoint sources (e.g., roads, streets, parking lots, farm fields, construction sites, etc.) are especially difficult to monitor, regulate and treat. Natural and anthropogenic pollutants emitted from a given source are defined as primary pollutants.

A number of primary pollutants can undergo changes owing to reactions with other pollutants or with some components of the environment. In this way, new compounds can be formed, which are often of increased toxicity; they are known as secondary pollutants. Both primary and secondary pollutants occur in all the environmental media, i.e., atmosphere, hydrosphere and soil [13].

Another important problem results from photochemical smog that occurs as a direct consequence of secondary atmospheric reactions driven by solar energy. One of the most important of these reactions is the formation of ozone by molecular and atomic oxygen, the latter being obtained from splitting O_2 or NO_2 by UV radiation.

About half the incoming solar radiation is absorbed or reflected by clouds or the atmosphere itself. Different components of the atmosphere react selectively with incoming radiation. For example, visible light is absorbed by stratospheric ozone, which provides protection for the Earth's inhabitants. The substances which are the most dangerous with regard to stratospheric ozone are CFCs, the presence of which causes ozone-layer depletion. On the other hand, so-called greenhouse gases (CO_2, CH_4, N_2O, NO_2, tropospheric O_3 and CFCs), i.e., substances that absorb in the spectral range where thermal energy radiated from Earth is at its maximum, represent a major challenge, in particular to industrialized societies, to prevent global warming.

Some major conventional pollutants are listed in Table 1.

Current environmental issues of adsorption

The subject of the use of modern adsorption technologies is of tremendous environmental, economic and legal importance and it constitutes a serious challenge to all industrialized societies with strong potential for further

Table 1 Major conventional pollutants. Chlorofluorocarbons are compounds such as $CFCl_3$, CF_2Cl_2, and others Volatile organic compounds comprise methane, benzene, formaldehyde, vinyl chloride, phenol, chloroform, trichloroethylene and gasoline ingredients. Organic sulfur includes methyl mercaptane, carbon disulphide, and dimethyl sulphide, among others. Metals comprise As, Cd, Cr, Cu, Se, Ni, Mo, Hg and Pb. Suspended particulate materials comprise soot, dust and ash

Species	Major sources	Average "half time" in atmosphere (days)
CO_2	Fossil fuel burning, biomass	2,500
CO	Fossil fuel burning, biomass	75
CH_4	Anthropogenic activities	3,600
Chlorofluorocarbons	Industry, overconsumption of fossil fuel	1–1,000
Volatile organic compounds	Anthropogenic activities	1–1,000
Volatile organic compound	Isoprene, terpenes from plants	<1
NO_x	Fossil fuel, burning biomass	4
N_2O	Fertilizers	60,000
SO_2, SO_4^{2-}	Fossil fuels, smelting	1–4
H_2S, organic sulfur	Anthropogenic activities, biogenic	1–900
Metals	Leaded gasoline, coal, industrial waste	1–30
Suspended particulate materials	Anthropogenic activities, fires, wind erosion	1–1,000

Table 2 Current typical environmental tasks related to adsorption science

Issue	Task
Local environmental problems	
Flue gas treatment	SO_x, NO_x and Hg emission removal
Solvent vapor recovery and solvent fractionation	Volatile organic compound recovery from work places, among them from ground water; adsorption methods are needed to prevent volatile organic compound emission into air by increasing emphasis on development/use of air purification and solvent vapor recovery processes.
Wastewater treatment	Removal of organics, nitrogen and phosphorus, i.e., removal/recovery nutrients from wastewater; removal of heavy metal traces from wastewater
Drinking water production	Deterioration of water sources, advanced treatment of wastewater, etc.
Desiccant dehumidification technology	Improvement of indoor air quality; removal of air pollutants; a number of microorganisms are removed/killed by desiccants owing to coadsorption by desiccant materials.
Global environmental problems	
Issue	Task
Global warming control	Emission control of greenhouse gases (CO_2, CH_4, N_2O, NO_x); utilization of CH_4
Ozone layer depletion	Emission control/recovery/replacement of chlorofluorocarbons as ozone-depleting gases used in refrigeration systems; chlorofluorocarbon-free refrigeration systems as alternative solutions to vapor-compression heat-chill pumps, which use valuable electric energy and polluting refrigerants (chlorofluorocarbons) that cause ozone depletion and the greenhouse effect
Defense applications	Removal of contaminants used by military, which comprised extremely toxic chemicals; in contrast to solvent vapor recovery processes, defense systems have to purify air only, but both applications are related to the environment with successful use of adsorption technology

intense development. The major important environmental tasks related to adsorption techniques are summarized in Table 2.

Adsorbents and their properties related to practical applications

The development and application of adsorption should be considered jointly with that of adsorbent technology as utilized on both laboratory and industrial scales [13, 14]. These adsorbents cover broad ranges of chemical specimens and geometrical surface structures, which is reflected by the range of their applications in industry and their importance for laboratory practice. This situation is compatible to that, as far as the variety of adsorbents is concerned, which could be found in various environmental applications [15].

In earlier adsorption processes, the choice of an adsorbent to respond to a given challenge was restricted to various types of activated carbons, silicas and aluminas [16, 17]. Natural zeolites were known as a scientific curiosity and have never been used as adsorbents on an industrial scale. The discovery of synthetic zeolites, also known as (one group of) molecular sieves, has broadened immensely the available range of adsorbents and has led to an enormous upsurge in the development of adsorption processes [18, 21, 22]. Hydrophobic clays also show excellent (selective) adsorption properties for different pollutants; therefore, the adsorption isotherms and the energetic characterization of adsorption on solid interfaces were investigated [23–27]. New theoretical approaches to adsorption phenomena and the emergence of numerous groups of novel adsorbents have generated new practical applications [2, 3, 30].

The main types of adsorbents of practical environmental importance are presented in Table 3.

In terms of pore-size engineering, a choice can be made from a wide class of nanoporous materials. These adsorbents have pore diameters in the range from several tenths of a nanometer to about 50 nm and they are of different basic pore geometry and chemical nature. Owing to their excellent adsorption, catalytic and thermal properties [19], many nanoporous adsorbents are very popular in science, fluid purification and separation process technology as well as in environmental protection.

The principal properties that relate to adsorbent applications are as follows [7, 10, 13]:

1. Adsorption characteristics.
2. Regeneration capability.
3. Physical characteristics (particle geometry, density, hardness, shaping ability, ash content, etc.).

Each of these properties can be important with regard to whether or not adsorption is considered to be a viable process option.

Adsorption characteristics are those with a direct effect on the cost–performance relation for the product in a given mode of application. They comprise:

Table 3 Basic types of adsorbents used for environmental purposes [12]

Carbon adsorbents	Mineral adsorbents	Other adsorbents
Active carbons	Silica gels	Synthetic polymers
Activated carbon fibers	Activated alumina	Metallorganic microporous and mesoporous materials
Carbon molecular sieves	Oxides of metals	Composite adsorbents: (complex mineral carbons, X–elutrilithe; X = Zn, Ca)
Mesocarbon microbeads	Hydroxides of metals	Mixed sorbents
Carbonaceous nanomaterials	Zeolites	
	Clay minerals	
	Pillared clays	
	Porous clay heterostructures	
	Inorganic nanomaterials	

1. Adsorption energy and its distribution.
2. Adsorption capacity.
3. Porosity and/or specific surface area for a given porosity type.
4. Material to extract.
5. Resistance to the pH value of fluid to be treated.

In general, the use of solid adsorbents in science and technology requires their manifold characterization [12]. For achieving this goal, a spectrum of various modern physicochemical analytical techniques provide useful and very specific information. In particular, these techniques comprise X-ray diffraction, X-ray spectroscopy, small-angle X-ray spectroscopy, solid-state NMR, atomic emission spectroscopy, Raman spectroscopy, temperature-programmed desorption, etc. In addition, the utilization of classical adsorption/desorption measurements is unavoidable to collect information on the adsorption behavior of a solid with regard to a pollutant [20].

Configuration of adsorption systems and regeneration of adsorbents

The configuration of an adsorption system is defined in accordance with the basic unit operation, by which a fluid (gas or liquid) is contacted with a solid adsorbent. There are only three main configuration categories [11], viz., bath, fixed bed and moving bed, each of which can be operated with particular regard to the flow of fluid and solid, i.e., co- or countercurrently.

Irrespective of the configuration of the adsorption systems, a suitable regeneration mode is desired. The economical success of adsorption processes depends on the regenerability of the adsorbent. The regeneration mode corresponds, in general, with the types of fluid and solid adsorbent used.

For liquid–solid adsorption systems, the following regeneration techniques can be utilized [7, 13]:

1. Thermal (suitable for granular activated carbon, shaped zeolites, etc.).
2. Steam regeneration (for granular activated carbon, polymeric adsorbents).
3. Acid or base regeneration (activated carbon, polymeric adsorbents).
4. Solvent regeneration (polymeric adsorbents).
5. Biological regeneration (e.g., carbon regeneration by biological oxidation).

For cyclic gas and most liquid-phase adsorption processes (gas separation, solvent recovery, drying of gases and liquids, recovery of chemicals from industrial and vent gases) the following adsorbent regeneration methods are in use [21]:

1. Temperature swing adsorption (TSA) – bed regeneration by raising the temperature.
2. Pressure swing adsorption (PSA) – bed regeneration by alternate reduction of the total pressure in the adsorber beds at constant temperature.
3. Inert purge stripping – removal of adsorbate without changing the temperature or the pressure; the void in the bed is filled with an inert gas upon complete regeneration.

PSA/TSA and related technologies, such as vacuum swing adsorption (VSA), pressure VSA (PVSA) and rapid PSA (RPSA), are executed in many sophistical variants that have helped to improve separation efficiency and to decrease energy consumption. PSA/TSA and related adsorption processes can be considered as being very efficient and relevant toward environmental applications [28, 29].

Apart from PSA/TSA and related technologies, the following important adsorption-based methods are utilized for treating issues of environmental pollution [12, 31]: membrane processes, ion-exchange methods, process-scale chromatographic techniques, hybrid methods (e.g., adsorption catalysis, adsorption-membrane separation, coagulation, sedimentation and filtration).

Environmental analysis

Environmental analysis represents a discrete and sophisticated branch of analytical chemistry in which a series of various analytical techniques are applied. Of these, adsorption chromatography, adsorption-electrochemical analysis, spectroscopic and spectrophotometric methods are of great importance. Environmental analysis is the most applied trace analysis method, where pollutants are present in trace and ultratrace quantities. Thus, the fundamental requirement of environmental analysis is for a fast, modern and reliable method in order to detect pollutants in very small quantities, quickly and precisely [8]. Environmental analysis relevant to adsorption science should be exemplified as follows.

Electrochemical analysis of adsorbable surface-active substances

This type of analysis deals with the formation of electrode double layers as a result of adsorption of many chemical compounds of both organic and inorganic nature, at the solution–electrode interface. This process effects the properties of the electrode double layer in a measurable manner, and it forms the basis for an electrochemical analysis of adsorbable surface-active substances that are present in solution. From a historical point of view and first of all, polarographic adsorption analysis, which was introduced by Heyrovsky [32], should be mentioned. The method of tensammetry developed from polarography or the measurement of the differential capacity of the electrode double layer is closely related to the former and is widely utilized. From a practical point of view, with respect to recent voltammetry methods, it is necessary to point out procedures that are based on adsorptive accumulation of the analyte on the electrode surface:

- Adsorptive stripping voltammetry.
- Adsorptive stripping potentiometry.

The scope of application of these methods ranges from metal trace analysis to that of organic compounds and, in general, to environmental, biochemical, pharmaceutical, toxicological and other applications [33].

Chromatographic methods proceeded by sampling and sample preparations

Adsorption phenomena are widely applied for sampling many fluids, such as air, surface water and wastewater [8]. Sampling is realized together with enrichment of analytes. owing to the inherent strong selectivity of adsorption, pollutants of interest can be removed selectively from the bulk sample matrix; they can be preconcentrated, cleaned up, separated into individual substances and analyzed by gas and liquid adsorption chromatography, or by related chromatographic techniques (high-performance liquid chromatography, thin-layer chromatography, etc.).

Among the adsorption methods that are used for isolating analytes from liquid matrices and for their preconcentration, increasing practical importance has been shown for the solid-phase extraction (SPE) technique. The idea of SPE consists of retention of analytes from a large sample volume on a small bed of adsorbent, followed by elution of analytes with a small volume of solvent. The proper selection of appropriate parameters of adsorbents and solvents is a crucial condition for successful employment of this method [34].

Challenges

The future development of adsorption science represents a major challenge to industry and environmental fields because of its low cost, regenerability of most of the adsorbents used and potential ability to design and manufacture novel materials which are specific to a given need.

The following comments can be made with regard to developments in environmental adsorption.

1. New applications of adsorption methods will emerge as new types of adsorbents specific to a given need are developed [30, 35]. Additional advantages will emerge if novel microporous and mesoporous adsorbents, like MCM-41 materials, metallorganic molecular sieves, inorganic ceramics or other solids with nanoporosity, are manufactured cost-effectively and if certain practical constraints are overcome.
2. Environmental applications of adsorption phenomena have to be developed in conjunction with new theoretical approaches and molecular simulations of industrial and environmental processes [36]; the same is true for adsorption techniques, in particular, for those with new classes of adsorbents, catalysts, membranes and ion exchangers [37].
3. Another field in continuous progress is that of the expansion of new separation and purification techniques for environmental purposes. In this context, the improvement and further development of chromatographic techniques demonstrates huge practical importance. A similar challenge also exists for various types of membrane separation methods that present very advanced and innovative molecular separation units with great potential for applications in all sectors of environmental protection; a number of environmental goals can be approached by hybrid processes that combine the processes of adsorption,

membrane separation, catalysis, filtration, chromatography, etc [38].

4. The prospective demands toward adsorption and related fields are based on a growing concern for environmental control, in order to maintain and increase the quality of life. Many of the aspects aimed at this concept will require both the uses of novel adsorbents, membranes, catalysts and other materials and the development of ecologically friendly, low-energy adsorption technologies. In this context, the PSA, PVSA, RPSA, TSA and other purification and separation techniques will prove to be more favorable in terms of computer simulation and experiment research [22, 39].

Conclusions

Only a fraction of the environmental roles played by adsorption, have been touched upon here. Numerous examples of practical applications in environmental protection are presented in various monographs and conference proceedings [2, 3, 14, 16, 21, 40, 41].

The development and utilization of molecular modeling to understand fundamentals of adsorption phenomena and to synthesize novel classes of adsorbent materials is crucial and becomes extremely valuable if used in conjunction with experimental techniques. On the other hand, new theoretical approaches and novel groups of adsorbents stimulate the development of new practical environmental applications, among other issues.

It is widely known that well-understood adsorption science has gained a dominating role in modern industry with regard to environmental, economic and energy aspects. Doubtlessly, adsorption technologies have improved rapidly and they were adapted to contemporary tasks of mankind. Nowadays, only such technologies which provide the possibility of sustainable development of people and societies are justified. In this context, adsorption and related domains are held in high esteem as important environmental technologies of this century.

References

1. (1996) Stud Surf Sci Catal 99
2. (1999) Stud Surf Sci Catal A 120
3. (1999) Stud Surf Sci Catal B 120
4. Suzuki M (1996) In: LeVan MD (ed) Fundamentals of adsorption, vol 5. Kluwer, Boston, p 4
5. Pokrovskiy VA, Bogillo VJ, Dabrowski A (1999) Stud Surf Sci Catal B 120:571
6. Bolt BA, Barcicki J, Kozak Z, Pawlowski L (1984) In: Pawlowski L, Verdier AJ, Lacy WJ (eds) Chemistry for protection of the environment. Elsevier, Amsterdam, p 5
7. Thomas WJ, Crittenden B (1998) Adsorption technology and design. Butterworth-Heinemann, Oxford
8. Bladek J, Neffe S (1999) Stud Surf Sci Catal B 120:3
9. Gosh TK, Hines AL (1999) Stud Surf Sci Catal A 120:879
10. DallBauman LA, Fin JE (1999) Stud Surf Sci Catal B 120:455
11. Slejko FL (ed) (1985) Adsorption technology. Dekker, New York
12. Dabrowski A (2001) Adv Colloid Interface Sci 93:4
13. Ge Y, Murray P, Hendershot WH (2000) Environ Pollut 107:137
14. Suzuki M (1990) Adsorption engineering. Elsevier, Amsterdam
15. Cohen Y (ed) (1995) AIChE Symp Ser 91
16. Oscik J (1982) Adsorption. Horwood, Chichester
17. Mantell CL (1951), Adsorption. McGraw-Hill, NewYork
18. Breck DW, Eversole WG, Milton RM, Reed TB, Thomas TL (1956) J Am Chem Soc 78:5963
19. Kaneko T (1999) Stud Surf Sci Catal B 120:635
20. Jaroniec M (1995) In: Pinnavaia TJ, Thorpe MF (eds) Access in nanoporous materials. Plenum, New York
21. Yang RT (1987) Gas separation by adsorption processes. Butterworth, Boston
22. LeVan MD (1998) In: Meunier F (ed) Fundamentals of adsorption, vol 6. Elsevier, Amsterdam, p 19
23. Patzko A, Dekany I (1993) Colloids Surf A 71:299
24. Regdon I, Kiraly Z, Dekany I, Lagaly G (1994) Colloid Polym Sci 272:1129
25. Marosi T, Dekany I, Lagaly G (1994) Colloid Polym Sci 272:1136
26. Kiraly Z, Dekany I, Klumpp E, Lewandowski H, Narres HD, Schwuger MJ (1996) Langmuir 12:423
27. Dekany I, Farkas A, Kiraly Z, Klumpp E, Narres HD (1996) Colloids Surf A 119:7
28. Liu Y, Subramanian D, Ritter JA (1999) Stud Surf Sci Catal B 120:213
29. Ritter JA, Liu Y, Subramanian D (1998) Ind Eng Chem Res 37:1970
30. Bülow M, Fitch FR (1997) Filtration and separation. October 1997, 5:839
31. Bülow M, Lutz W, Suckow M (1999) Stud Surf Sci Catal A 120:301
32. Heyrovsky J, Kuta J (1965) Principles of polarography. Academic, New York
33. Stulik K, Kalvoda R (eds) (1996) Electrochemistry for environmental protection, UNESCO Venice Office, Venice
34. Raisglid M, Burke S (1999) Stud Surf Sci Catal B 120:137
35. Ruthven DM (2000) Ind Eng Chem Res 39:2127
36. Thomas P (1999) Simulation of industrial processes for control engineers. Butterworth–Heinemann, Boston
37. Bülow M, von Gemmingen U, Izumi J, Pullumbi P, Sherman J, Coe C, Suzuki M, Baron G (1998) In: Meunier F (ed) Fundamentals of adsorption, vol 6 Elsevier, Amsterdam, p 47
38. (1997) Proceedings of the Topical Conference on Separation Science and Technology. AIChE Annual Meeting November1997, Los Angeles, Calif
39. Ruthven DM, Farooq D, Knaebel KS (1994) Pressure swing adsorption. VCH, New York
40. Inui T, Anpo M, Izu K, Yanagida S, Yamaguchi T (eds) (1998) Stud Surf Sci Catal 114
41. (2000) Proceedings of the Topical Conference on Energy and the Environment. AIChE Annual Meeting November 2000, Los Angeles, Calif

Progr Colloid Polym Sci (2001) 117: 76–79

T. Textor
T. Bahners
E. Schollmeyer

Organically modified ceramics for coating textile materials

T. Textor (✉) · T. Bahners
E. Schollmeyer
Deutsches Textilforschungszentrum
Nord-West e.V., Adlerstrasse 1,
47798 Krefeld, Germany
e-mail: info@dtnw.de
Tel.: +49-2151-8430
Fax: +49-2151-843143

Abstract Fabrics made of poly(ethylene terephthalate) (PET) and glass fibres were coated with different modified inorganic–organic hybrid polymers based on 3-glycidyloxpropyl trimethoxysilane. In the case of PET the focus was to affect the surface specific properties especially the hydrophobic, oleophobic and hydrophilic properties. The glass fibre fabric was finished with a composite that improves the wear resistance.

Key words Poly(ethylene terephthalate) · Glass fibre · Organically modified ceramic · Surface energy · Wear resistance

Introduction

Increasing demands for functional and highly specialized textiles, for example, technical textiles, can be observed worldwide. Intense research is presently aimed at new methods for surface modification in order to establish improved or new properties.

An innovative method for textile finishing is the modification of fibre material with a thin coating of organically modified ceramics that combine the advantages of organic polymers and ceramics [1–3].

The application of ceramic materials, primarily established as thin ceramic layers, offers far-reaching possibilities for permanent surface modification in modern material development. Primarily because of very high processing temperatures but also owing to huge technical effort textile materials have not been accessible to these technologies so far. A new perspective is offered on the background of recent work in the sol–gel technique [4, 5]. Organically modified ceramics, which are made by sol–gel-processing, combine qualities of ceramics and synthetic polymers and have an immense potential for creative modifications of surface properties with low technical effort at moderate temperatures. These materials are derived from silica alkoxides that are modified with one organic group. This group consists of, for example, a hydrocarbon chain with functional epoxy, metacrylic or thiol groups. In the presence of certain amounts of water, under basic or acidic conditions these alkoxides undergo a hydrolysis reaction and partly condense to form sols. In a following curing step, the condensation of the hydrolysed silica alkoxides can be completed by simultaneously cross-linking the functional groups. The sols can be modified by mixing them with other metal alkoxides or dispersing nanosized metal oxides. Therefore the resulting three-dimensional networks are build from organic and inorganic domains. The sols can be applied by common methods, for example, dipping, spraying or knife coating, and in contrast to ceramic processing the curing temperatures are very moderate. Depending on the organic modification the curing can be carried out with UV radiation as well.

Experimental

Materials

The coating experiments were carried out with a technical polyester fabric and a glass fibre fabric. Prior to the treatment, the fabrics were extracted with ethanol and petrol ether.

Chemicals

As organically modified alkoxysilane 3-glycidoxypropyl trimethoxysilane (GTPMS) (98%, Aldrich) and for catalysing the cross-linking reaction of the epoxy-groups 1-methylimidazol (purity above 97%, Fluka) were used. For hydrophobic coatings the sols were modified with tridecafluoro-1,1,2,2-tetrahydrooctyltriethoxysilane (ABCR) and hydrophilic modifications were carried out by adding a surfactant (Tween80, Fluka). Aluminium oxide (aluminium C, Degussa, particle size 5–20 nm) and bisphenol A (Fluka) were used to modify the sols for wear-resistant coatings of glass fibre fabrics. Water-repellent aluminium-modified glass fibre fabrics were made using *n*-propyl trimethoxysilane (Merck).

Analytical methods

The effects on the hydrophobic and hydrophilic properties were examined by measuring the contact angles of water with a Krüss G40 measuring system and by the investigation of the drop penetration behaviour of distilled water following the TEGEWA test [6].

The oil repellency was analysed following AATCC (118–1972). The wear resistance was tested using the Martindale abrasive test (DIN EN ISO 12947). The climatized fabric samples were scrubbed radially over a defined testing fabric. After 5,000 cycles the samples were investigated visually.

Preparation of the sols

Most of the sols used were based on GPTMS. GPTMS (1 mol) was prehydrolysed with 1.5 mol water (0.01 M hydrochloric acid) and stirred for at least 2 h to build the basis sol. The particular sols were produced by appropriate modifications of the GPTMS sol. The catalyst for the cross-linking reaction of the epoxy group – 1-methylimidazol – was added just before the coating process.

For the hydrophilic coatings two routes were followed. One way is to convert the epoxy group of the GPTMS to a glycol group, which is carried out by refluxing it for 3 h with a certain amount of diluted sulfuric acid and subsequent removal of the sulfate ion by ion exchange. Alternatively hydrophilic coatings were yielded by stirring in a surfactant 5 min before the coating process, followed by addition of the catalyst. The same procedure can be used for hydrophobic coatings by stirring in tridecafluoro-1,1,2,2,-tetrahydro-octyltriethoxysilane. For wear-resistant coatings the GPTMS sols were modified with 10 wt% Aluminium C. The aluminium oxide was dispersed in acidic water with an ultrasonic generator. Bisphenol A in a ratio of 4:1 was additionally dissolved in these sols. Finally the complete sol was diluted with ethanol to a concentration of 10 vol%.

To prepare a hydrophobic surface for a wear-resistant surface in a second coating step an *n*-propyl trimethoxysilane sol was made by partly hydrolysing 1 mol silane with 1.5 mol hydrochloric acid (0.01 M) for at least 5 h.

Coating process

The coatings were carried out by a padding process with a laboratory padder. The coated fabrics were tightened on a pin tenter heat setter and dried in an oven for about 1 h. The drying temperature was adjusted to 130 °C.

Results

General effects on the textile properties

The coatings are clear but have a distinct yellow shade. The tensile strength of the coated fabrics is slightly, but significantly, increased. The fabrics show increased stiffness and a decreasing wrinkle recovery angle (DIN 53890). The degree of these effects depends on the modification of the different sols. Scanning electron microscopy micrographs (Fig. 1) show that the coatings, which cover the fibres, are stuck together by the coating material at the points of intersection.

Hydrophilic and hydrophobic coatings

By converting the epoxy ring of the GPTMS into a glycol group the number of hydroxyl groups rises; therefore the surface energy of the coating material is increased. The measured contact angle of distilled water for this coating is 39° and for the unmodified GPTMS the contact angle is about 66°, which is a little lower than contact angle for uncoated polyester (76°). Although the surface energy for the coated fibre increases, the drop penetration time for the coated fabrics rises from 135 s to more than 1 h and the drop shows only slight spreading. The prepara-

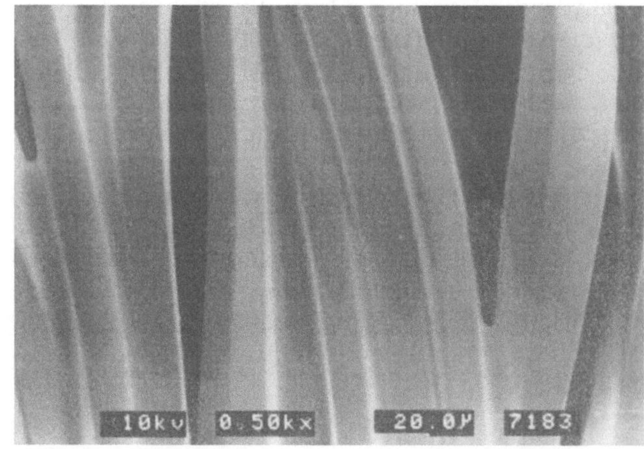

Fig. 1 Scanning electron microscopy micrograph of a polyester fabric coated with a 3-glycidoxypropyl trimethoxysilane sol

Table 1 Contact angles of distilled water for differently coated polyester surfaces

Surface	Contact angle (degrees)
Polyester	76
Polyester coated with 3-glycidoxypropyl trimethoxysilane sol	66
Polyester coated with 3-glycidoxypropyl trimethoxysilane sol with hydrolysed epoxy function	39
Polyester coated with tenside-modified 3-glycidoxypropyl trimethoxysilane sol	≈0
Polyester coated with fluorine-modified 3-glycidoxypropyl trimethoxysilane sol	95

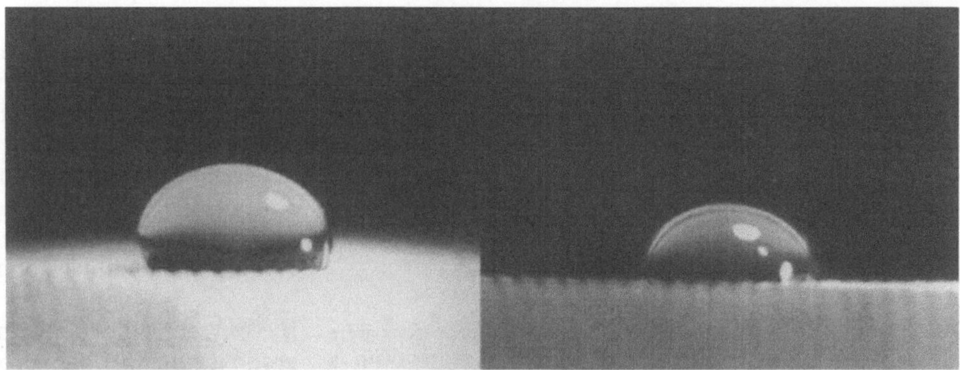

Fig. 2 A drop of distilled water on a polyester fabric with a hydrophobic coating. The *left photograph* is taken directly after the application, the *right photograph* 2 h later

tion of a hydrophilic coating by incorporating 5% of a nonionic surfactant – in this case Tween 80 – leads to a very hydrophilic surface. The contact angle of water is close to 0° and the coated fabrics soak up water quickly, the drop penetration time is only 85 s and is even lower than for the uncoated fabric.

In order to establish very hydrophobic and oleophobic surfaces a sol was prepared from GPTMS and a highly fluorinated trialkoxysilane, tridecafluoro-1,1,2,2-tetrahydrooctyl triethoxysilane. By applying this sol, surfaces can be created with contact angles of up to 95°. The experiments showed that 1–2% of the highly fluorinated silane is necessary to yield such low-energy surfaces. Higher amounts do not lead to higher contact angles.

The TEGEWA test shows no water penetration into the fabrics, the drop remains on the material for several hours, and even the oil test showed good results. The oil repellency is "satisfactory" (grading 4) compared to the unmodified fabric, which was unsatisfactory (grading 0). The high water repellency of the hydrophobized fabric is illustrated in Fig. 2.

Wear-resistant coating

One of the disadvantages of glass fibre fabrics is their insufficient wear resistance; therefore, they must be protected for many applications. By coating glass fibres with aluminium oxide modified sol fabrics can be prepared that withstand at least 5,000 cycles of a wear resistance test, whereas the unmodified fabric is destroyed completely under these conditions. This is demonstrated impressively by the photographs in Fig. 3.

Owing to the hydrophilic character of the aluminium oxide particles the resulting samples have relatively high surface energies, which is disadvantageous for many applications, for example, for textile architecture. Good water repellency can be achieved in a second coating step with the previously mentioned fluorine modified sol or alternatively with prehydrolysed and diluted *n*-propyl trimethoxysilane.

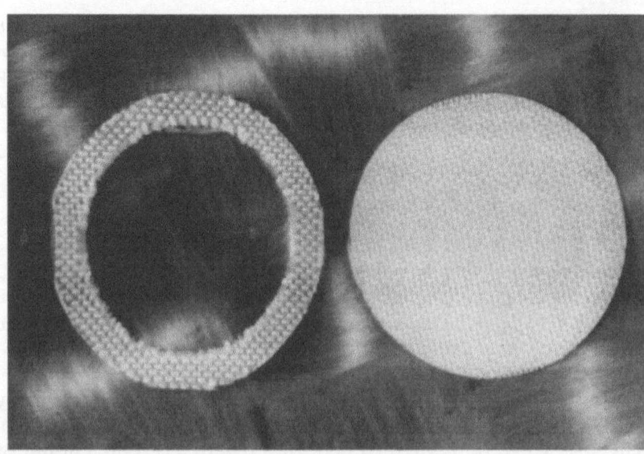

Fig. 3 Photographs taken after the abrasive test from an unmodified (*left*) and a modified (*right*) glass fibre fabric

Conclusion

Coatings based on organically modified ceramics were intensively investigated in the last years and several applications are already used commercially. For textile materials the composites promise far-reaching possibilities for creative surface design, not only with regard to hydrophilic and hydrophobic effects or to improve the wear resistance of glass fibres.

Additional interesting properties are conceivable by suitable modifications of the coating materials. Examples are sun-protection properties, self-adapting colorations or magnetic features. By adequate coatings trans-dermal-therapy-systems could be created by incorporating drugs that are set free over an adjustable period and in adjustable amounts into the amorphous organically modified ceramic networks.

Acknowledgements We thank the Forschungskuratorium Textil e.V. for funding this research project (AiF nos.10954 N and 12000 N). The project was funded with financial resources of the Bundesministerium für Wirtschaft und Technologie (BMWi) with a grant from the Arbeitsgemeinschaft industrieller Forschungsvereinigungen "Otto von Guericke" e.V. (AiF).

References

1. Nass R, Schmidt HJ (1990) J Non-Cryst Solids 121:329–333
2. Schmidt HJ (1994) J Non-Cryst Solids 178:302–312
3. Nass R, Arpac E, Glaubitt W, Schmidt H (1990) J Non-Cryst Solids 121:370–374
4. Knittel D, Textor T, Bahners T, Schollmeyer E (1998) UMIST Conference: Textiles engineered for performance, 20–22 April 1998, Manchester, UK, pp 1–13
5. Textor T, Bahners T, Schollmeyer E (1999) Melliand Textilber 80:847–848
6. Anonymous (1987) Melliand Textilber 68:581ff

Progr Colloid Polym Sci (2001) 117: 80–87

K. Esumi
K. Torigoe

Preparation and characterization of noble metal nanoparticles using dendrimers as protective colloids

K. Esumi (✉) · K. Torigoe
Department of Applied Chemistry
and Institute of Colloid
and Interface Science,
Science University of Tokyo,
Kagurazaka, Shinjuku-ku,
Tokyo 162-8601, Japan
Tel.: +81-3-32604272

K. Torigoe
Department of Industrial Chemistry,
Faculty of Engineering,
Science University of Tokyo, Kagurazaka,
Shinjuku-ku, Tokyo 162-8601, Japan

Abstract Noble metal nanoparticles such as gold, platinum, and silver have been prepared by reduction of their metal salts in the presence of poly(amidoamine) dendrimers with various surface functional groups. In aqueous solutions of dendrimers with surface amino groups, stable nanoparticles of gold and platinum are obtained by adsorbing the surface amino groups of the dendrimers in which the particle size of the gold nanoparticles is considerably affected by the generation of the dendrimer as well as the dendrimer concentration. On the other hand, in nonaqueous solutions of dendrimers with surface methyl ester groups and hydrocarbon chains dendrimer-encapsulated noble metal nanoparticles are obtained. Thus, the size of the noble metal nanoparticles and the morphology of noble metal–dendrimer composites can be controlled in aqueous or nonaqueous solutions by using dendrimers with various surface groups.

Key words Noble metal nanoparticles · Dendrimers · Reduction · Encapsulation of nanoparticles

Introduction

The preparation of metal nanoparticles has been studied intensively because metal nanoparticles have been applied in electrooptical devices, electronic devices, imaging materials, catalysis, and so on. Fabrication of nanoparticles becomes one of the important topics in nanotechnology. Accordingly, for that purpose it is very important to be able to control the particle size, shape, and size distribution of metal nanoparticles. In the wet method, various metal nanoparticles have been prepared in the presence of polymers or surfactants as a protective colloid [1].

Recently, dendrimers having unique structures and properties have been attracting increasing attention [2–5]. Generally, dendrimers of lower generation tend to exist in relatively open forms, while higher generation dendrimers take a spherical three-dimensional structure, which is very different from linear polymers adopting random-coil structures. Dendrimers might provide reaction sites including their interior or periphery. In addition, we can also prepare water-soluble dendrimers or organic-solvent-soluble ones. Accordingly it is expected that nanoparticles prepared in the presence of dendrimers in aqueous or nonaqueous solutions are affected by the generation of dendrimers and show different behavior from those prepared using conventional linear polymers. In fact, some groups [6–9] have characterized the metal nanoparticles obtained in the presence of poly(amidoamine) dendrimers with various surface groups, such as amino, carboxyl or hydroxyl, and have shown various applications using such nanosized materials.

In this article, we describe the physicochemical properties of noble metal nanoparticles prepared in aqueous solutions or in nonaqueous solutions using poly(amidoamine) dendrimers with various surface functional groups.

Experimental

Materials

Poly(amidoamine) dendrimers were synthesized, involving exhaustive Michael addition to an ethylenediamine core with methyl acrylate and exhaustive amidation of the resulting esters with a large excesses of ethylenediamine [10]. The ester-terminated dendrimers were hydrolyzed by addition of NaOH to obtain dendrimers with surface carboxyl groups. In addition, a hydrophobically modified dendrimer was also synthesized [11] from the reaction of epoxy-1,2-dodecane and poly(amidoamine) dendrimer with surface amino groups, and is referred as GH(4). The dendrimers with surface amino groups are referred to as G(3–5) and those with surface carboxyl groups as G(3.5–5.5). The ester-terminated dendrimers are referred to as GE(1.5–5.5). Metal salts were kindly supplied by Tanaka Kikinzoku Kogyo. The water used in this study was purified through a Milli-Q Plus system. The other chemicals were of analytical grade. The structures of the dendrimers are shown in Scheme 1.

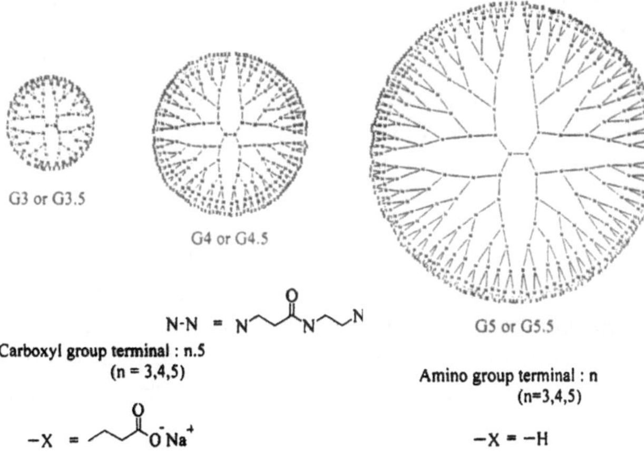

Methods and measurements

The metal salts and dendrimers were separately dissolved in deionized water or organic solvents. Then, both solutions were mixed with deionized water or organic solvents to make the desired concentrations of the metal salts and dendrimers. The solutions were stirred vigorously during the addition of excess reductant.

UV–vis spectra of the solutions before and after the reduction of the metal salts were measured with a UV spectrophotometer (Hewlett-Packard 8452A). The average particle sizes of the metals obtained were determined from several negative films (approximately 100 particles) taken using a transmission electron microscope (Hitachi H-800). The samples were prepared by mounting a drop of the solution on carbon-coated Cu grids and allowing the drop to dry in air.

Fourier transform IR (FT-IR) spectra for dendrimers and metal particles were obtained by forming a thin transparent KBr pellet containing the samples.

Hydrogenation of cyclohexene was carried out at 40 °C. The amount of catalyst used was 0.1 mmol dm^{-3} in 50 ml ethanol. Before addition of cyclohexene, hydrogen gas was bubbled in the ethanol for 20 min. The hydrogenation from cyclohexene to cyclohexane was determined by gas chromatography from the analysis of the reactant collected from the reaction vessel.

Results and discussion

Preparation of metal nanoparticles in aqueous solutions

Gold, platinum, and silver nanoparticles were prepared in aqueous solutions containing dendrimers by reduction of the respective metal salts with $NaBH_4$.

Transmission electron microscopy (TEM) micrographs and the particle size distribution of the gold particles are shown in Fig. 1. Here, the concentration of dendrimer means the concentration of the surface functional group. One can see that the average particle size of gold using G5 dendrimer is smaller than that using G3 dendrimer. To obtain the relationship between the particle size of gold and the generation of the dendrimer, the average diameter of the gold particles is plotted against the ratio of the concentrations of dendrimer to Au^{3+} for G3–G5 dendrimers in Fig. 2. The average diameter decreased rapidly and then gradually with increasing ratio of the concentration of dendrimer to Au^{3+} for G3–G5 dendrimer, where the average diameter for G3 dendrimer was greater than those for G4 and G5 dendrimers at the same ratio of the concentration of

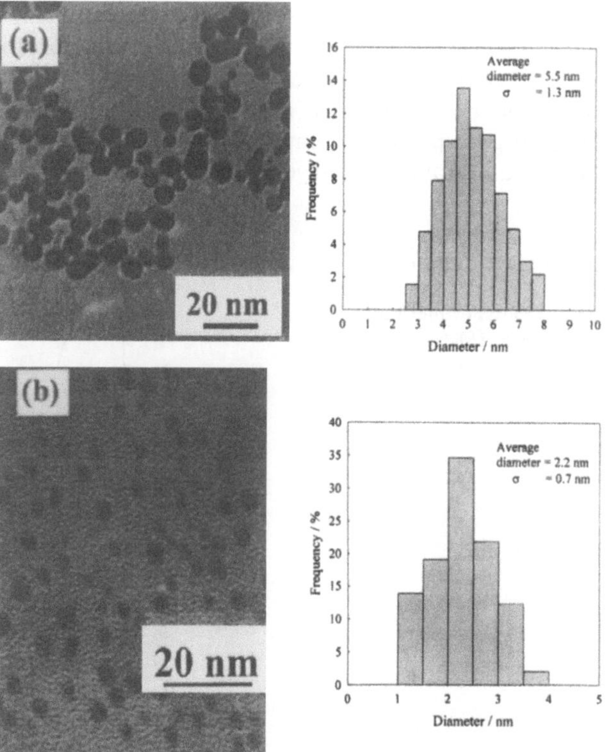

Fig. 1 Transmission electron microscopy (*TEM*) and particle size distribution of gold nanoparticles obtained in the presence of dendrimer: **a** $[HAuCl_4] = 0.2$ mmol dm^{-3}, $[G3] = 0.4$ mmol dm^{-3}, $[NaBH_4] = 2$ mmol dm^{-3}; **b** $[HAuCl_4] = 0.2$ mmol dm^{-3}, $[G5] = 0.4$ mmol dm^{-3}, $[NaBH_4] = 2$ mmol dm^{-3}

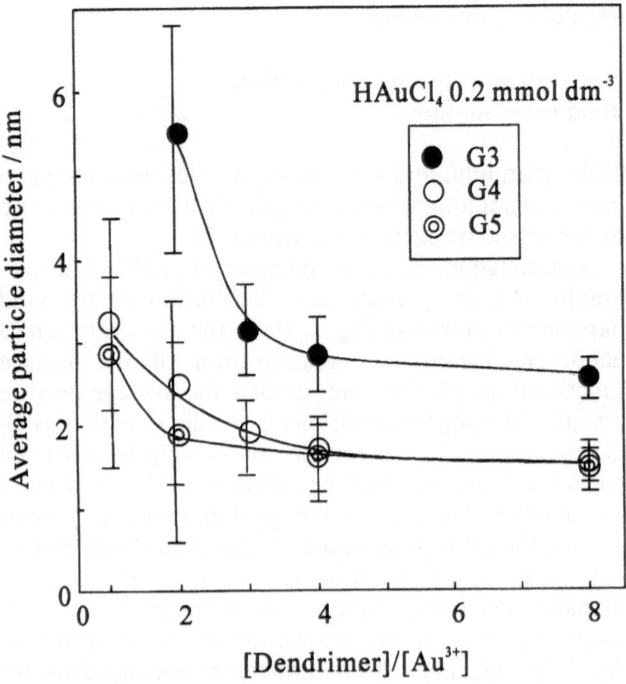

Fig. 2 Change in average particle size of gold nanoparticles with ratio of [dendrimer]/[Au^{3+}]

dendrimer to Au^{3+}. Thus, it is found that the dendrimers operate as a very effective protective colloid for the preparation of gold particles since only a very small amount of the dendrimer is required to obtain nanometer-sized gold particles compared to other linear polymers. In addition, it is suggested that the size and the shape of the dendrimers play an important role for the preparation of gold nanoparticles. A similar trend in the average particle size of gold with the generation has been reported for the reduction of Au^{3+} with UV radiation [6].

To confirm the interaction between gold nanoparticles and the dendrimers, FT-IR spectra of G5 dendrimer alone and gold nanoparticles obtained from dendrimer solution were measured (Fig. 3). The amide bands at 1,630 and 1,540 cm^{-1}, which are characteristic of the dendrimer branches, are very similar for both Fig 3, curves a and b. This may suggest that gold nanoparticles adsorb on the exterior of the dendrimers.

In the preparation of platinum particles, many aggregates of platinum colloids were observed below the ratio of the concentration of G3 to that of Pt^{4+} of 40:1 and stable platinum colloids could not be obtained. When the ratio increased, stable platinum nanoparticles were obtained. As one example, TEM micrographs of platinum colloids obtained in the presence of G4 and G5 dendrimers and their size distributions are given in Fig. 4. It is apparent that the size distributions are relatively small, and the average particle sizes are about 2.4 nm for both systems. In the case of G3 dendrimer, the average size was about 3.0 nm at the ratio of the concentration of dendrimer to that of Pt^{4+} of 80:1. Thus, these results indicate that the particle size of platinum is insensitive to the size and the shape of the dendrimers above some ratio of the concentration of dendrimer to Pt^{4+}. In addition, the fact that high dendrimer concentrations are required for stable platinum colloids compared to that for stable gold colloids may suggest that the interaction between platinum

Fig. 3 Fourier transform IR spectra of G5 dendrimer alone (*a*) and G5 dendrimer/gold nanoparticles (*b*)

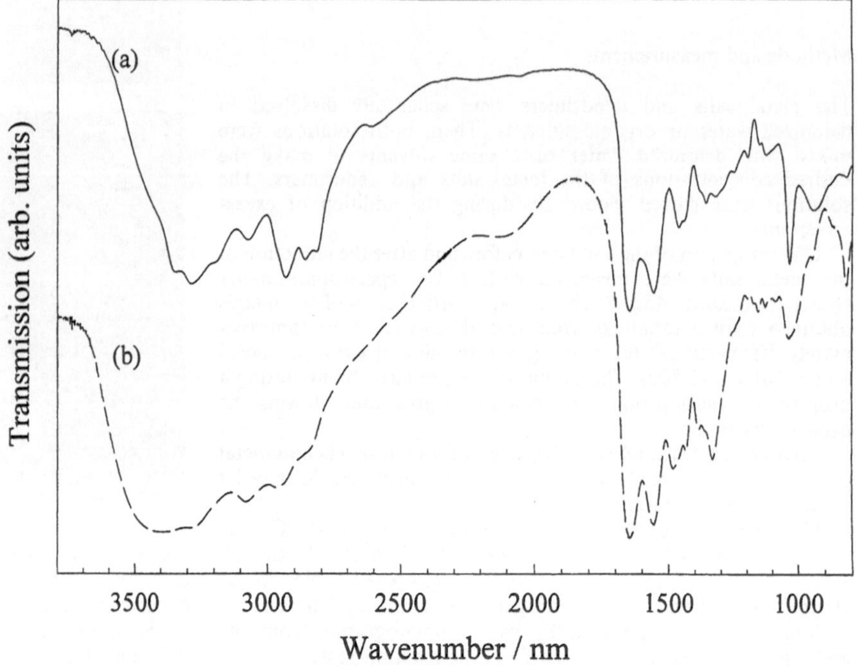

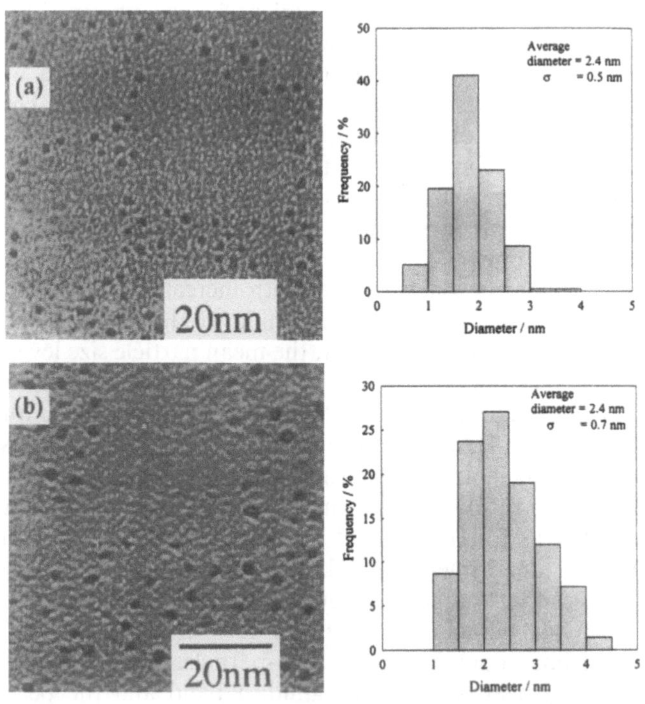

Fig. 4 TEM and particle size distribution of platinum nanoparticles obtained in the presence of dendrimer: **a** $[H_2PtCl_6] = 0.1$ mmol dm^{-3}, $[G4] = 8.0$ mmol dm^{-3}, $[NaBH_4] = 1$ mmol dm^{-3}; **b** $[H_2PtCl_6] = 0.1$ mmol dm^{-3}, $[G5] = 8.0$ mmol dm^{-3}, $[NaBH_4] = 1$ mmol dm^{-3}

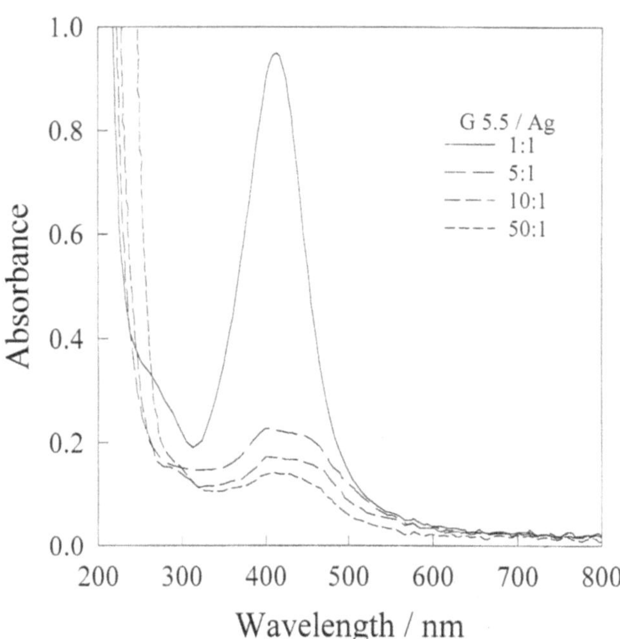

Fig. 5 UV–vis spectra of $AgNO_3$ in the presence of G5.5 dendrimer after reduction with $NaBH_4$

colloids and the dendrimers is weaker than that between gold colloids and the dendrimers. In fact, FT-IR spectra of the platinum colloids prepared in the presence of G5 dendrimer show that the absorption band due to the interaction between platinum colloids and amino groups of the dendrimer is very weak.

Since it is expected that Ag^+ ions adsorb strongly on dendrimers with surface carboxyl groups through electrostatic attractive forces, the reduction of $AgNO_3$ by addition of $NaBH_4$ in the presence of G3.5, G4.5, or G5.5 dendrimer was carried out. In the absence of the dendrimers we only obtained very unstable silver particles, which sedimented rapidly. When Ag^+ ions were reduced with $NaBH_4$ in the presence of the dendrimers, the color of the silver colloids changed with increasing concentration of G5.5; yellow at the ratio of the concentration of G5.5 to that of Ag^+ of 1:1, but turned to orange at higher ratios. In the case of G3.5 and G4.5 dendrimers, a similar result was observed as for G5.5 dendrimer. It is interesting to note that the time required for the color change to yellow or orange after addition of $NaBH_4$ increases with increasing concentration of the dendrimers because of the repulsive action of BH_4^- with the negatively charged dendrimers that carry reacting Ag^+ ions. Such a delay in the reduction of Ag^+ ions has been observed in the presence of polyacrylate [12]. Unfortunately, above the ratio of the concentration of G5.5 to that of Ag^+ of 5:1 we could not observe any silver particles by TEM because the particle size was too small to detect under this condition. The absorption spectra of $AgNO_3$ in the presence of G5.5 after reduction with $NaBH_4$ (Fig. 5) show that a typical plasmon band of silver colloids is observed at 380 nm at the ratio of the concentration of G5.5 to Ag^+ of 1:1 and with an increase in the ratio another band appears at 450 nm. A similar absorption band at 440 nm for silver cluster has been reported when the reduction of Ag^+ on polyacrylate in aqueous solution is carried out by γ irradiation [12].

Preparation of metal nanoparticles in nonaqueous solutions

Two nonaqueous systems have been investigated to prepare nanosized gold [13], platinum [14], and silver particles. One is the reduction of metal salts with dimethylamineborane in ethyl acetate in the presence of dendrimers with surface methyl ester groups. The other is that metal ions are extracted from aqueous solution to organic solvents containing a hydrophobically modified dendrimer, followed by reduction of metal ions in organic solvents with dimethylamineborane [15].

When $HAuCl_4$ in ethyl acetate was added to dendrimers in ethyl acetate, the formation of a dendrimer–Au^{3+} complex was observed (Fig. 6). As shown in Fig. 6a, $HAuCl_4$ in ethyl acetate shows an intense absorption at 324 nm, which can be assigned to the ligand-to-metal charge transfer band of $AuCl_4^-$. On the other hand, GE5.5 dendrimer shows a small absorption

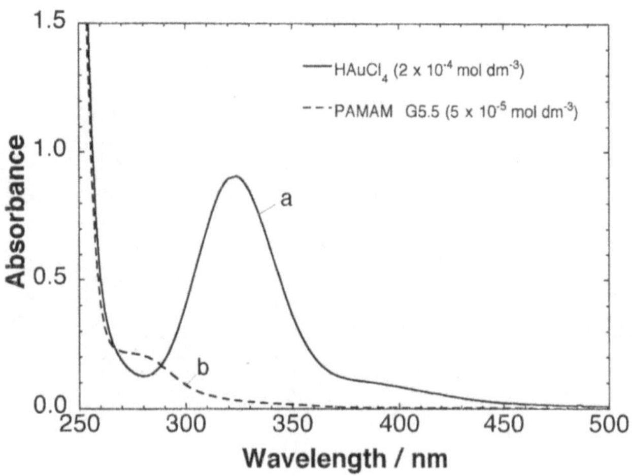

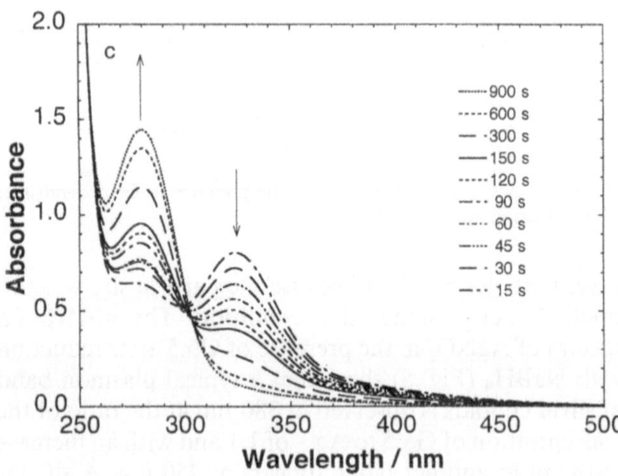

Fig. 6 UV–vis spectra of $HAuCl_4$–dendrimer in ethyl acetate: **a** $HAuCl_4$; **b** GE5.5; **c** mixture of $HAuCl_4$ and GE5.5 with running time

maximum at 278 nm, as shown in Fig. 6b. On mixing Au^{3+} with GE5.5 dendrimer, the absorption spectra are changed as shown in Fig. 6c. A new band develops at 280 nm at the expense of the 324 nm band and the latter completely disappears in 15 min of stirring. In addition, an isobestic point is found at 304 nm, indicating that a ligand substitution from $AuCl_4^-$ to another Au^{3+} complex occurs through the participation of GE5.5 dendrimer. To confirm the complex formation of GE5.5 dendrimer and Au^{3+}, the FT-IR spectra of GE5.5 dendrimer were measured in the absence or presence of Au^{3+}. It was found that the bands for the surface methyl ester group at 1,044, 1,200, 1,438, and 1,736 cm^{-1} do not change by introducing Au^{3+}, while the amide I band at 1,647 cm^{-1} for interior amide groups undergoes a high-frequency shift to 1,649 cm^{-1} and the single peak for the amide II band splits into two peaks on addition of Au^{3+}. These results suggest that the interior amide groups of GE5.5 dendrimer are responsible for the coordination to Au^{3+}. A similar result for the formation of a complex of dendrimers and Au^{3+} was obtained for GE1.5 and GE3.5 dendrimers. Au^{3+} ions in the presence of dendrimers in ethyl acetate were readily reduced by adding an excess amount of dimethylamineborane. The TEM images of Au nanoparticles prepared at various dendrimer concentrations for a fixed Au^{3+} concentration (0.2 mmol dm^{-3}) are shown in Fig. 7. One can see that although large particles are produced when the concentration of GE5.5 dendrimer is low, the particle size of gold becomes smaller with increasing dendrimer concentration and above an amide to $HAuCl_4$ concentration ratio of 10 or greater, the mean particle size levels off at 3.5 + 0.2 nm. Furthermore, it is important to note that the maximum particle size is in the range 7–8 nm for an amide to $HAuCl_4$ concentration of 10 or greater. This size is in good agreement with the hydrodynamic diameter of GE5.5 dendrimer in ethyl acetate. A similar result was found for gold nanoparticles obtained with GE3.5 dendrimer ($d_H = 4.8$). These results strongly suggest that the gold nanoparticles are encapsulated in the dendrimer molecules. The FT-IR spectrum for GE5.5 dendrimer containing Au nanoparticles also indicates that the peak shift of the amide I band and the peak splitting of the amide II band remain after reduction of Au^{3+} to Au, while only a small peak shift is observed for the stretching band of the surface methyl ester group. However, since d_{max} is larger than d_H in the case of GE1.5 dendrimer, encapsulation of gold nanoparticles by GE1.5 dendrimer may not occur.

Platinum nanoparticles were also prepared in ethyl acetate in the presence of dendrimers with a surface methyl ester group. Similar to the preparation of gold nanoparticles, it is important to control the time elapsed after mixing H_2PtCl_6 and the dendrimer in ethyl acetate. For example, the reduction of $PtCl_6^{2-}$ ions in the presence of GE5.5 dendrimer by addition of an excess amount of dimethylamineborane (90 times that of $PtCl_6^{2-}$) after 5 min of elapsed time provides large and small platinum particles, as shown in Fig. 8. It is very interesting to visualize by TEM that small platinum particles are encapsulated by GE5.5 dendrimer, while large platinum particles are covered with GE5.5 dendrimer. When the time elapsed increased to 1 day, it was found that relatively uniform platinum nanoparticles which are encapsulated by GE5.5 dendrimer were obtained; however, in the case of GE1.5 dendrimer, platinum nanoparticles having diameters of 8.8–10.4 nm were obtained; the particle size was greater than the diameter of GE1.5 dendrimer, indicating that GE1.5 dendrimer is a protective colloid rather than an encapsulating agent. Similarly, silver nanoparticles were prepared in ethyl acetate using GE5.5 dendrimer.

Gold nanoparticles were prepared through the extraction of Au^{3+} from aqueous solution to toluene or chloroform containing GH4 dendrimer and reduction by addition of dimethylamineborane. The transfer ratio of

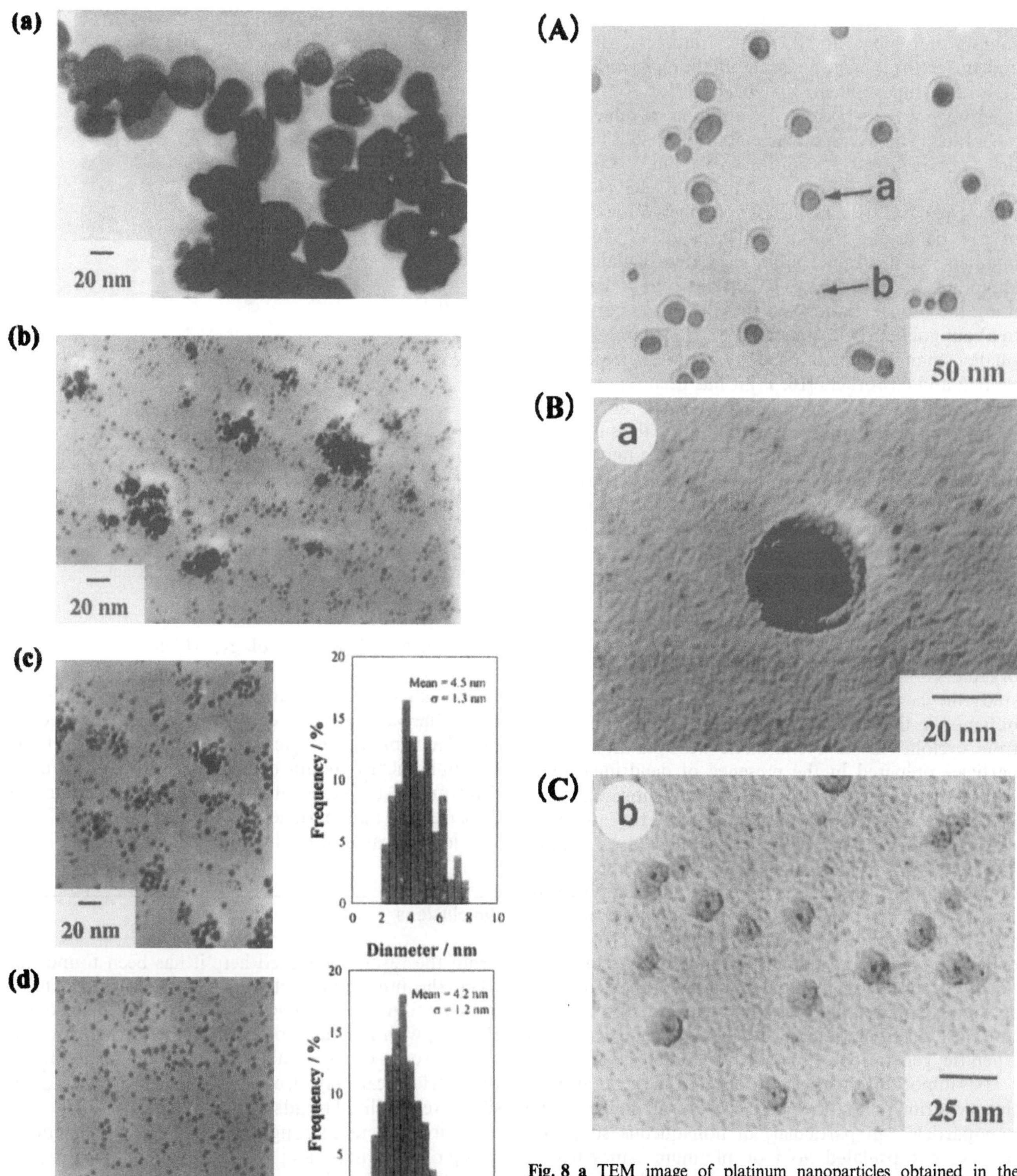

Fig. 7 TEM and size distribution of gold nanoparticles prepared at various GE5.5 dendrimer concentrations in ethyl acetate: **a** [amide]/[Au^{3+}] = 1.5; **b** [amide]/[Au^{3+}] = 5; **c** [amide]/[Au^{3+}] = 10; **d** [amide]/[Au^{3+}] = 15

Fig. 8 a TEM image of platinum nanoparticles obtained in the presence of GE5.5 dendrimer after 5 min elapsed time. **b** Enlargement of the particle in **a**. **c** Enlargement of the particle in **b**

Au^{3+} from aqueous solution to toluene containing GH4 dendrimer was as follows: 30% at 0.01 g dm^{-3}, 40% at 0.03 g dm^{-3}, and 42% at 0.05 g dm^{-3} dendrimer, where the initial concentration of Au^{3+} in aqueous solution was kept constant at 0.2 mmol dm^{-3}. Since Au^{3+} is not soluble in toluene at all, Au^{3+} transferred from the aqueous to the toluene phase should be incorporated into GH4 dendrimer. Then, the reduction of Au^{3+} in the dendrimer was carried out by addition of dimethylamineborane. The average diameters of the gold nanoparticles obtained was 2.5 nm for 0.01 g dm^{-3}, 3.0 nm for 0.03 g dm^{-3}, and 3.4 nm for 0.05 g dm^{-3} GH4 dendrimer, respectively, the diameters being below the diameter of G4 dendrimer (4.5 nm). The average Au^{3+} ion numbers per dendrimer molecule were calculated to be about 300–80 from 0.01 to 0.05 g dm^{-3} GH4 dendrimer using the transfer ratios and these numbers correspond to about gold particle of diameter 2.1–1.4 nm, which are smaller than those obtained by TEM. According to Balogh and coworkers [16, 17] it has been reported that noble metal particles consisting of small numbers of atoms or molecules dispersed within or on the surfaces of the dendrimers are often amorphous. This may suggest that the the gold particles formed in the GH4 dendrimer are composed of many gold clusters. This idea can explain the gold particle size obtained by TEM observation. Gold nanoparticles were also obtained in the chloroform–Au^{3+}–GH4 dendrimer system; their diameters ranged between 3.0 and 4.0 nm.

Polymer-stabilized noble metal particles have been used as catalysts for the hydrogenation of unsaturated organic compounds. Accordingly, it is interesting to study the catalytic activity of noble metal nanoparticles obtained in the presence of dendrimers. The conversion from cyclohexene to cyclohexane with platinum nanoparticles prepared in the presence of dendrimers with surface methyl ester groups in ethanol is shown in Fig. 9 [18]. It is clearly demonstrated that the platinum nanoparticles obtained in the presence of GE1.5 dendrimer have much higher catalytic activity than those obtained in the presence of GE3.5 and GE5.5 dendrimers. At the reaction time of 24 h, the conversion was about 86% for GE1.5 dendrimer and 48% for GE3.5 dendrimer and GE5.5 dendrimer, respectively. This result may be derived from a unique structure of the dendrimer so that cyclohexene molecules are less able to penetrate into the interior platinum nanoparticles of the dendrimer when the generation of the dendrimer increases.

It is important to discuss the effect of the generation of the dendrimer for the preparation of noble metal nanoparticles, in particular, in nonaqueous solutions, because encapsulated gold or platinum nanoparticles were obtained in ethyl acetate, toluene, or chloroform. In the early generation of the dendrimers, relatively large nanoparticles are obtained in which the dendrimers adsorb on the nanoparticles and act as protective colloids. This behavior is very similar to that of linear polymers because the earlier generation dendrimers have an open structure or are less compact. On the other hand, for the later generations, the dendrimers consist of a more dense, spheroid topology, which can incorporate metal ions in the interior of the dendrimers. As a result, encapsulated noble metal nanoparticles are obtained in which the particle size is controlled by the generation of the dendrimers. In particular, in the case of the hydrophobically modified dendrimer–Au^{3+}–toluene or dendrimer–Au^{3+}–chloroform systems, relatively monodisperse gold nanoparticles are obtained since Au^{3+} ions should be located in the interior of the dendrimer.

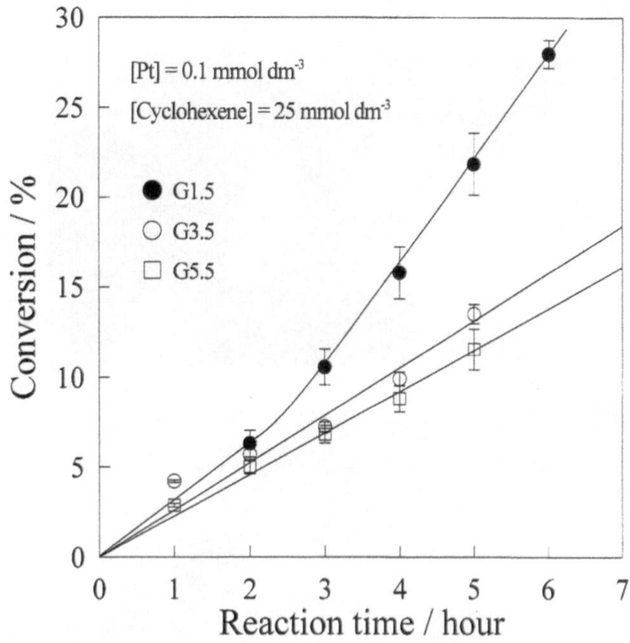

Fig. 9 Hydrogenation of cyclohexene by platinum nanoparticles prepared in the presence of GE dendrimers in ethanol

Conclusions

From the results presented here it has been found that when the interaction between metal ions and exterior functional groups of dendrimers is strong, for example, $HAuCl_4$ with poly(amidoamine) dendrimer with surface amino groups or Ag^+ with poly(amidoamine) dendrimer with surface carboxyl groups, nanoparticles of gold and silver are stabilized by adsorption of the dendrimer, while metal ions interact strongly with the interior functional group of dendrimers and encapsulated nanoparticles are often formed. In addition, an increase of the generation as well as of the concentration leads to the formation of smaller nanoparticles. Thus, dendrimers play a very interesting role for the preparation of nanoparticles and is different from conventional linear polymers. The hydrogenation of cyclohexene is considerably affected by platinum nanoparticles prepared by different generation dendrimers.

References

1. Bradley JS (1994) In: Schmid G (ed) Clusters and colloids: from theory to applications. VCH, Weinheim, pp 459–544, and references therein
2. Tomalia DA, Naylor AM, Goddard WA III (1990) Angew Chem Int Ed Engl 29:138–175
3. Frechet JM (1994) Science 263:1710–1715
4. Bosman AW, Janssen HM, Meijer EW (1999) Chem Rev 99:1665–1688
5. Zeng F, Zimmerman SC (1997) Chem Rev 97:1681–1712
6. Esumi K, Suzuki A, Aihara N, Usui K, Torigoe K (1998) Langmuir 14:3157–3159
7. (a) Esumi K, Hosoya T, Suzuki A, Torigoe K (2000) Langmuir 16:2978–2980; (b) Esumi K, Suzuki A, Yamahira A, Torigoe K (2000) Langmuir 16:2604–2608
8. (a) Carcia ME, Baker LA, Crooks RM (1999) Anal Chem 71:256–258; (b) Zhao M, Crooks RM (1999) Adv Mater 11:217–220; (c) Chechik V, Zhao M, Crooks RM (1999) J Am Chem Soc 121:4910–4911; (d) Chechik V, Crooks RM (2000) J Am Chem Soc 122:1243–1244
9. He J-A, Valluzzi R, Yang K, Dolukhanyan T, Sung C, Samuelson L, Balogh L, Tomalia DA (1999) Chem Mater 11:3268–3274
10. Tomalia DA, Baker H, Dewald J, Hall M, Kallos G, Martin S, Roeck J, Ryder J, Smith P (1986) Macromolecules 19:2466–2468
11. Sayed-Sweet Y, Hedstrand DM, Spinder R, Tomalia DA (1997) J Mater Chem 7:1199–1205
12. Ershov BG, Henglein A (1998) J Phys Chem B 102:10667–10671
13. Torigoe K, Suzuki A, Esumi K (in press)
14. Esumi K, Nakamura R, Suzuki A, Torigoe K (2000) Langmuir 16:7842–7846
15. Esumi K, Hosoya T, Suzuki A, Torigoe K (2000) J Colloid Interface Sci 229:303–306
16. Balogh L, Valluzzi R, Laverdure KS, Gido SP, Hagnauer GL, Tomalia DA (1999) J Nanoparticle Res 1:353–368
17. Balogh L, Tomalia DA, Hagnauer GL (2000) Chem Innov 30:19–26
18. Esumi K, Satoh K, Suzuki A, Torigoe K (2000) J Jpn Soc Colour Mater 73:434–437

Progr Colloid Polym Sci (2001) 117: 88–93

K. Mogyorósi
J. Németh
I. Dékány
J. H. Fendler

Preparation, characterization, and photocatalytic properties of layered-silicate-supported TiO_2 and ZnO nanoparticles

K. Mogyorósi · J. Németh
I. Dékány (✉)
Department of Colloid Chemistry, University of Szeged and Nanostructured Materials Research Group of the Hungarian Academy of Science, Aradi v.t. 1, 6720 Szeged, Hungary
e-mail: i.dekany@chem.u-szeged.hu
Tel.: +36-62-544210
Fax: +36-62-544042

J. H. Fendler
Center for Advanced Materials Processing, Clarkson University, Potsdam, NY 13699-5814, USA

Abstract TiO_2 nanoparticles were prepared in 2-propanol and in water from titanium alkoxide, and ZnO nanoparticles were formed from aqueous $ZnCl_2$ either in situ in sodium montmorillonite and in synthetic hectorite or, alternatively, were incorporated into the silicate lamellas. Transmission electron microscopy (TEM), X-ray diffraction, and Brunauer–Emmett–Teller measurements were used to establish the incorporation of semiconductor nanoparticles into the interlamellar space of the clays. The average particle diameter of the nanoparticles, determined by TEM, ranged between 4 and 14 nm. The photocatalytic activity of TiO_2, intercalated into montmorillonite, for salicylic acid photooxidation was determined to be better than that elicited by TiO_2 alone.

Key words TiO_2 · ZnO nanoparticles · Montmorillonite · Intercalation · Photooxidation

Introduction

The preparation and photochemical properties of semiconductors has attracted increasing attention from a wide range of scientific interests [1, 2]. Photofunctional semiconductors play an important role in the photocatalytic degradation of organic pollutants in water and soil [3, 4]. In order to control the photocatalytic properties we also have to control the size, the dispersity, and the stability of the nanoparticles. This aim can be achieved by synthesizing the nanoparticles in droplets of microemulsions [5], in micellar systems [6], in solid–liquid interfacial adsorption layers as nanophase reactors, or in the interlayer spacings of clay minerals [7–16]. The intention of the present work was the preparation of TiO_2 and ZnO photocatalysts without any organic solvents or surfactants. The synthesis occurred in aqueous suspensions of layered silicates (montmorillonite and synthetic hectorite).

Experimental

Materials

TiO_2 nanoparticles were prepared using the sol–gel method by the hydrolysis of titanium(IV) ethoxide (tetraethyl orthotitanate, Merck, pro anal.) and titanium(IV) isopropoxide (tetraisopropyl orthotitanate, Fluka Chemika, pract.) in Milli-Q water (18 MΩcm deionized water, Millipore Co., MilliQ system). For reference materials we chose P 25 Degussa TiO_2. The sodium montmorillonite (Wyoming montmorillonite, USA, 85 mEq/100 g clay cation-exchange capacity) was used in the 2-propanol (Reanal, a.r.) medium as a stabilizing agent and support. The basal distance in the air dry state is 1.2 nm. We also used other layered silicates, such as Optigel SH (Süd-Chemie, synthetic hectorite). For the synthesis of nanosol hydrochloric acid (Reanal, pro anal.) and for the neutralization of the sols NaOH (Reanal, puriss) were used.

Zinc(II) chloride (Reanal), sodium hydroxide (Reanal) and ZnO (Reanal) were used as received. Sodium montmorillonite was purified by removing particles that have a diameter greater than 2 mm by sedimentation for 12 h. Smaller montmorillonite particles were removed by suction from the coarse settled minerals.

Sedimentation was repeated twice and the sodium montmorillonite nanoparticles were isolated by centrifugation (15 min at 5,000 rpm).

The water used in all the preparations was purified by a Milli-Q system (Millipore), resulting in a resistivity of 18 MΩcm.

Methods

Sample preparation

TiO_2 sol preparation method

Titanium dioxide nanoparticles were prepared by the hydrolysis of titanium(IV) alkoxides and subsequent heat treatment. In a typical preparation 500 ml 5.3 vol% titanium(IV) tetraisopropoxide–2-propanol mixture was added drop by drop to purified 4000 ml Milli-Q water under vigorous stirring over 45 min to yield a milky white dispersion. Subsequent to the addition of 27.2 ml concentrated hydrochloric acid the dispersion was stirred at 50 °C for 12 h. During the stirring the dispersion became a transparent homogeneous sol. After calcination (at 400 °C for 4 h) the average particle size was 5.4 nm (standard deviation 1.1 nm) (sample: TiO_2/0/S/100%, see sample notation in the Appendix).

Preadsorption of alkoxides

In these experiments we adsorbed titanium(IV) ethoxide and titanium(IV) isopropoxide onto Wyoming sodium montmorillonite and onto synthetic hectorite. In a typical preparation we dispersed 1.0 g montmorillonite in 50 ml 2-propanol, then we added a 24 ml solution of titanium(IV) isopropoxide in 2-propanol, $c_{alkoxide}$ = 17 v/v%. After 20 min in the adsorption equilibrium 33.5 ml 40 v/v% water–2-propanol mixture was added drop by drop to the suspension under vigorous stirring at ambient temperature. The system was allowed to stand for 2 h, centrifuged and after this dried at 60–70 °C (sample TiO_2/M/P/50%, since the TiO_2 content is approximately 50%). This dispersion was repeated a number of times with different amounts of alkoxide (0.9–13.3 mmol alkoxide/g montmorillonite, samples TiO_2/M/P/7%, TiO_2/M/P/50%) and with the other silicate support – at constant alkoxide–water ratio (1:56 molar ratio).

Heterocoagulation method

The TiO_2 nanosol (pH 1.5, 0.1 wt%) was neutralized with 1 M NaOH solution to pH 4. We washed and centrifuged the precipitated material with Milli-Q water three times. The TiO_2 sediment was redispersed into Milli-Q water and later the pH was adjusted at pH = 4.0 with 0.1 M HCl solution. In this state the sol has positively charged particles which coagulate in a montmorillonite suspension wherein the platelets are negatively charged. The hydrolysis product, titanium(IV) oxide/hydroxide, was calcined in air at 320 °C for 4 h (TiO_2/M/P/7%, TiO_2/M/P/50%), at 400 °C for 4 h (hydrolyzed from alkoxide TiO_2/0/A/100%, heterocoagulated montmorillonite/TiO_2 sample TiO_2/M/H/33%), or at 450 °C for 6 h (preadsorption method, Hec/TiO_2 nanocomposite TiO_2/Hec/P/50%).

Montmorillonites intercalated by ZnO nanoparticles were prepared by heterocoagulation of previously prepared $Zn(OH)_2$ sols with negatively charged platelets of the clay mineral. The following method was used for the heterocoagulation of ZnO montmorillonite complexes. $ZnCl_2$ (1.5 g) was dissolved in 1.5 ml distilled water and subsequent to dissolution 4.0 ml aqueous 1.0 M NaOH was quickly introduced to produce $Zn(OH)_2$. The crude precipitate was homogenized by stirring and ultrasonication and was diluted with distilled water to 1,600 ml and after further ultrasonication was characterized using a particle charge detector. The surface potential of the sol particles was 429 mV, the pH was 6.7, and the $Zn(OH)_2$ content was 0.2 g in 1,600 ml of the sol. The $Zn(OH)_2$ sol was added to 100 (sample: ZnO/M/H/14%) and 10 ml (sample: ZnO/M/H/61%) 1% montmorillonite suspension. The samples were centrifuged, dried, and calcined at 400 °C for 3 h.

Measurement methods

The formation of the anatase and the thermostability of the clay carriers were established by thermoanalytical measurements using a MOM Derivatograph Q-1500 D instrument.

X-ray diffraction (XRD) measurements were performed on solid samples using a Philips 1800 X-ray diffractometer (Cu Kα, 1.5418 nm, 40 kV, 35 mA). The basal distances (d_L) were calculated from the first (001) Bragg reflections by using the PW 1877 automated powder diffraction software.

Surface areas were determined by N_2 adsorption (Brunauer–Emmett–Teller, BET) experiments using a Gemini 2735 (Micromeritics) sorptometer at 77 ± 0.5 K. Prior to the measurements the samples were pretreated in a vacuum at 393 K for 2 h. The sample vessel was loaded with about 0.1–0.7 g sample. Step-by-step (cumulative) measurements were controlled by the Gemini software.

The surface-potential examinations of the sol particles were carried out using a Mutek PCD 02 particle charge detector with Au electrodes.

Transmission electron microscopy (TEM) images were taken using a Philips CM-10 electron microscope, using an accelerating voltage of 100 kV. Powder samples were suspended in ethanol (at a concentration of approximately 0.01%) and the aliquots were dropped on 2-mm diameter Formvar-coated copper grids. The particle size distributions were determined by using the UTSSCSA Image Tool program.

The analysis of the samples after photooxidation was carried out using a UVIKON 930 UV–vis dual-beam spectrophotometer. The degradation of model pollutants (salicylic acid and surfactants) was monitored by UV–vis measurements.

Results and discussion

Nanoparticle growth and incorporation into layer silicates

The calcination of $Ti(OH)_4$ is a multistep polycondensation process. By thermogravimetric measurements it was established that the transformation of $Ti(OH)_4$ into TiO_2 involved the endothermic loss of water up to 200–250 °C and the subsequent exothermic formation of the anatase crystals between 300 and 400 °C. The optimum calcination temperature is between 400–450 °C, i.e., below the dehydroxylation temperature of the silicates. The formation of the anatase crystal structures manifested itself in the broadening of the XRD peaks at $2\Theta = 25.3°$, which permitted the assessment of the nanoparticle diameters possible by using the Scherrer equation.

The XRD measurements showed that heterocoagulation method results in an intercalated structure of montmorillonite (Fig. 1). The basal distance ranges from 2.5 to 5.5 nm in the case of TiO_2/montmorillonite nanocomposites. In the case of the ZnO nanoparticles, Fig. 2 shows the XRD patterns of the nanocomposites prepared by heterocoagulation synthesis. Since pillaring of the montmorillonite by ZnO is incomplete, the peak

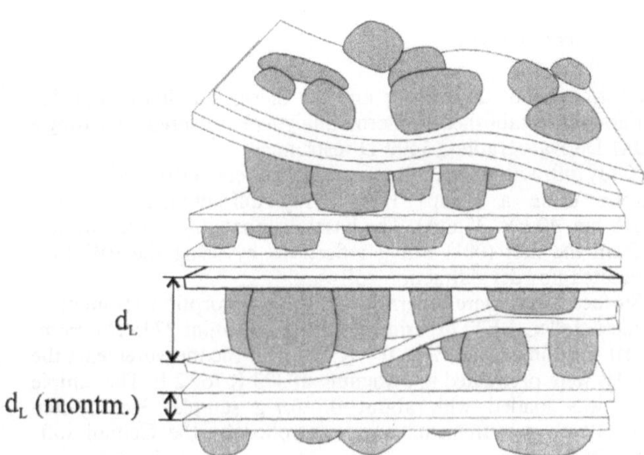

Fig. 1 Heterocoagulated structure: semiconductor particles on montmorillonite platelets

due to the basal spacing of the clay lamellae ($d_L = 0.96$ nm) appears along with that due to the intercalated ZnO ($d_L = 5.6$–6.9 nm, Table 1). Increasing the amounts of ZnO diminishes, however, the reflection due to the montmorillonite basal spacing (in the absence of any intercalating compounds).

Representative reflections of the ZnO wurtzitic structure can be seen between 30 and 40°. Sample ZnO/M/H/14%, which contains 10 times more stabilizing montmorillonite than sample ZnO/M/H/61%, shows lower and broader peaks, indicating the presence of smaller ZnO particles (see TEM measurements).

The XRD patterns of ZnO/montmorillonite complexes also include an intercalation peak which shows the formation of the pillared structure. The exact structural parameters can be seen in Table 1.

Using the preadsorption method the TEM images indicated the presence of TiO_2 particles of 2–3-nm diameter between the lamellae, but particles are much larger (10–14 nm) without the support. By the heterocoagulation method the average diameter of the TiO_2 (anatase) particles is 6.5 nm. The TiO_2 particles are approximately 5.4 nm by the sol method and 7.5 nm by normal hydrolysis of alkoxide in 2-propanol after calcination.

The electron micrographs clearly show the presence of the nanoparticles. These images can also be used for the construction of particle size distribution functions. A representative picture and size distribution function of ZnO nanoparticles can be seen in Figs. 3 and 4.

The reference ZnO (Reanal) has an average particle diameter of 700 nm. Samples ZnO/M/H/14% and ZnO/M/H/61% (Figs. 3, 4) were prepared from the same $Zn(OH)_2$ sol by the heterocoagulation method. An important difference between these composites is that the amount of the stabilizing montmorillonite is 10 times greater in sample ZnO/M/H/14% than in sample ZnO/M/H/61%. The greater amount of stabilizing agent results in smaller nanoparticles and better monodispersity as can be seen in the pictures showing the size distribution functions. A possible explanation is that the increasing number of the negatively charged montmorillonite lamellae can better take apart the positively charged, aggregated secondary particles of the $Zn(OH)_2$-sol. The TEM images demonstrated the particles to be in the nanosize range and that their size and dispersity can be controlled by adding different amounts of the stabilizing montmorillonite.

The data collected in Table 1 show that incorporation of nanoparticles results in a strong effect on the specific surface area of sodium montmorillonite. The BET specific surface areas of samples containing anatase varied between 50 and 270 m^2/g as determined by N_2 adsorption measurement. These data are in the

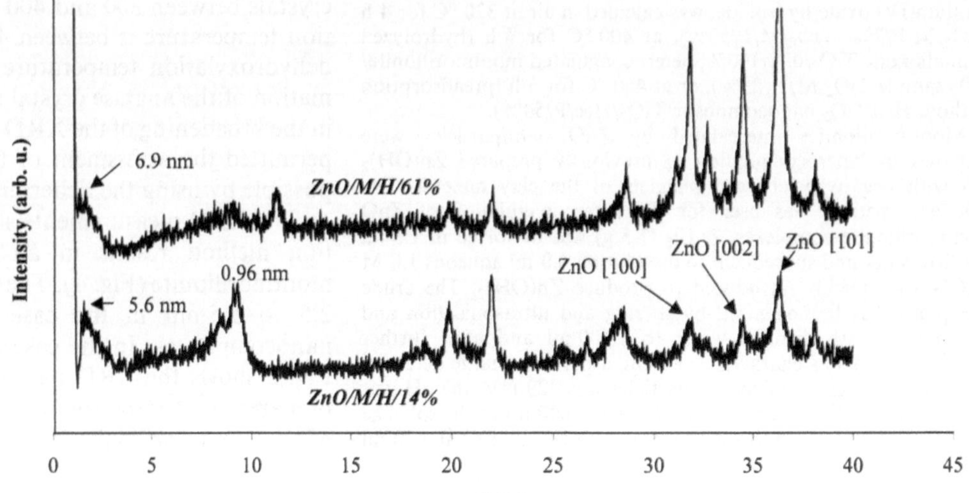

Fig. 2 X-ray diffraction patterns of calcined (400 °C for 3 h) ZnO montmorillonite complexes synthesized by the heterocoagulation method, ZnO/M/H/14%: 2 mmol ZnO g montorillonite, ZnO/M/H/61%: 20 mmol ZnO/g montmorillonite. There are intercalation peaks in the low-angle range, and in the 30–40° range the representative reflections of the wurtzitic ZnO structure show the different rates of the crystallinity

Table 1 Structural prpoperties of TiO_2 and ZnO layer silicate nanocomposites

Name of the sample	Semiconductor content		Basal distance (nm)	Specific surface area (m^2/g)	Mean particle diameter (nm)
	(wt%)	(mmol/g)			
Sodium montmorillonite calcined at 300 °C	–	–	0.96	4.9	–
TiO_2 (P 25)	100	–	–	50.0	22
TiO_2/0/S/100%	100	–	–	105.8	5.4
TiO_2/M/P/7%	6.7	0.9	3.9(0.96)	55.0	–
TiO_2/M/P/50%	51.6	13.3	4.0	142.8	5.7
TiO_2/Hec/P/50%	51.4	13.2	4.08	267.7	10.3
TiO_2/M/H/33%	33	8.3	6.3	130.3	4.3
ZnO (Reanal)	100	–	–	2.3	700
ZnO/M/H/14%	13.8	2	5.6	31.2	6.3
ZnO/M/H/61%	61.5	20	6.9	42.4	13.4

range 30–40 m^2/g after calcination in the case of ZnO intercalated montmorillonites, while the montmorillonite support without nanoparticles and reference ZnO show only values under 5 m^2/g. This phenomenon verifies the higher degree of dispersity of nanocomplexes rather than that of the bulk materials, and higher dispersity ensures a better possibility of using them as catalysts.

As can be observed, on increasing the semiconductor content by adding more zinc or tin hydroxide sol, the

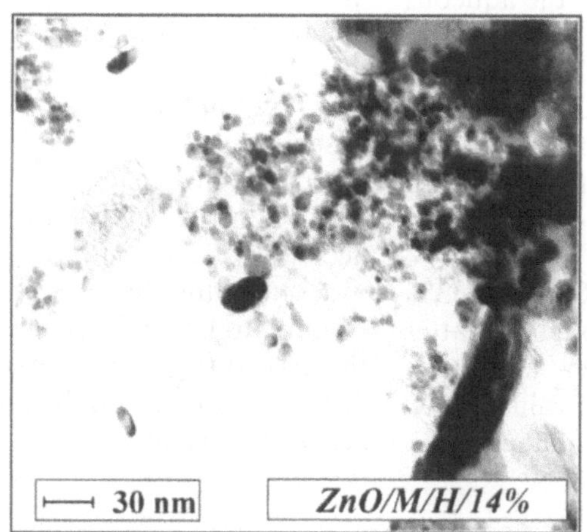

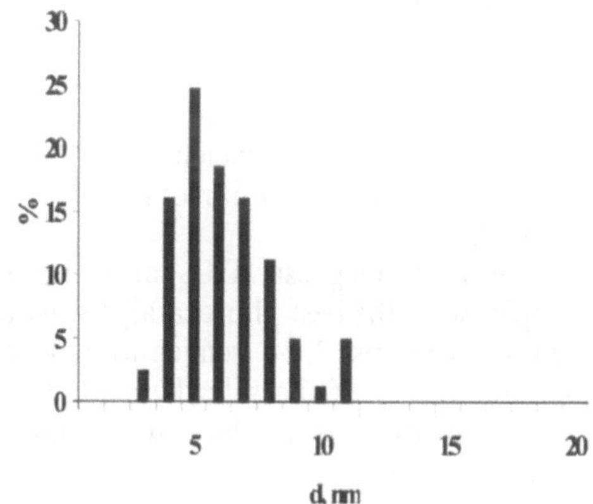

Fig. 3 Transmission electron microscopy (*TEM*) image of ZnO/M/H/14% ZnO/montmorillonite nanocomposite and its size distribution. The ZnO content is 2 mmol ZnO/g montmorillonite. The mean particle diameter is 6.3 nm

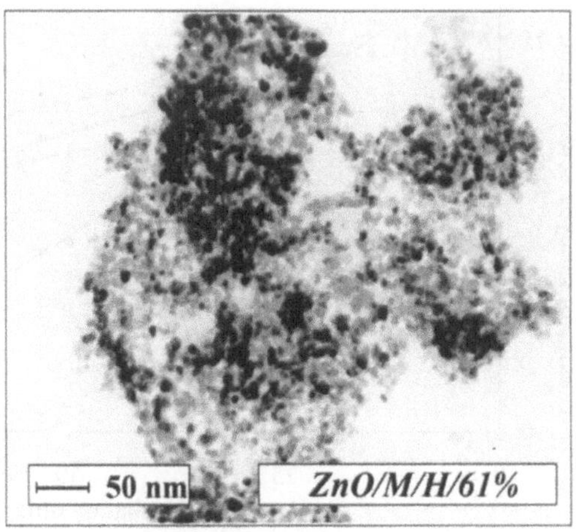

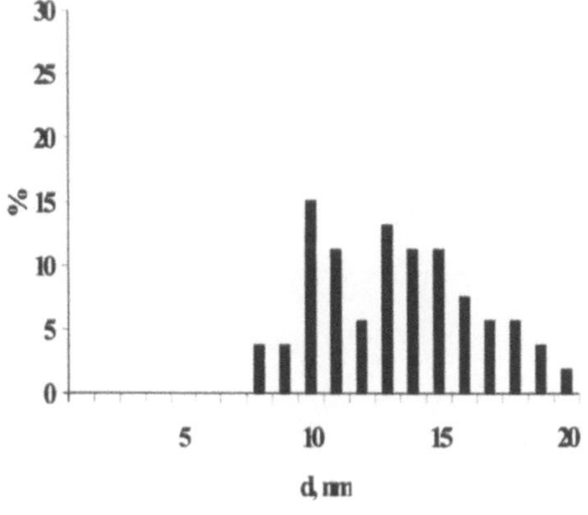

Fig. 4 TEM image of ZnO/M/H/61% ZnO/montmorillonite nanocomposite and its size distribution. The ZnO content is 20 mmol ZnO/g montmorillonite. The mean particle diameter is 13.4 nm

Table 2 The photocatalytic efficiency of the catalysts using a Pyrex reactor (λ = 310–390 nm) or a quartz reactor

Name of the sample	Semiconductor content (wt%)	Amount of pollutant removed (%)	Amount of pollutant removed (mmol/g semiconductor)
TiO_2 (P 25)	100	83.0	0.125
TiO_2/0/S/100%	100	51.8	0.078
TiO_2/M/P/50%	51.6	42.1	0.122
TiO_2/M/H/33%	33	49.1	0.223
ZnO (Reanal)[a]	100	88.4	0.133
ZnO/M/H/14%[a]	13.8	34.3	0.372
ZnO/M/H/61%[a]	61.5	46.2	0.113

[a] λ = 250–390 nm

specific surface area also increases, indicating the incorporation of the pillaring sols.

Photocatalytic properties of nanocomposites

The photocatalytic efficiency of the synthesized and the reference materials was tested by the degradation of salicylic acid in a Pyrex and a quartz UV photoreactor with an inner radiation-type high-pressure mercury arc lamp. The concentration of salicylic acid in the aqueous solution was 0.15 mm. The catalysts were employed in suspension form with a concentration of 0.1 wt%. The concentration of the remaining salicylic acid was determined by measuring the absorbance of the solution at 296 nm (Fig. 4.). The efficiencies of the catalysts, which were calculated after 60 min of irradiation, can be seen in Table 2.

The P 25 Degussa TiO_2 and the TiO_2/M/H/33% samples were the best photocatalysts. Although the latter sample contains 33% semiconductor even so it can degrade 49.1% of the salicylic acid. These values show that by incorporating the semicondutor nanoparticles into the sodium montmorillonite we can improve the photodegradation rate of salicylic acid compared to the reference materials. In the case of ZnO/M/H/14% nanocomposite the improvment is much greater than for the ZnO/M/H/61% nanocomposite because the semiconductor content of the ZnO/M/H/14% catalyst is only 13.8%, but still it degrades 34.3% of the pollutant, while ZnO/M/H/61% with 61.5% semiconductor content can oxidize 46.2% of the salicylic acid. This can be explained by the fact that the average particle diameter of ZnO in sample ZnO/M/H/14% is 6.3 nm but in sample ZnO/M/H/61% it is 13.4 nm. The semiconductor particles have higher dispersity in the first sample than that in ZnO/M/H/61%, which results in a better adsorption of the organic pollutant on the surface of ZnO nanoparticles, bécause the geometrical surface area of ZnO in the catalyst increases if the particle diameter decreases.

Controlling the particle size and stability by incorporation into layered silicates has a great effect not only on the structural parameters (specific surface area, basal distance) but also on the photocatalytic properties of the nanocomposites.

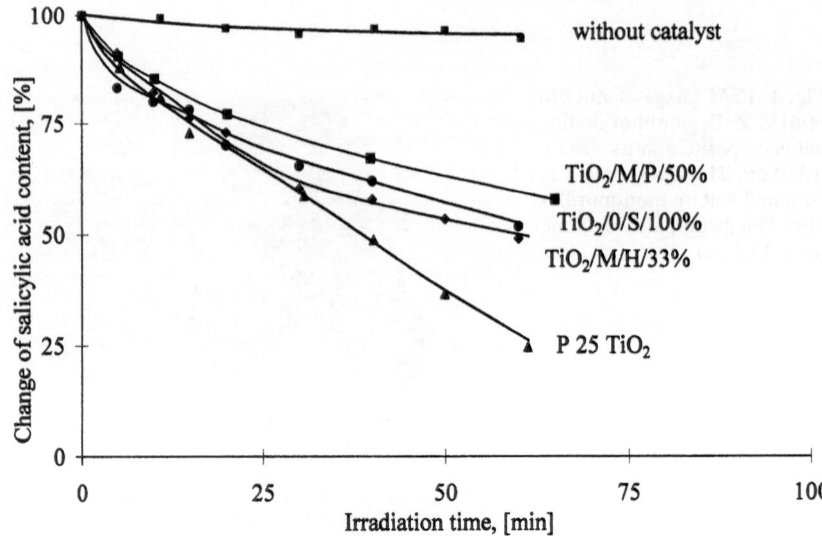

Fig. 5 Photodegradation of salicylic acid by different catalysts containing anatase

Conclusions

TiO_2 and ZnO nanoparticles were formed in aqueous media and were heterocoagulated with sodium montmorillonite suspension to maintain the particle size and the stability. Independent measurement methods (XRD, TEM, BET experiments) indicated the incorporation of the nanoparticles into the interlamellar space of montmorillonite. Nanoparticles stabilized by the layered silicates showed an increased photocatalytic activity for the degradation of salicylic acid model pollutant than their unstabilized counterparts. The present inexpensive photocatalysts, prepared by an environmentally friendly method, may be very profitable for industrial applications in a wide range of photocatlytic pollutant destructions.

Acknowledgements We thank OTKA (TO25392), FKFP (0402/1999) and NATO SfP 972652 for support of this work.

Appendix: notation of the samples

Type of semiconductor	TiO_2
Type of support	ZnO
	Hec (hectorite)
	M (montmorillonite)
	0 (no support)
Preparation method	P (preadsorption method)
	H (heterocoagulation method)
	A (alkoxide method)
Semiconductor content of catalyst	0–100%

References

1. Ogawa M, Kuroda K (1995) Chem Rev 95:399–438
2. Fujishiro J, Uchida S, Sato T (1999) Int J Inorg Mater 1:67–72
3. Zhuang J, Rhusu CN, Yates JT, Jr. (1999) J.Phys Chem 103:6957–6967
4. Kormann C, Bahnemann DW, Hoffmann MR (1988) Environ Sci Technol 22:798–806
5. Song KC, Kim JH (1999) J Colloid Interface Sci 212:193–196
6. Sato H (1996) PhD dissertation. Osaka University, p 34
7. Németh J, Dékány I (2000) Colloid Polym Sci 278:211–219
8. Kiricsi I, Pálinkó I, Tasi G, Hannus I (1994) Mol Cryst Liq Cryst 244:149–154
9. Dékány I, Turi L, Tombácz E, Fendler JH (1995) Langmuir 11:2285–2292
10. Dékány I, Fendler JH (1996) In: Pelizetti E (ed) NATO Advanced Study Institute Series, vol 18, pp 443–455
11. Dékány I, Túri L, Homonnay Z, Vértes A, Burger K (1996) Colloids Surf A 119:195–203
12. Dékány I (1996) NATO Advanced Study Institute Series 18. High technology. Kluwer, Dordrecht, pp 293–322
13. Dékány I, Túri L, Vankó Gy, Juhász G, Vértes A Burger K (1996) NATO Advanced Study Institute Series 18. High technology. Kluwer, Dordrecht, pp 555–568
14. Dékány I, Nagy L, Turi L, Király Z, Kotov NA, Fendler JH (1996) Langmuir 12:3709–3715
15. Dékány I, Túri L, Szûcs A, Király Z (1998) Colloids Surf A 141:405–417
16. Dékány I, Túri L, Galbács G, Fendler JH (1999) J Colloid Interface Sci 213:584–591

Progr Colloid Polym Sci (2001) 117: 94–100

S. Papp
I. Dékány

Growth of nearly monodisperse palladium nanoparticles on disaggregated kaolinite lamellae

S. Papp · I. Dékány (✉)
Nanostructured Materials Research Group of the Hungarian Academy of Science, Aradi Vértanúk tere 1, 6720 Szeged, Hungary
e-mail: i.dekany@chem.u-szeged.hu
Tel.: + 36-62-544211
Fax: + 36-62-544042

I. Dékány
Department of Colloid Chemistry
University of Szeged,
Aradi Vértanúk tere 1, 6720 Szeged, Hungary

Abstract Palladium nanoparticles were prepared by in situ growth between kaolinite layers. Expansion of the interlayer space kaolinite was effected by intercalation of dimethyl sulfoxide (DMSO) at 65 °C. The basal spacing increased from 0.72 to 1.12 nm during this process. After washing with methanol (partial displacement of DMSO), nonionic poly(vinylpyrrolidone)/kaolinite, cationic poly(diallyldimethylammonium)/kaolinite and octylammonium/kaolinite complexes were generated by adsorbing the polymer or the surfactant from solution. Kaolinite samples containing palladium nanoparticles were prepared by hydrazine and $NaBH_4$ reduction of palladium ions adsorbed from a solution of H_2PdCl_4 (pH 4). The size of the palladium particles formed was determined by transmission electron microscopy measurements which showed that nearly spherical, nearly monodisperse particles of size of about 1 nm were generated.

Key words Palladium nanoparticles · Kaolinite · Intercalation · Polymer stabilization

Introduction

Several procedures were devised for the preparation of nanoparticles in the 1–50-nm size range [1]. In these procedures an important role is assigned to stabilizing agents which protect the nanoparticles formed against aggregation, making possible the preparation of nanoparticles with diameters of a few nanometers [2, 3]. Polymers and surfactants are most often used as stabilizing agents for sols of subcolloidal size. Particles of controlled size can also be prepared within the internal space of micelles and microemulsions [4, 5]. Silicate minerals and layer double hydroxides are excellent supports for the preparation of semiconductor and transition-metal particles with diameters of a few nanometers on the surface of these layered compounds but also in the interlamellar space. Clay minerals (montmorillonite, hectorite, etc.) are especially suitable for this purpose because they swell readily in aqueous media and, therefore, provide a large (500–800 m^2g^{-1}) internal surface area [6, 7].

Pd, Pt and Au nanosols stabilized by poly(vinylpyrrolidone) (PVP) in aqueous media have been synthesized using alcohol and H_2 for reduction [8–10]. Chen and Akashi [11] obtained Pt sols from H_2PtCl_6 in water–ethanol mixtures of a volume fraction of 0.6 with respect to ethanol and showed the formation of particles in the polymer-stabilized system.

Transition-metal particles are most conveniently grown within the interlamellar space of clay minerals by displacing the exchangeable by precursor transition-metal cations and by subsequent reduction. Au and Ag clusters were obtained in laponite by Aihara et al. [12], with hydrazine hydrate, sodium borohydride and UV irradiation as reducing agents.

In our earlier studies the adsorption layer at the solid/liquid interface was employed as a "nanophase reactor" for the generation of nanocrystalline semiconductor particles (CdS, ZnS, TiO_2) and for their stabilization [6, 7]. To summarize briefly, the procedure consists of adsorbing the precursor ions of the nanocrystalline material in the interfacial adsorption layer of solid particles (i.e. about 1-nm thick lamellae) dispersed in a liquid phase. The synthesis is carried out in the adsorption layer by introducing the appropriate reagent. [6, 7]. The nanoparticles are formed attached to the surface in a well-controllable number and size between the silicate layers. Király et al. [13, 14] reported the synthesis of Pd nanoparticles in organic suspensions on hydrophobized montmorillonite by alcohol reduction of Pd acetate. We now present new procedures for the preparation of nanocrystalline palladium in the interlamellar space of kaolinite, a clay mineral that does not swell in water. The large specific surface area necessary for nanoparticle growth is created by breaking the hydrogen bonds between the kaolinite lamellae, i.e. by delamination of the mineral particles. According to Weiss et al. [15], silicate lamellae develop strong dipole–dipole interactions with, for example, dimethyl sulfoxide (DMSO), and hydrogen bonds with formamide, acetamide and hydrazine. The study of the intercalation of kaolinite with similar molecules is well known [15–20]. Gábor et al. [18] obtained the intercalation complex of kaolinite with hydrazine and potassium acetate. The basal spacing of kaolinite is 0.72 nm, that of the kaoline–hydrazine complex is 1.05 nm and that of the kaoline–potassium acetate complex is 1.41 nm. The maximum degree of intercalation was shown to increase with the time of reaction and the concentration of the guest compound but was found to be unaffected by changes in temperature. In the case of hydrazine, the reaction was 94% complete within 1 h, whereas with potassium acetate the reaction never exceeded 86% completion, even after prolonged reaction times.

Komori et al. [19] cleaved interlamellar hydrogen bonds in kaolinite by *N*-methylformamide ($d_L = 1.08$ nm) and in methanol ($d_L = 1.11$ nm) and their PVP was incorporated ($d_L = 1.24$ nm). Intercalation of alkylamines (C = 6–18) and water was also accomplished by the same authors [20]. The layers were expanded by *N*-methylformamide, washed with methanol and the methanol-wet product was treated with alkylamine; alternatively, methanol was displaced by treatment with water. The basal spacing increased with the alkyl chain length in a linear way in the range 2.69–5.75 nm with a slope of 0.225 nm per carbon atom.

We chose direct intercalation of DMSO for expanding the kaolinite structure. The excess DMSO was displaced from the kaolinite by washing with methanol before the polymer and surfactant were adsorbed. When palladium ions were added to the suspension, they were also adsorbed on the external and internal surfaces and the Pd^{2+} ions were reduced by addition of a reducing agent. Thus, the previously adsorbed polymer or surfactant promotes the adhesion of newly formed nanoparticles on the surface of the lamellae.

Experimental

Materials

Supports

Kaolinite from Zettlitz (Germany, particle diameter: 10–20 μm) was used as a support for the preparation of nanoparticles. The basal spacing was 0.72 nm. The specific surface area as determined by N_2 adsorption measurement was 14 m^2/g.

Reagents

The reagents were used as received without further purification. The metal precursor $PdCl_2$ (purity 99%) was obtained from Aldrich. PVP (K-30, average molecular weight 40,000, Fluka), a 20% aqueous solution of poly(diallyldimethylammonium chloride) (PDDA, average molecular weight 40,000–50,000, Aldrich), and octylamine (purity 99%, Fluka) were used as protective agents for the Pd nanoparticles. Methanol (Reanal, Hungary) was of analytical purity. DMSO (analytical purity, Reanal, Hungary) was used for delamination of the kaolinte particles. The hydrazine hydrate reducing agent was a 55% aqueous solution (Carlo Erba).

Methods

Preparation of Pd^0 nanoparticles

The kaolinite was expanded by the intercalation of DMSO at 65 °C. The excess DMSO was removed from the sample by several washings with methanol over 5 days. Pd/kaolinite complexes were prepared directly by reduction of Pd^{2+} ions (0.7 mM aqueous solution) previously adsorbed in the methanol/DMSO/kaolinite system and by applying a polymer and surfactant to ensure binding to the lamellae and steric stabilization. Macromolecules were adsorbed on the support from a methanol or aqueous solution and this was followed by adsorption and reduction of Pd^{2+} ions. Intercalation complexes of nonionic PVP/kaolinite and cationic PDDA/kaolinite (0.02–0.4 g polymer/g kaolinite) were prepared by this method in systems containing various (0.1–2.0%) concentrations of methanol/PVP, PDDA or water/PVP, PDDA by polymer adsorption. Intercalation complexes of octylammonium/kaolinite (0.12–1.2 g/g kaolinite) were prepared at pH 4.0 in systems containing various (0.5–0.05 M) concentrations of water/octylamin. Methanol/DMSO/kaolinite intercalation compound (1 g) was dispersed in 20 ml polymer or surfactant solution, and the mixtures were stirred at room temperature for 24 h. Kaolinite samples containing different amounts of palladium were obtained by (0.9–3.6 ml, 0.1 M) hydrazine or (1.8 ml, 0.1 M) $NaBH_4$ reduction of the palladium ions. The metal content of the products was 0.45–1.4% (Table 1). The schematic diagram of the synthesis is presented in Fig. 1.

X-ray diffraction experiments

X-ray diffraction measurements were made using a Philips PW 1820 diffractometer (Cu Kα radiation, 40 kV, 35 mA). The basal spacing was calculated from the (001) Bragg reflections using the PW 1877 automated powder diffraction software.

Electron microscopy

Transmission electron microscopy (TEM) images were made using a Philips CM-10 transmission electron microscope with an

Table 1 Basal spacings of intercalated kaolinite derivatives and particle diameters of the Pd^0 nanoparticles produced

Kaolinite samples	Pd content (w/w%)	Poly(vinylpyrrolidone)/poly(diallyldimethylammonium)/C8N content (w/w%)	d_L (nm) from X-ray diffraction	d_{ave} (nm) from transmission electron microscopy
Kaolinite	0.00	0.00	0.72	–
Dimethyl sulfoxide/kaolinite	0.00	0.00	1.12	–
Methanol/dimethyl sulfoxide/kaolinite(K)	0.00	0.00	1.12	–
Rehydrated kaolinite	0.00	0.00	0.72	–
Poly(vinylpyrrolidone)/methanol/kaolinite	0.00	1.90	1.12	–
Poly(vinylpyrrolidone)/H_2O/kaolinite	0.00	1.90	3.64[a], 0.72[b]	–
Poly(vinylpyrrolidone)/H_2O/kaolinite	0.00	1.90	3.92[a], 0.74[b]	–
C8N/kaolinite	0.00	56.50	2.48[a], 1.28[b]	–
Palladium kaolinite samples Prepared in methanol, reduced by N_2H_4				
PdK	0.47	0.00	7.53	3.25
PVPPdK1	0.47	3.80	2.21[a], 1.00[b]	2.24
PVPPdK2	0.95	3.80	4.56[a], 0.72[b]	2.45
PVPPdK3	0.95	17.00	4.24[a], 0.71[b]	2.45
PVPPdK4	0.95	28.00	0.84	2.59
PVPPdK5	1.41	3.80	0.74	2.10
PDDAPdK1	0.95	3.80	3.52[a], 0.82[b]	6.33
Palladium kaolinite samples Prepared in methanol, reduced by $NaBH_4$				
PVPPdK6	0.95	3.80	3.78[a], 0.84[b]	4.16
PDDAPdK2	0.95	3.80	5.07[a], 0.83[b]	2.67
Palladium kaolinite samples Prepared in water, reduced by N_2H_4				
PVPPdK7	0.95	3.80	0.72	2.81
PVPPdK8	1.90	3.80	4.36[a], 0.72[b]	5.83
PDDAPdK3	0.95	3.80	4.24[a], 0.72[b]	4.90
PDDAPdK4	1.90	3.80	4.69[a], 0.72[b]	9.44
C8NPdK1	0.95	11.60	0.73	2.17
C8NPdK2	0.95	20.10	4.42[a], 0.73[b]	2.04
C8NPdK3	0.95	40.00	2.65	1.95
C8NPdK4	0.95	56.50	5.95[a], 0.72[b]	2.22
Palladium kaolinite samples Prepared in water, reduced by $NaBH_4$				
PVPPdK9	0.95	3.80	4.62[a], 0.72[b]	4.66
PDDAPdK5	0.95	3.80	6.44[a], 0.72[b]	3.52

[a]Basal spacings of intercalated kaolinite
[b]Basal spacings of nonintercalated or partially intercalated kaolinite

accelerating voltage of 100 kV. Aliquots of the ethanol suspensions of the samples were dropped on copper grids (diameter 2 mm) covered with Formvar foil and were then left to stand for 3–40 min and then transferred to the microscope. The particle size distribution was determined using the UTHSCSA Image Tool program.

Results and discussion

In order to create the space necessary for the synthesis of the nanoparticles, the hydrogen bonds tightly interlinking the kaolinite layers must be broken. As indicated by XRD experiments, the interlamellar expansion was nearly 100% after the formation of the DMSO/kaolinite intercalation complex. The basal spacing increased from 0.72 to 1.12 nm within 24 h at 65 ° (Fig. 2a, b). The interlamellar distance was unchanged after removal of excess DMSO (Fig. 2c). The appearance of a peak at 7.53 nm on the XRD pattern of the Pd/kaolinite complex prepared by reduction of Pd^{2+} ions previously adsorbed in the methanol/DMSO/kaolinite system and the disappearance of the (001) reflection of kaolinite demonstrate the particle growth in the interlayer space (Fig. 2d). For the synthesis, macromolecules were first adsorbed on the support from a methanol or aqueous solution (0.1–2.0% w/v). The adsorption of the amount of polymer necessary for successful particle synthesis (2–20% w/w) did not increase the basal spacing (Fig. 3a). When palladium particles (0.47–1.41% w/w) were formed, the (001) reflection of 2.21 nm proves the incorporation of the nanoparticles (Table 1). This peak was much more prominent in sample PVPPdK1 than in the other preparations (Fig. 3b). The particle size calculated from the position of the new reflection (2.21–0.72 = 1.49 nm) is

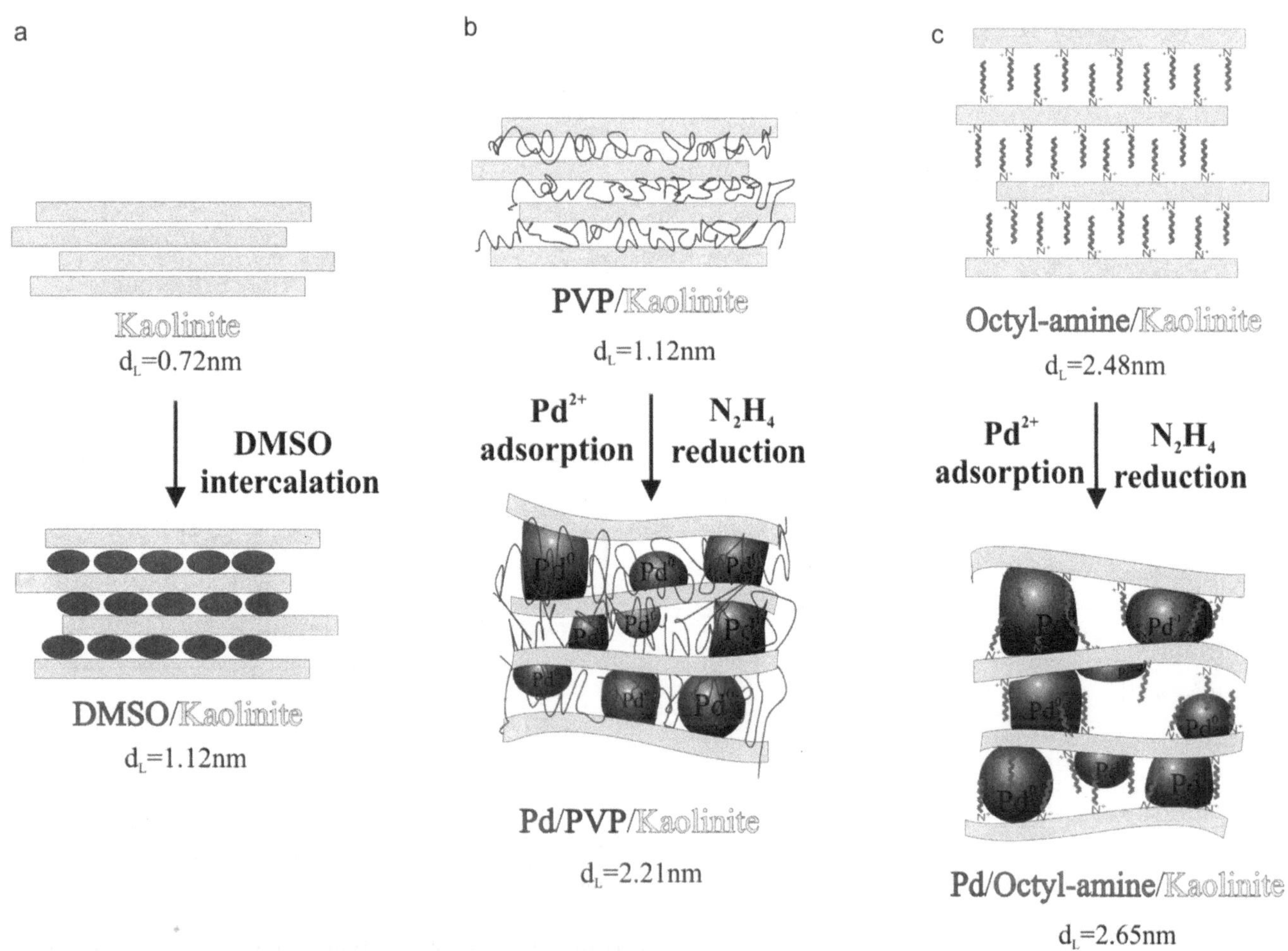

Fig. 1a–c Schematic illustration of preparation of Pd nanoparticles on kaolinite. **a** Intercalation of kaolinite, **b** adsorption of poly(vinylpyrrolidone) (*PVP*) and **c** octylammonium ions on dimethyl sulfoxide (*DMSO*)/methanol intercalated kaolinite

in good agreement with the results of TEM measurements (see later).

When the aqueous PVP solution was contacted with kaolinite, the peak characteristic of the intercalation compound of kaolinite was not observed (Fig. 4a). Rather, the initial (001) reflection reappeared and a broadened peak of $d_L \sim 3.6$ nm was observed (Fig. 4b). This indicates that in contact with the aqueous PVP solution some of the kaolinite layers reaggregate, whereas the remaining layers adsorb PVP. The formation of metal clusters in the interlamellar space was indicated by a shift of the reflection from $d_L = 3.6$ to $d_L = 4.6$ nm (Fig. 4c). When sodium borohydride is used for reduction, the peak appears at higher lamellar distances of 5–6 nm.

Kaolinite containing palladium was also prepared using octylamine in acidic media (pH 4.0) to separate the silicate layers. The amount of octylamine was varied between 0.12 and 1.2 g/g kaolinite. The reflection at $d_L = 2.48$ nm indicates the intercalation of octylammmonium ions (Fig. 5a). In the course of the reduction by hydrazine following the adsorption of palladium ions, a further shift of the (001) reflection was observed (2.65 nm, Fig. 5b). In the case of PVPPdK457 and C8NPdK1 Pd nanoparticles only formed on the external surfaces of the kaolinite particles.

The TEM picture of the PdK sample shows spherical particles with diameters of 2–4 nm, separately, without aggregation attached to the lamellae (Fig. 6a). Electron micrographs of the complexes containing nonionic PVP display relatively small, nearly monodisperse particles (Fig. 6b) which tend to stick together when the polymer concentration is increased. The polymer concentration did not have an effect on the Pd particle size (Table 1). In samples containing cationic PDDA as a stabilizer, the size distribution of the particles formed was more polydisperse. The Pd^0/kaolinite samples prepared in the aqueous system yielded comparatively large palladium clusters of polydisperse size distribution (2–5 nm) densely covering the lamellae (Fig. 6c). The larger particle size deduced from the shift of the (001) reflection in the presence of PDDA is supported by the TEM pictures

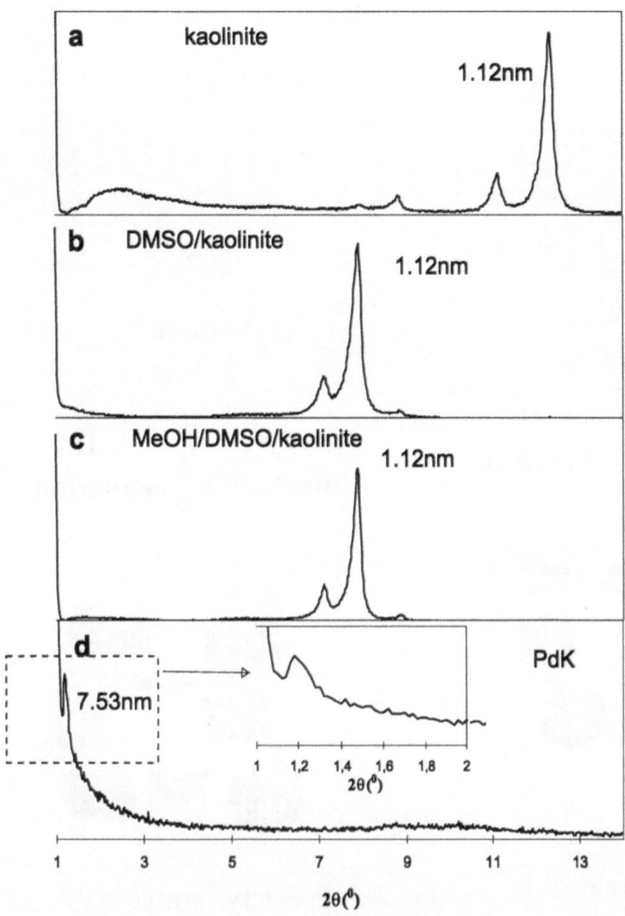

Fig. 2 X-ray diffraction (*XRD*) diagrams: **a** kaolinite; **b** DMSO-intercalated kaolinite; **c** methanol-washed DMSO/kaolinite; **d** palladium nanoparticles between disaggregated kaolinite layers

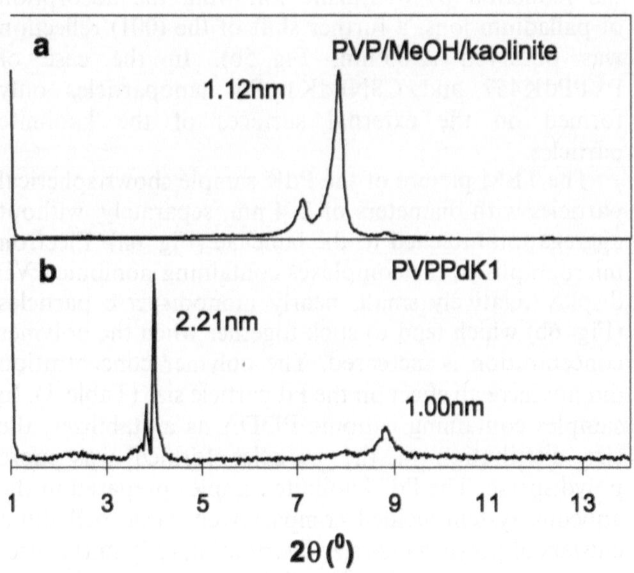

Fig. 3 XRD diagrams: **a** methanol/DMSO kaolinite sample with adsorbed PVP; **b** palladium/PVP/kaolinite

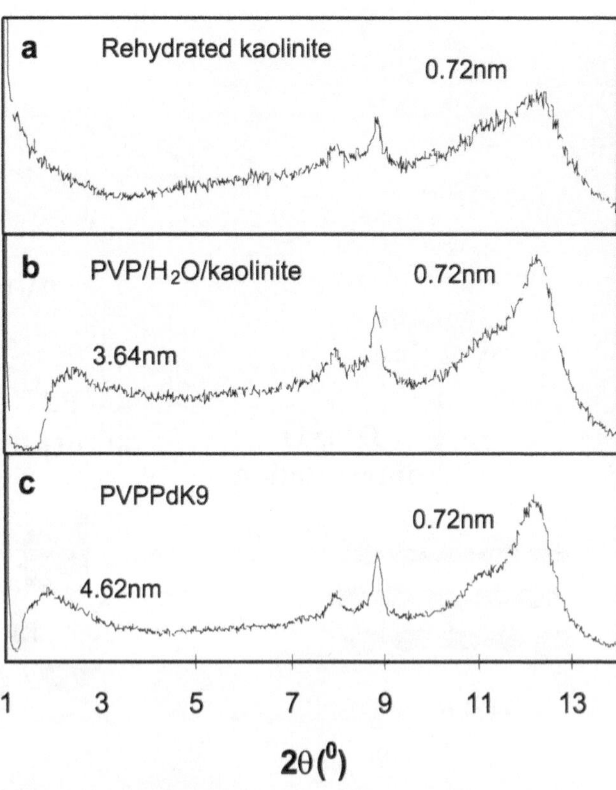

Fig. 4 XRD diagrams: **a** rehydrated methanol/DMSO kaolinite; **b** rehydrated methanol/DMSO kaolinite with adsorbed PVP; **c** palladium/PVP/kaolinite

(Fig. 6d, e; Table 1). Increasing the palladium content not only increased the particle size but also broadened the polydispersity. With octylamine the particle size was

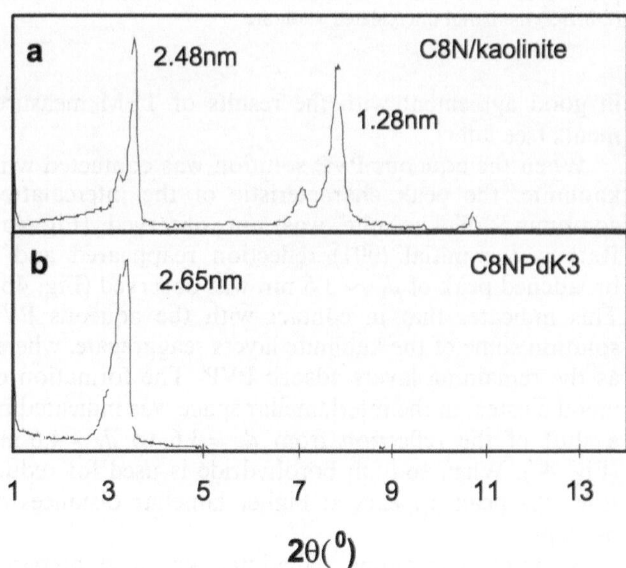

Fig. 5 XRD diagrams: **a** rehydrated methanol/DMSO kaolinite with adsorbed octylamine; **b** palladium/octylammonium/kaolinite

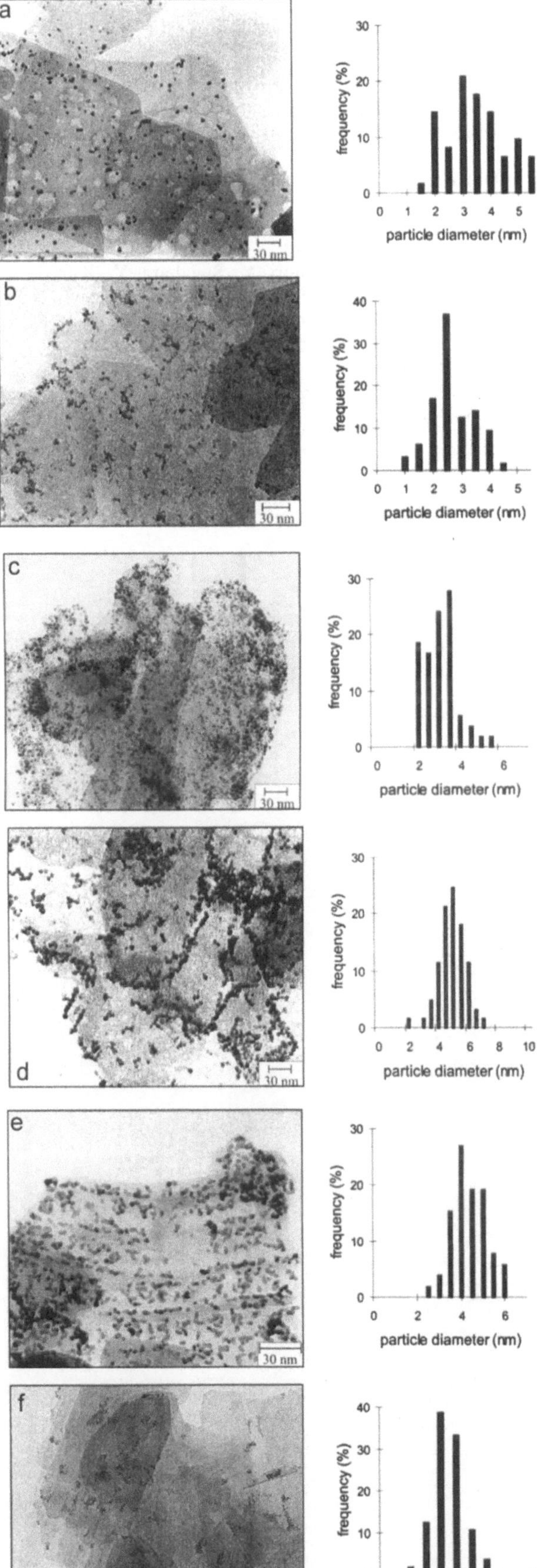

◄

Fig. 6 **a** Transmission electron microscopy (*TEM*) image and particle size distribution of sample PdK (0.47%Pd, reduced by 0.1 M N_2H_4 solution). **b** TEM image and particle size distribution of sample PVPPdK3 (0.95%Pd, 17% PVP, methanol treatment, reduced by 0.1 M N_2H_4 solution). **c** TEM image and particle size distribution of sample PVPPdK7 (0.95%Pd, 3.8% PVP, prepared in aqueous suspension, reduced by 0.1 M N_2H_4 solution). **d** TEM image and particle size distribution of sample PVPPdK6 (0.95%Pd, 3.8% PVP, methanol treatment, reduced by 0.1 M $NaBH_4$ solution). **e** TEM image and particle size distribution of sample PVPPdK9 (0.95%Pd, 3.8% PVP, prepared in aqueous suspension, reduced by 0.1 M $NaBH_4$ solution). **f** TEM image and particle size distribution of sample C8NPdK2 (0.95%Pd, 20.1% C8 N, methanol treatment, reduced by 0.1 M N_2H_4 solution)

similar to that in the presence of polymers; however, even at the lowest amine concentrations studied the particles aggregated up to form small islets (Fig. 6f) and, as the amount of surfactant was increased, large clusters developed. In each case, even at high metal contents, the palladium particles remained noncrystalline without distinct crystal shapes.

Conclusions

Well-crystallized kaolinite intercalated by DMSO served as an excellent support for the production of metal nanoparticles. After the adsorption of palladium ion precursors, metal nanocrystals were formed by reduction with hydrazine or sodium borohydride. The interlayer spaces acted as nanoreactors. The intercalation of nanoparticles into the kaolinite structure was proven by the XRD measurements. This procedure makes possible the steric control and restriction of nanoparticle growth. The stability of the nanoparticles was further enhanced by the addition of polymers (PVP, PDDA) and surfactants (alkylammonium salts) which were also intercalated into the kaolinite. TEM measurements proved the presence of the particles and yielded reliable particle sizes and size distributions.

Acknowledgements The authors wish to express their thanks for the financial support of the FKFP 0402/1999 project of the Hungarian Ministry of Education and the National Scientific Research Foundation (OTKA T 13/034430).

References

1. Siegel RW (1994) In: Fujita FE (ed) Springer Series in Materials Sciences, vol 27. Springer, Berlin Heidelberg New York, p 65
2. Henglein A, Guiterrez M, Ber Bunsenges (1983) Phys Chem 87:474
3. Ueno A, Kakuta N, Park KH, Finlayson MF, Bard AJ, Champion A, Fox MA, Webber SE, White JM (1985) J Phys Chem 89:3828
4. Trickot YM, Fendler JH (1984) J Am Chem Soc 106:7359
5. Your H-C, Baral S, Fendler JH (1988) J Phys Chem 92:6320
6. Dékány I, Nagy L, Turi L, Király Z, Kotov NA, Fendler JH (1996) Langmuir 12:3709
7. Dékány I, Túri L, Galbács G, Fendler JH (1997) J Colloid Interface Sci 195:307
8. Teranishi T, Miyake M (1998) Chem Mater 10:594
9. Liu H, Mao G, Meng S (1992) J Mol Catal 74:275
10. Yu W, Liu H (1998) Chem Mater 10:1205
11. Chen C-W, Akashi M (1997) Langmuir 13:6465
12. Aihara N, Torigore K, Esumi K (1998) Langmuir 14:4945
13. Király Z, Dékány I, Mastalir Á, Bartók M (1995) Mag Kem Foly 101:539
14. Király Z, Dékány I, Mastalir Á, Bartók M (1996) J Catal 161:401
15. Weiss A, Thielepape W, Orth H (1963) Z Anorg Allg Chem 320:183
16. Wada K (1961) Am Min 46:78
17. Poyato-Ferrera J, Becker H-O, Weiss A (1977) Proceedings of the 3th European Clay Conference, Oslo, p 148
18. Gábor M, Tóth M, Kristóf J, Komáromi-Hiller G (1995) Clays Clay Miner 43:223
19. Komori Y, Sugahara Y, Kuroda K (1999) Chem Mater 11:3
20. Komori Y, Sugahara Y, Kuroda K (1999) Appl Clay Sci 15:241

Progr Colloid Polym Sci (2001) 117: 101–103

Katharina Landfester

Quantitative considerations for the formulation of miniemulsions

Abstract The polymerization of styrene in miniemulsions stabilized with anionic sodium dodecyl sulfate or nonionic Lutensol AT50 results in stable polymer dispersions with particle diameters between 30 and 480 nm and narrow particle size distributions. Steady-state miniemulsification results in a system "with critical stability", i.e. the droplet size is the product of a rate equation of fission by ultrasound and fusion by collisions, and the minidroplets are as small as possible for the timescales involved. The droplet growth by monomer exchange, or the τ_1 mechanism, is effectively suppressed by addition of a very hydrophobic material, whereas droplet growth by collisions, or the τ_2 mechanism, is subject to the critical conditions. The growth of the critically stabilized miniemulsion droplets is usually slower than the polymerization time; therefore, in ideal cases, a 1:1 copy of droplets to particles is obtained, and the critically stabilized state is frozen.

Key words Miniemulsion · Critical stability · Polymerization · Osmotic pressure · Hydrophobe

K. Landfester
Max Planck Institute of Colloids and Interfaces, Am Mühlenberg, 14424 Potsdam/Golm, Germany
e-mail: landfester@mpikg-golm.mpg.de
Tel.: +49-331-5679509
Fax: +49-331-5679502

Introduction

Regular miniemulsions can be defined as aqueous dispersions of surfactant-stabilized oil droplets within a size range of 50–500 nm prepared by shearing a system containing oil, water, a surfactant, and a strong hydrophobe. Recently, a quantitative approach for the formulation of a miniemulsion was given [1]. The hydrophobe acts as an osmotic agent which stabilizes the system against Ostwald ripening. The growth of the droplets by collision is controlled by the density of the surfactant layer:freshly prepared miniemulsions are "critically stabilized" and show slow, but pronounced growth, whereas a miniemulsion in equilibrium exhibits constant particle size on longer time scales. Polymerization of the oil droplets of such miniemulsions turned out to be very promising and extends the possibilities of classical emulsion polymerization. Owing to the fact that the polymerization time is usually shorter than the growth of the droplets by collisions, polymerization in carefully prepared miniemulsions results in latex particles which have about the same size as the initial droplets, as shown by a combination of small-angle neutron scattering, surface tension measurements, and conductometry [2]. In the case of appropriately formulated miniemulsions where polymerization is initiated in each droplet and the solubility of the monomer in the continuous phase is low, the ideal, limiting case of a 1:1 copy of the droplets to the particles can be obtained; therefore, each droplet can be considered as a small reactor in which polymerization reactions takes place. On the basis of the quantitative understanding of miniemulsions, the process allows new structures in particles to be created by the polymerization, for example, by unusual combinations of monomers or by the incorporation of materials which are not soluble in the continuous phase [3].

Results and discussion

Creating a miniemulsion starts with the homogenization step. Oil, a hydrophobe, and an aqueous solution of emulsifier are first mixed by vigorous stirring of droplets to obtain droplets in the micrometer range. Then, high shear, such as ultrasonication, is applied in order to obtain small monodisperse droplets in the hundred-nanometer range. With increasing time of sonication, the droplets shrink and the polydispersity decreases till the miniemulsion has reached a steady state. In principle, there are two possibilities for the degradation of emulsions:the droplets can grow by diffusion (Ostwald ripening, τ_1 mechanism) or the droplets can grow by collisions (τ_2 mechanism). The first growth process can be efficiently suppressed by the addition of a hydrophobe, which prevents the droplets from Ostwald ripening by generating an osmotic pressure in the droplets. This osmotic pressure counteracts the Laplace pressure. Directly after sonication, the osmotic pressure is usually smaller than the Laplace pressure. Growth by the τ_1 mechanism is not observed; however, it was found that growth by collision can not be fully suppressed. Therefore the droplets grow very slowly till the Laplace pressure and the osmotic pressure are equal. This scenario of equal pressure can also be reached intentionally by adding an adequate additional amount of emulsifier.

The size of the droplets in freshly prepared miniemulsions and after polymerization of the resulting polymer particles mainly depends on the amount of emulsifier used. As an example, the dependence of the amount of the anionic emulsifier sodium dodecyl sulfate on the size of polystyrene particles is shown in Fig. 1. Particle sizes between 180 and 30 nm can be obtained. As a comparison, the nonionic emulsifier Lutensol AT50 is shown, where particle sizes between 480 and 70 nm are realized. The domain of the miniemulsion is in the low surfactant load. In all cases, the droplets and the final particles are incompletely covered by surfactant molecules. At a surfactant amount of more than about 50 %, one leaves the domain of miniemulsions since full coverage of the particles is obtained.

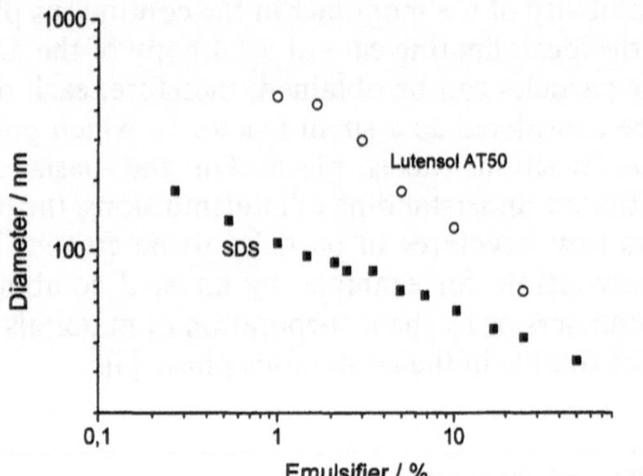

Fig. 1 The size of polystyrene particles obtained in miniemulsion polymerization by using different amounts of anionic sodium dodecyl sulfate or nonionic Lutensol AT50

Since each droplet behaves as a nanoreactor, many application are realized, and only three examples can be given here:

- The miniemulsion polymerization process is not limited to radical polymerizations. Owing to the fact that each droplet can be handled as a minireactor, there are a lot of new syntheses which can be obtained in miniemulsions but not in conventional emulsion polymerization processes. We could successfully perform polyaddition reactions with diamines and diepoxides in the droplets [4]. Depending on the type and the amount of emulsifier, particle sizes between 60 and 400 nm can be obtained. The excess of one component allows the synthesis of particles with surface reactivity.
- It was shown that the principle of aqueous miniemulsions can be transferred to nonaqueous media. In direct miniemulsions using polar media such as formamide or glycol instead of water and hydrophobic monomer, the miniemulsion stability was obtained by a hydrophobic agent, which prevents the droplets from Ostwald ripening. In the case of inverse systems, hydrophilic monomers were miniemulsified in a nonpolar medium, such as cyclohexane. In order to provide osmotic-stabilized droplets, a lipophobe has to be added to the monomer phase. This allows the synthesis of homogeneous structured particles containing water-soluble monomers [5].
- The miniemulsion polymerization enables the formation of nanocrystalline particles consisting of pure polyacrylonitrile (PAN) in the size range between 100 and 180 nm. Since the polymer is insoluble in the monomer, the polymer formed precipitates during the polymerization within the droplets, and large polymer nanocrystals (about 10 nm) are formed. Pure PAN latexes have a crumpled appearance [6].

Conclusion

Steady-state dispersed miniemulsions are osmotically stable but are critically stabilized with respect to their colloidal stability. The interface energy between the oil and water phase in a miniemulsion is greater than zero. The surface coverage of the droplets by surfactant molecules is incomplete. The formation of a miniemulsion requires high mechanical agitation to reach a steady state given by a rate equilibrium of droplet fission and fusion. The osmotic stability of miniemulsion droplets results from an osmotic pressure in the droplets, which controls the solvent or monomer evaporation. The osmotic pressure results from the addition of a hydrophobe, which has an extremely low water solubility. During polymerization each droplet behaves like a nanoreactor. This allows the formation of particles which are not accessible by other types of heterogeneous polymerization.

References

1. Landfester K, Bechthold N, Tiarks F, Antonietti M (1999) Macromolecules 32:5222
2. Landfester K, Bechthold N, Förster S, Antonietti M (1999) Macromol Rapid Commun 20:81
3. Bechthold N, Tiarks F, Willert M, Landfester K, Antonietti M (2000) Macromol Symp 150:549
4. Landfester K, Tiarks F, Hentze HP, Antonietti M (2000) Macromol Chem Phys 201:1
5. Landfester K, Willert M, Antonietti M (2000) Macromolecules 33:2370
6. Landfester K, Antonietti M (2000) Macromol Rapid Commun 21:820

Progr Colloid Polym Sci (2001) 117: 104–109

L. Vékás
D. Bica
I. Potencz
D. Gheorghe
O. Bălău
M. Raşa

Concentration and composition dependence of rheological and magnetorheological properties of some magnetic fluids

L. Vékás (✉) · D. Bica · I. Potencz
Centre for Fundamental and Advanced Technical Research, Romanian Academy, Timişoara Branch,
Bd. Mihai Viteazul no. 24,
1900 Timişoara, Romania

D. Gheorghe · O. Bălău · M. Raşa
Institute for Complex Fluids
Politehnica University of Timişoara
Bd. Mihai Viteazul, no. 1
1900 Timişoara, Romania

Abstract Rheological and magnetorheological behaviour of magnetic fluids with nonpolar and polar carrier liquids were investigated, using capillary and rotary viscometers, the measuring cells being completed with specially designed electromagnets. In the absence of a magnetic field the relative viscosity–hydrodynamic particle volume fraction dependence is well fitted especially by the formulas of Vand, Krieger–Dougherty and Chow. The fitted values of the Chow interaction coefficient are close to the theoretical value, except at low temperatures and close packing of magnetic nanoparticles. The effective viscosity of various types of magnetic fluids in an applied magnetic field is strongly dependent on microstructural characteristics. A relatively large, almost 1 order of magnitude, increase in the relative viscosity was measured for medium-concentration water-based magnetic fluids, with large secondary agglomerates. The magnetoviscous effect was found to be rather small, below 10%, for highly stable magnetic fluid samples on decahydronaphathalene and pentanol carriers.

Key words Magnetic fluids (ferrofluids) · Rheological properties · Magnetorheological properties · Colloidal stability · Chemical composition

Introduction

Magnetic fluids are a special category of colloids. Magnetic nanoparticles (Fe_3O_4, γ-Fe_2O_3, $CoFe_2O_4$, Co, Fe, etc.) of about 7–10-nm mean physical diameter are dispersed using suitable surfactants and by applying various stabilisation procedures in nonpolar and polar carrier liquids, like hydrocarbons, synthetic oils, water or alcohols, to mention only a few possible carrier liquids. The most important feature of these magnetic colloids is that they maintain their high degree of colloidal stability even in intense and very nonuniform magnetic fields, i.e. from a macroscopic point of view magnetic fluids are quasihomogeneous magnetisable liquid media Many details of the preparation, properties, magnetohydrodynamics and applications of magnetic fluids may be found in Ref. [1].

The rheological and magnetorheological properties of magnetic fluids are strongly dependent on the microstructural characteristics; therefore, the macroscopic treatment of the rheological behaviour is not adequate. In a recent investigation conducted in the absence of a magnetic field [2] we showed that composition details, for example, the nature and chemical purity of the surfactants used, as well as the hydrodynamic volumic fraction of the nanoparticles, strongly influence the flow properties of magnetic fluids.

In this article we present experimental results concerning the rheological and magnetorheological behaviour of several nonpolar and polar magnetic fluids, as well as fits of the data to appropriate semiempirical and theoretical formulae concerning the dependence of the dynamic viscosity on the volumic concentration of the nanoparticles and the details of their surface coverage (nature and quality of the surfactant/s, effective thickness of the surfactant layer), in the absence or the presence of an applied magnetic field.

Magnetic fluid samples and experimental methods

High-concentration magnetic fluid samples were prepared up to the upper limit of the hydrodynamic volume fraction of magnetic nanoparticles, $\varphi_h \cong 0.6$. The Fe_3O_4 nanoparticles were synthesised by chemical coprecipitation of Fe^{2+} and Fe^{3+} ions in the presence of excess 25% NH_4OH at 80–82 °C [3].

Depending on the nonpolar or polar character of the carrier, for example, transformer oil, decahydronaphthalene (DHN), dioctylsebacate, alcohol, and water, the dispersion of magnetic nanoparticles was performed by a monolayer or a double-layer steric stabilisation procedure [3, 4]. The first chemisorbed surfactant layer, which consists of technical grade oleic acid (TOA) or chemically pure oleic acid (POA), ensures the stability of nonpolar magnetic fluids. The free oleic acid was eliminated from the magnetic fluid. In order to ensure the dispersion of oleic acid covered Fe_3O_4 nanoparticles in polar carriers, a suitable secondary surfactant has to be used [dodecyl benzene sulfonic acid, poly(isobutylene succinanhydride), etc.], which is physically adsorbed on the oleic acid covered Fe_3O_4 particles. The stabilisation of magnetic nanoparticles, in particular, in strongly polar carriers, like in medium-chain and short-chain alcohols or water, is a difficult task [4–6]. The preparation of very high concentration C_3–C_{10} alcohol based magnetic fluids was reported recently [7]. The pentanol (C_5) based samples of this series were thoroughly investigated in this work. The excellent colloidal stability of this type of magnetic fluid was shown in Refs [8, 9] by applying various physical methods of analysis of microstructural properties, in particular, sedimentation velocity analysis and electrophoresis. Several results of magnetic, electronic micrography, rheological, magnetorheological and magnetooptical methods applied to microstructural characterisation of various type of magnetic fluids may be found in Refs. [10–12].

These investigations evidenced the effects of composition and particle concentration on the agglomerate formation process and implicitly on the macroscopic behaviour of different samples. For the investigations performed in this work the following nonpolar and polar magnetic fluids were selected: magnetic fluid/DHN (300 G), magnetic fluid/pentanol (830 G) and magnetic fluid/H_2O (300 and 420 G), as well as a series of dilute samples of the concentrated magnetic fluid/pentanol liquid.

The rheological behaviour of the samples was investigated with an Ubbelohde-type capillary viscometer and with a Couette-type rotary rheometer (Rheotest 2). The experimental cells were completed with a magnetic-field source, an electromagnet with plane-parallel pole pieces (Fig. 1a; $B_{max} \cong 0.3$ T) and a specially designed electromagnet (Fig. 1b; $B_{max} \cong 0.1$ T), for the magnetorheological measurements.

Experimental results and discussion

As was shown [2, 10, 11], the magnetic, rheological, magnetorheological and magnetooptical properties of magnetic fluids strongly depends on the microstructural characteristics, in particular, on aggregate formation. Particle agglomeration processes may be induced, among

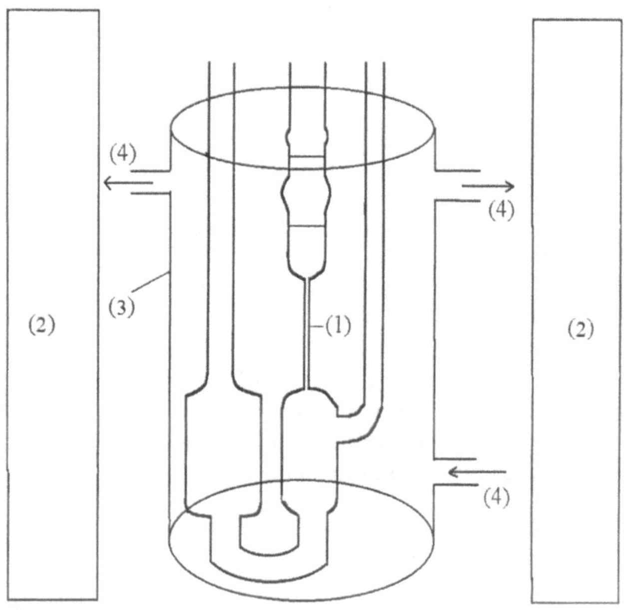

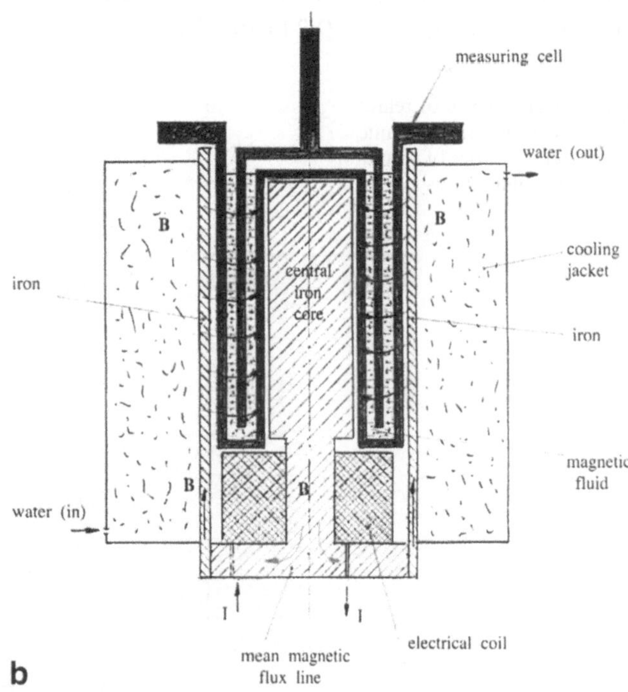

Fig. 1 **a** Ubbelohde-type capillary viscometer (*1* – capillary tube, *2* – polar pieces, *3* – support cylinder, *4* – water entrance/exit), **b** Couette-type rotary rheometer

others, by insufficient covering of particles owing to the low chemical purity of the surfactant (e.g. with a large fraction of shorter chain length carboxylic acids in TOA), by the reduction of the mean distance between magnetic dipole particles owing to the increase in the particle volume fraction or by the application of a strong magnetic field.

The mixed stabilisation mechanism, double-layer steric and electrostatic, appropriate for pentanol-based magnetic fluids [8] and for other polar magnetic fluids prepared by the same method [4], allowed *n*-pentanol magnetic fluid samples of very high solid volume fraction to be obtained ($\varphi_p \cong 19.4\%$ and with correspondingly high saturation magnetisation, $M_s = 800$–900 G, which are the highest reported up to now [5, 7]).

The dependence of relative viscosity, η/η_o, where η is the dynamic viscosity of the magnetic fluid and η_o is the dynamic viscosity of the pentanol carrier liquid, on the solid particle volume fraction, T, was determined at various temperatures in the −10–70 °C interval. In order to determine the hydrodynamic volume fraction, φ_h, i.e. the effective volume fraction of the double-layer-covered Fe_3O_4 nanoparticles, the data were fitted to the Vand formula using the fit parameter $p = \varphi_h/\varphi_\rho$:

$$\eta/\eta_o = \exp[(2.5p\phi_\rho + 2.7p^2\phi_\rho^2)]/(1 - 0.609p\phi_\rho) \ . \quad (1)$$

The fitted $\eta/\eta_o = f(\varphi_s)$ curves at various temperatures are presented in Fig. 2 together with the detailed results of the fits and the Vand formula [13]. Using the fitted p values, the effective mean surfactant layer thickness may be obtained, $\delta = (p^{1/3} - 1)D/2$ (Table 1).

The values of the maximum hydrodynamic volume fraction, φ_m, were determined by fitting the data to the well-known two-parameter formula of Krieger and Dougherty [14],

$$\eta/\eta_0 = (1 - \varphi_h/\varphi_m)^{-[\eta]\varphi_m} \ , \quad (2)$$

where $[\eta]$ is the intrinsic viscosity (Table 1).

Note that the $[\eta]$ values obtained are not far from that corresponding to spherically shaped particles, $[\eta] = 2.5$. Also, it may be observed that p and consequently δ, φ_m and $[\eta]$ are slightly temperature dependent. Indeed, the effective surfactant layer thickness, δ, depends on the temperature and has greater values at lower temperatures, especially because the physically adsorbed secondary surfactant layer is more influenced by the thermal motion.

In the Refs. [15, 16] Chow performed a thorough theoretical analysis of concentrated suspensions, taking into account the contribution of many-body particle interactions on the effective viscosity. On the basis of a liquid lattice model, the low-shear limiting viscosity resulted as

$$\frac{\eta}{\eta_0} = \exp\left(\frac{2.5\varphi_h}{1 - \varphi_h}\right) + \frac{A\varphi_h^2}{\left(1 - A\varphi_h^2\varphi_m\right)} \ , \quad (3)$$

where A is the coupling coefficient. Without considering dipole–dipole type interactions between particles, the theoretical value of A was determined to be 4.67.

The theoretical formula (Eq. 3) obtained by Chow proved to be well fitted by viscosity data for various magnetic fluids at 20 °C [2]. The fitted values of A are

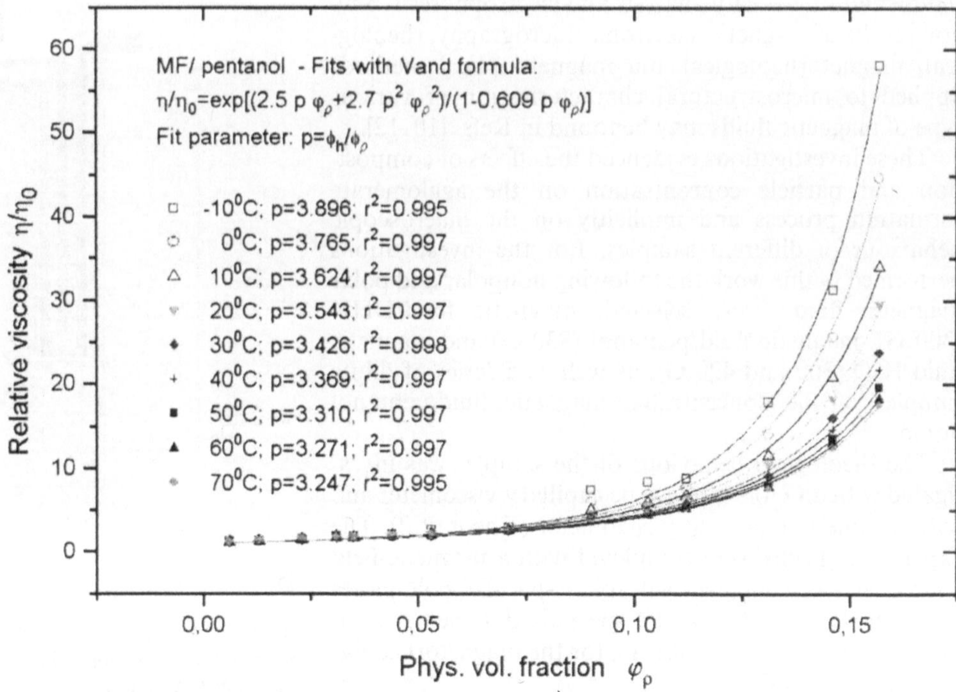

Fig. 2 Dependence of relative viscosity on physical volume fraction for a series of pentanol-based magnetic fluids

Table 1 Fit parameters from Vand, Krieger–Dougherty (*K-D*) and Chow relations (pentanol magnetic fluids)

T (°C)	p (Vand)	δ (nm)	φ_m (K-D)	$[\eta]$ (K-D)	A (Chow) with $\varphi_m^{K\text{-}D}$
−10	3.896	2.466	0.670	2.858	3.317
0	3.765	2.389	0.654	2.886	3.663
10	3.624	2.305	0.637	2.925	3.975
20	3.543	2.255	0.625	2.953	4.170
30	3.426	2.182	0.614	2.971	4.356
40	3.369	2.146	0.607	2.980	4.460
50	3.310	2.108	0.600	2.988	4.552
60	3.271	2.083	0.596	2.997	4.599
70	3.247	2.967	0.593	3.000	4.627

A_{theor} (Chow) = 4.67

close to the theoretical one for $\varphi_h \leq 0.45$. At higher values of the hydrodynamic volume fraction, the resulting A value is lower, evidencing the role of dipolar interactions at close packing.

The results of a fit of all the viscosity data corresponding to the temperature interval −10–70 °C and up to the highest hydrodynamic volume fraction, $\varphi_m \sim 0.6$, are shown in Fig. 3. The experimental data are well fitted by the Chow formula (Eq. 3), the resulting overall coupling coefficient being 3.78. This fit and the previous one [2] show that A is smaller than the theoretical value of 4.67 at lower temperatures and at close packing of particles when the role of dipolar interactions cannot be neglected, even in the case of a magnetic fluid of very high degree of colloidal stability. Note, that at $T > 50$ °C, the fitted values of A approach the theoretical one, the increased thermal motion diminishing the influence of dipolar interactions.

Under the influence of a magnetic field, magnetic fluids increase their effective viscosity owing to the supplementary dissipation due to the motion of particles relative to the surrounding carrier liquid.

The magnetorheological effect is described by the model developed by Shliomis [17], but its validity is limited to a very low volumic concentration of particles even if we take into account the size distribution of the particles and the Shliomis diameter, as in the generalised formula proposed in Ref. [10].

The intensity of the magnetorheological effect, will be shown, is strongly dependent on various microstructural processes, especially on those relating to agglomerates. Incomplete surfactant covering of particles initiates the formation of agglomerates, which significantly change the behaviour of magnetic fluids in a magnetic field. This is well illustrated by the experimental data represented in Fig. 4 for two decahydronaphthalene-based magnetic

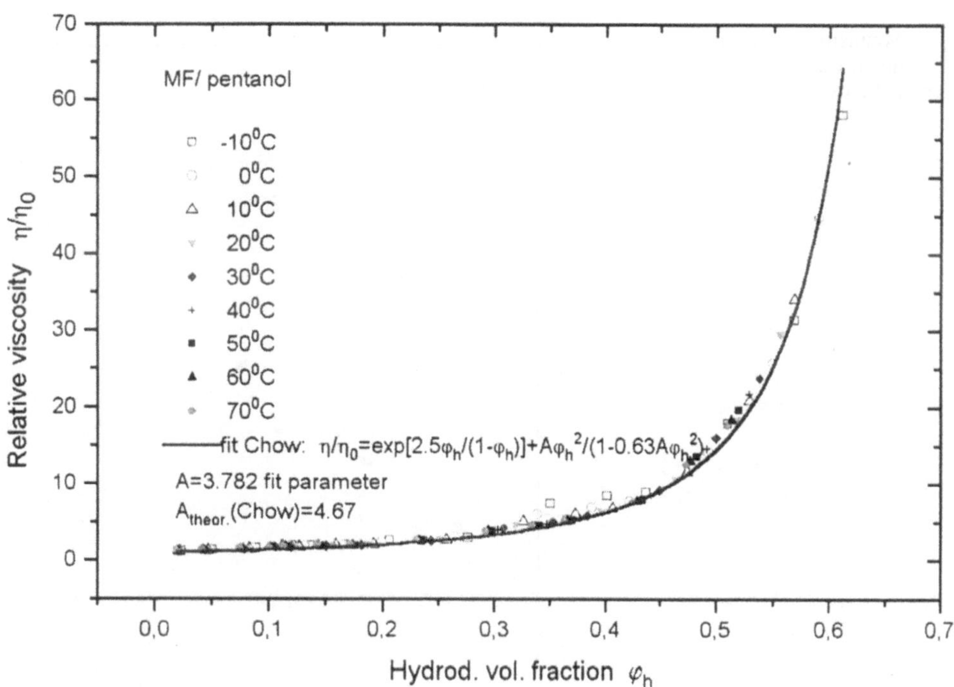

Fig. 3 Dependence of relative viscosity on hydrodynamic volume fraction for a series of pentanol-based magnetic fluids

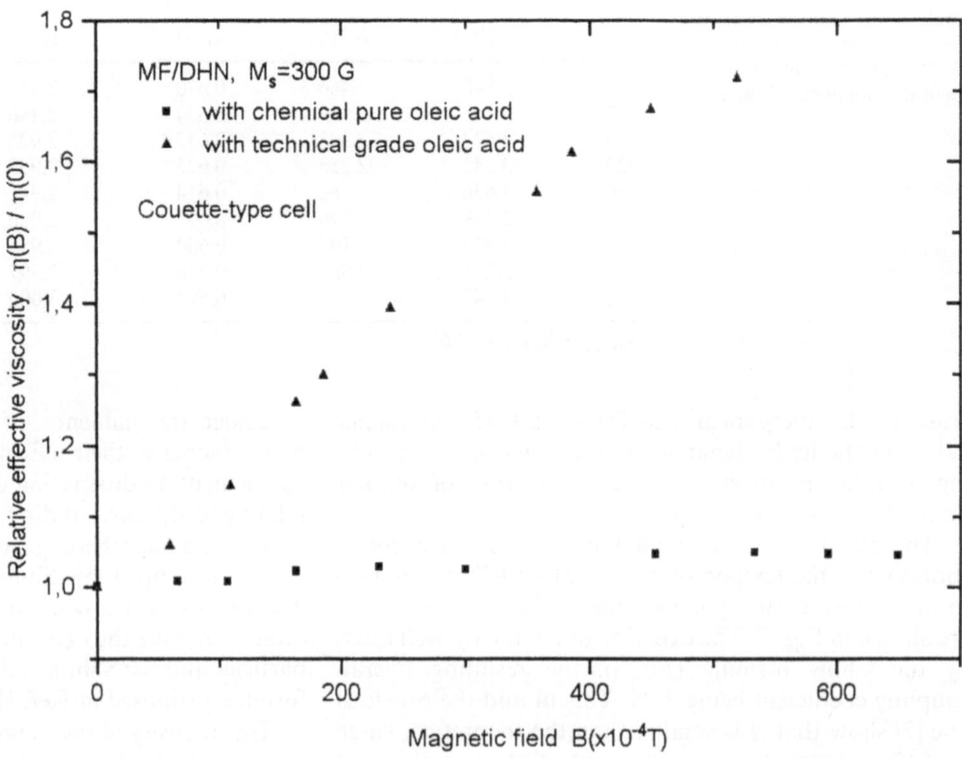

Fig. 4 The influence of the degree of purity of chemisorbed surfactant on the magnetoviscous effect

fluid samples of the same magnetisation, but stabilised with technical grade and chemically pure surfactant (TOA and POA, respectively).

Water-based magnetic fluids, except those with low particle volume fraction, $\varphi_p \sim 0.05$, present a relatively large fraction of primary agglomerates, which develop further into secondary, large agglomerates under the influence of an applied magnetic field of moderate or even low magnetic induction ($B \cong 0.01$ T). The process of agglomerate formation in water-based samples pre-

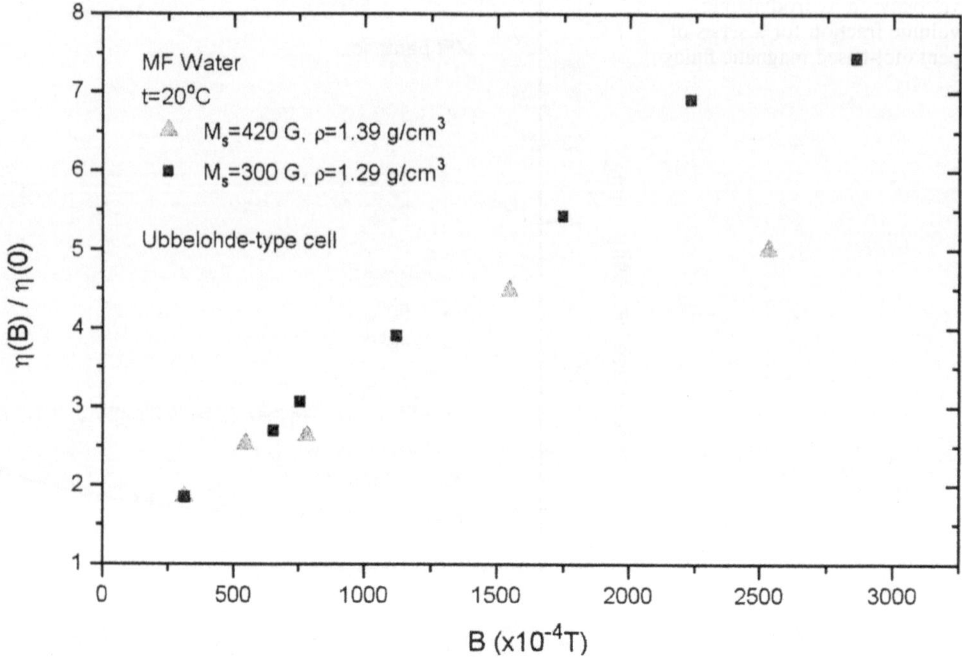

Fig. 5 Magnetorheological behaviour of water-based magnetic fluids

pared according to the procedure given in Ref. [6] was recently investigated by magnetic [18] and magnetooptical [11] methods. Magnetic-induced phase condensation occurs, i.e. the magnetic fluid becomes a biphasic system in an external magnetic field. The microstructural changes, in particular, the formation of condensed-phase droplets, significantly modify the magnetisation curve, especially in the low-field region [18], as well as the magnetic dichroism effect [11], compared to other highly stable magnetic fluids.

The significant, almost 1 order of magnitude, increase of the relative viscosity, $\eta(B)/\eta(0)$, may be observed in the case of two medium-concentration water-based magnetic fluids in Fig. 5. The large effects observed illustrate the high sensitivity of magnetorheological measurements concerning microstructural changes in magnetic colloids.

Conclusions

- Rheological and magnetorheological properties of magnetic fluids are strongly dependent on the nature and quality of surfactant(s), as well as on the volumic concentration of magnetic nanoparticles.
- The relative viscosity–volumic concentration dependence, at various temperatures, is well fitted, especially by the formulae of Vand, Krieger–Dougherty and Chow. This is well illustrated by the data for a series of pentanol-based magnetic fluids. The fitted values of the Chow interaction parameter, A, are very close to the theoretical one (4.67). Somewhat lower values resulted at lower temperatures ($T \leq 20$ °C) and at close packing of magnetic nanoparticles ($\varphi_h \cong 0.6$).
- The magnetorheological behaviour of magnetic fluids is very sensitive to microstructural processes, in particular, to agglomerate formation induced by low-quality stabilisation. Water-based magnetic fluids, with large secondary agglomerates, show an almost 1 order of magnitude increase of relative viscosity in a magnetic field.

Acknowledgements The permanent support of Prof. I. Anton (Romanian Academy – Tourizoara Branch) is gratefully acknowledged. We are indebted to A. Philipse (University of Utrecht) for many helpful discussions on the stability and microstructural charactersitics of polar magnetic fluids. The present work was partly supported by a grant of the Romanian Academy (contract no. 64/2000). The financial support of the Romanian Agency for Science, Technology and Innovation (contract no. 91/1996) is also gratefully acknowledged.

References

1. Berkovski B, Bashtovoi V (eds) (1996) Magnetic fluids and applications handbook. Begell House, New York
2. Vékás L, Bica D, Gheorghe D, Potencz I, Raşa M (1999) J Magn Magn Mater 201:159–162
3. Bica D (1995) Rom Rep Phys 47: 265–272
4. Bica D, Vékás L (1994) Magn Gidrodin (Magnetohydrodynamics), 30: 194–196
5. Fujita T, Miyazaki T, Nishiyama H, Jeyadevan B (1999) J Magn Magn Mater 201:14–17
6. Bica D (1985) Romanian Patent RO 90078
7. Bica D, Vékás L (2000) Concentrated magnetic fluids on various alcohol carrier liquids, "last-minute poster". International conference "Adsorption and nanostructures – from theory to application", 25–28 September 2000, Budapest
8. Donselaar LN, Philipse AP, Surmond J (1997) Langmuir 13:6018
9. Donselaar L (1998) PhD thesis. University of Utrecht
10. Vékás L, Raşa M, Bica D (2000) J Colloid Interface Sci 231:247–254
11. Socoliuc V, Bica D (2001) Prog Colloid Polym Sci 117:131–135
12. Schott M, Vékás L, Bica D (1995) Rom Rep Phys 47:411–436
13. Vand V (1984) J Phys Colloid Chem 52:277–321
14. Barnes HA, Hutton JF, Walters K (1989) An introduction to rheology. Elsevier, Amsterdam, p 199
15. Chow TS (1993) Phys Rev E 48:1997
16. Chow TS (1994) Phys Rev E 50:1274
17. Shliomis MF (1972) J Exp Theor Phys 61:2411
18. Vékás L, Raşa M, Bica D (2000) CAS 2000 Proceedings of the 23rd International Semiconductor Conference, 10–14 October 2000, Sinaia, Romania, vol 2. Institute of Electrical and Electronics Engineers, pp 495–498

Progr Colloid Polym Sci (2001) 117: 110–112

F. Tiarks
M. Willert
K. Landfester
M. Antonietti

The controlled generation of nanosized structures in miniemulsions

F. Tiarks · M. Willert · K. Landfester (✉)
M. Antonietti
Max Planck Institute of Colloids and Interfaces, Research Campus Golm, 14424 Potsdam, Germany
e-mail: landfester@mpikg-golm.mpg.de
Tel.: +49-331-5679509
Fax: +49-331-5679502

Abstract In this article the effective encapsulation of carbon black with polymers by cosonication of a carbon black dispersion and a miniemulsion is decribed. The inverse miniemulsion process is shown to be an excellent method for the preparation of particles consisting of hydrophilic polymers.

Key words Miniemulsion polymerization · Encapsulation · Carbon black · Cosonication · Polar monomers

Introduction

Regular miniemulsions are aqueous dispersions of relatively stable oil droplets with a size between 50 and 500 nm prepared by shearing a system containing oil, water, a surfactant, and a hydrophobe. The polymerization of miniemulsions extends the possibilities of the emulsion polymerization and provides advantages with respect to copolymerization reactions of monomers with different polarity, incorporation of hydrophobic materials, or the stability of the latexes formed [1, 2]. In this article the effective encapsulation of carbon black with polymers by cosonication of a carbon black dispersion and a miniemulsion polymerization recipe is described. Another subject of this article is the preparation of polymer particles from inverse miniemulsions. Inverse miniemulsions consist of a hydrophilic monomer, like acrylic acid, dispersed in a nonpolar continuous phase, for example, cyclohexane and a surfactant [3]. To stabilize the droplets against Ostwald ripening a strong lipophobe, like sodium chloride, has to be added. It could be shown that miniemulsions can also be made in organic systems with monomers, like acrylamide, styrene sulfonic acid sodium salt, or hydroxyethyl methacrylate, and show the typical characteristics of miniemulsions, i.e. incomplete coverage of the particles with surfactants and preservation of droplet number.

Results

Encapsulation of carbon black

In previous experiments encapsulated carbon black particles were obtained by first mixing carbon into the monomer and then formulating a miniemulsion of the monomer/carbon black dispersion in water [1]. Owing to aggregation of carbon black clusters and the coupled viscosity problems, the amount of carbon black in the monomer can be only as high as 10 wt%. To enable any ratio between the monomer and carbon black, another approach has to be developed. Both components have to be independently dispersed in water and mixed afterwards in the right proportion. The controlled fission/fusion process in the miniemulsification destroys the aggregates and liquid droplets, and only hybrid particles remain owing to their higher stability. This process can be realized by high-energy ultrasound or high-pressure homogenization.

By screening experiments, it was shown that a hydrophobic polyurethane is a favorable choice as the osmotic agent: It is not volatile in film formation processes, and the interaction with carbon black is strong, leading to a homogeneous and well-coupled polymer layer on the carbon. In addition, it diminishes the specific surface area of the carbon (presumably by

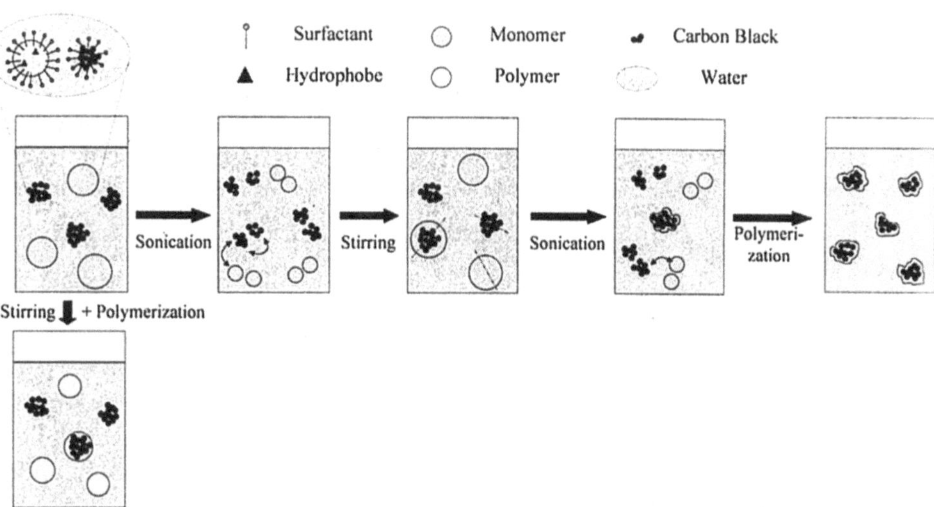

Fig. 1 Principle of encapsulation of carbon black

Table 1 Comparison between stirred and sonicated emulsions. FW18 was used as carbon black and FTPUR1 as the optimized hydrophobe. The miniemulsions comprised 1.0 g styrene, 0.025 g 2,2′-azobis(isobutyronitrile), 7.0 g water, and 0.03 g FTPUR1. A carbon black dispersion (10.0 g, γ = 43.5) was added

Latex	Surfactant	(g)	γ_{mini} (mNm^{-1})	γ_{carbon} (mNm^{-1})	US	% polystyrene–%C (thermogravi metric analysis)	Particle diameter (nm)	Poly-dispersity
FTC88					No	27:73	128	0.35
FTC89	Sodium dodecyl sulfate	0.033	48.7	43.5	Yes	62:38	70	0.49

"plugging" cavities and the inner surface), which results in a lower surfactant demand. For optimization of the technical parameters, the weight ratio of carbon to monomer was kept at 50:50, and the influence of sonication was studied.

The characteristics of the dispersion with and without sonication treatment are summarized in Table 1. The surface tension of the final dispersion is above the minimal surface tension of sodium dodecyl sulfate, indicating the absence of micelles and incomplete coverage of the polymer–carbon hybrid particles. As studied by sedimentation experiments in a preparative ultracentrifuge, the sample without sonication treatment consists of different species: a white fraction composed of carbon-black-free homopolystyrene particles and a black fraction with densities higher than 1.13 g/l. In the case of sonication treatment (FTC89) only one particle fraction, with a narrow density distribution (1.2 g/l), is detected, indicating that all the particles formed show a very similar carbon/polymer composition or degree of encapsulation. The absence of pure polystyrene particles or pure pigment aggregates is clearly proven. This composition heterogeneity is also observed by electron microscopy. In the transmission electron microscopy picture no pure polystyrene particles are detected (not shown). The carbon particles are well separated, which means that they are coated with a thin layer of polymer which screens the carbon–carbon interaction.

Table 2 Characteristics of latexes consisting of different polymers obtained by inverse miniemulsion polymerization in cyclohexane (25 g)

Monomer	Amount of monomer (g)	Emulsifier (mg)	Water/NaCl	Particle diameter (nm)
Acrylamide	3	255	4	142
Acrylic acid	1	250	1	145
Styrene sulfonic acid sodium salt	1	204	5	206
Acrylic acid-(3-sulfopropylester) potassium salt	1	200	1	205

Inverse miniemulsions

Inverse miniemulsions allow the synthesis of particles consisting of polar polymers. For the dispersion of polar monomers in nonpolar dispersion media, surfactants with low hydrophile-lipophile balance values are required. The

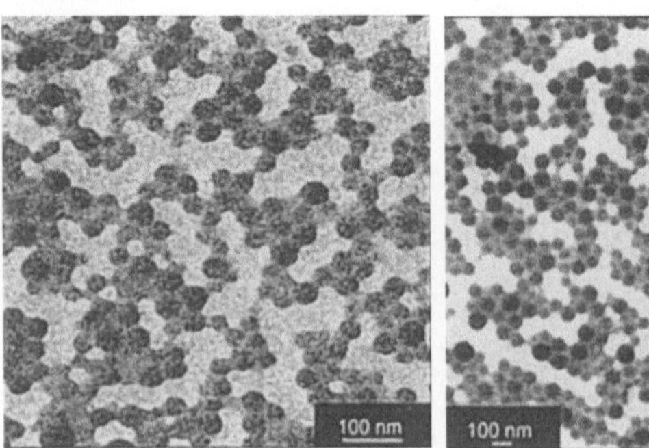

Fig. 2 Transmission electron micrograph of particles consisting of **a** poly(acrylic acid) and **b** polyacrylamide

concept of using a hydrophobe to stabilize the disperse droplets has to be adapted by adding a very lipophobic substance. A number of surfactants were screened. The nonionic block copolymer stabilizer poly(ethylene-*co*-butylene)-*b*-poly(ethylene oxide) (KLE3729) and sorbitan monooleate (Span80) turned out to be the most efficient surfactants for inverse miniemulsions. To obtain stable miniemulsions, the strong "lipophobe" NaCl was mixed with the monomeric phase in order to build up an osmotic pressure in the droplets.

For polymerization in inverse miniemulsions, a large variety of hydrophilic monomers can be used, such as acrylamide, acrylic acid-(3-sulfopropylester) potassium salt, styrene sulfonate, or acrylic acid. For the formulation of the miniemulsions, the solid crystalline monomers have to be dissolved in water. NaCl was added as an additional lipophobic agent to increase the stability of the miniemulsions, and the continuous phase was cyclohexane. The polymerization was started by adding 100 mg 2,2′-azobis(isobutyronitrile) to the continuous phase after miniemulsification. The characteristics of the final latexes are shown in Table 2. In the case of acrylamide and acrylic acid, particle sizes of about 140 nm are found; the particles in the case of styrene sulfonic acid sodium salt and acrylic acid-(3-sulfopropylester) potassium salt are larger (205 nm).

Conclusion

It was shown that the cosonication of a carbon black dispersion and a miniemulsion leads to the effective encapsulation of the pigment particles. It is our model that the fusion/fission process induced by ultrasound is only effective for the monomer droplets, whereas the monomer-coated carbon is not splittable. In this way, monomer redistribution takes place until all droplet fragments are heteronucleated onto the carbon to form a monomer film; the thickness of the film depending on the amount of monomer.

The inverse miniemulsion polymerization enables the formation of homogeneous particles consisting of polar polymers such as acrylamide, acrylic acid, styrene sulfonic acid sodium salt, or acrylic acid-(3-sulfopropylester) potassium salt and extends the possibilities of inverse emulsion and precipitation polymerization in a meaningful way.

References

1. Bechthold N, Tiarks F, Willert M, Landfester K, Antonietti M (2000) Macromol Symp 151:549
2. Tiarks F, Landfester K, Antonietti M (2001) Macromol Chem Phys 202:51
3. Landfester K, Willert M, Antonietti M (2000) Macromolecules 33:2370

Progr Colloid Polym Sci (2001) 117: 113–116

Wolfram Härtl
Christian Beck

The glass transition and propagating transverse phonons in colloidal systems

W. Härtl (✉) · C. Beck
Physikalische Chemie,
University of Saarbrücken, Gebäude 9.2,
Im Stadtwald, 66123 Saarbrücken,
Germany
e-mail: w.haertl@mx.uni-saarland.de
Tel: + 49-681-3024881
Fax: + 49-681-3024759

C. Beck
Paul Scherrer Institute,
5232 Villingen, Switzerland

Abstract The dynamics of monodisperse, highly charged colloidal systems are investigated in the glassy phase. The Debye–Waller factors obtained are compared to calculations based on the mode-coupling theory. For small Q values, low-frequency oscillations appear in the time correlation functions. An interpretation of these oscillations is given in terms of long-wavelength transverse phonons, which are characteristic of the glassy state.

Key words Glass transition · Colloids · Dynamic light scattering

Introduction

Monodisperse colloidal particles are well-suited model systems to investigate the liquid–solid phase transition. The main advantage compared to atomic systems is the possibility to study the dynamics directly in the time domain over several orders of magnitudes.

Whereas colloidal systems, which interact with a hard-sphere potential, were investigated extensively during the last decade [1, 2] only few experiments on charged colloidal systems are known which deal with the glass transition [3, 4].

In the hard-sphere systems the phase diagram and the glass-transition point were determined from experiments and good agreement with theoretical predictions was obtained [5]. In case of charged colloidal particles, which interact via a Yukawa potential, the exact conditions for obtaining a glassy phase are still unknown, but a recent theoretical investigation in the framework of the mode-coupling theory (MCT) seems to support the experimental data [6].

Preparation and characterisation

For our investigations we prepared two different model systems. The starting material for the silica colloids was tetraethoxysilane (TEOS), which was hydrolysed according to the Stöber synthesis [7]. The silica particles obtained were functionalised with carboxylic acid groups according to a procedure which was described in detail recently [8].

The second model system was prepared by emulsion polymerisation of fluorinated poly(acrylic ester) monomers [9]. Both model systems were index-matched in a mixture of water/glycerol in order to suppress multiple scattering. Our light scattering setup consisted of a photon correlation spectrometer from ALV Laservertriebsgesellschaft, Germany. The correlation functions were obtained with an ALV-5000 E fast correlator. As light source we used an He–Ne laser with a power of 33 mW (Zeiss). The scattered light was focused on a monomode cable and transferred to the detection unit, which consisted of two photomultipliers. The digitalised signals were cross-correlated in order to eliminate dead-time effects. All the measurements were done at 25 °C and the dust in the samples was removed with membrane filters of 0.8-μm pore size.

Theory

The interpretation of dynamic light scattering data from glassy colloidal suspensions has to take into account the nonergodic character of the system. The obtained time correlation functions are no longer identical with the ensemble-averaged correlation functions; however, it has been shown by Pusey et al. [10] that it is possible to convert them to the ensemble-averaged correlation functions. In the case of the monomode fibre system, only one coherence area is detected and the ensemble-averaged field correlation function can be expressed as

$$f(Q,t) = 1 + \frac{I_T}{I_E}\left(\sqrt{g_T^2(Q,t) - g_T^2(Q,0) + 1} - 1\right) , \quad (1)$$

where I_T and I_E are the time- and ensemble-averaged scattering intensities and $g_T{}^2(Q,t)$ is the time-averaged intensity correlation function. I_E was obtained by rotating the sample at constant speed during the measurements. The main features of the glass transition in colloidal systems can be described by two distinct relaxation processes, α and β, with the relaxation times τ_α and τ_β. The α process, which exists only in the liquidlike state of the colloidal system, describes the decay of density fluctuations at long times. The β process exists both in the liquidlike and in the glassy state and can be interpreted as the particle motion within the neighbouring cages. The MCT makes precise predictions about the dynamics in the vicinity of the glass-transition point. For hard-sphere colloids the separation parameter, ε, is defined as

$$\varepsilon = \frac{\Phi}{\Phi_G} - 1 , \quad (2)$$

where Φ_G is the volume fraction at the glass-transition point. The idealised MCT approach leads to the following result for the intermediate scattering function:

$$f(Q,t^*) = f(Q,\infty) + h_Q\sqrt{\varepsilon}g_\pm(t^*) . \quad (3)$$

The rescaled time, t^*, is connected to ε and diverges at the glass-transition point. The amplitude function. h_Q, and the master function, $g_\pm(t^*)$, are independent of ε and were calculated for the hard-sphere system [11]. The Debye–Waller factors, $f(Q,\infty)$, can be obtained from the long-time plateaus of the intermediate scattering function and compared with the MCT predictions if one knows the static properties of the system. The only inputs in these calculations are the static structure factor, $S(Q)$, and the direct correlation function, $c(Q)$ [11].

Results and discussion

First, we present some results we obtained for the highly charged silica systems. The samples we investigated had a mean diameter of 272 nm and were index-matched in a water/glycerol mixture. For our system we obtained from MCT calculations a glass transition at a volume fraction, Φ, of 0.175 if one uses 550 effective charges in the rescaled mean spherical approximation (RMSA) calculations of the structure factor. This result is in good agreement with recent calculations of Wang and Lai [6] on a similar system. The structure factors of our samples at volume fractions $\Phi = 0.21$ and $\Phi = 0.28$ could be modelled by RMSA calculations with 550 effective charges. The time correlation functions of these samples showed the characteristic plateau values for a glassy state. The ensemble-averaged field correlation functions are constructed from Eq. (1) and the measured intensity correlation function. The $f(Q,\infty)$ values are extracted from the long-time limits of the field correlation functions. In a next step we calculated the $f(Q,\infty)$ function within the idealised MCT approach and used as an input the corresponding experimental structure factors of the samples [12]. A comparison of the experimental and theoretical values is shown in Fig. 1. The agreement is remarkable if one keeps in mind that there are no free parameters involved. The main difference to the Debye–Waller factors of hard-sphere colloid particles can be found at small Q vectors. Here the Debye–Waller factors are much smaller, which means that for charged systems the long-wavelength density fluctuations are less frozen in as in the case of hard-sphere systems.

To investigate the dynamics of glassy colloidal systems at even smaller Q values compared to the maximum of the structure factor we synthesised perfluorinated colloids of diameter 110 nm. After concentrating the suspension to a volume fraction of 0.25 and treating it with an ionic exchanger we obtained a glassy phase, which can be identified unambiguously by the plateau values in the time correlation function. The intensity correlation functions are shown in Fig. 2 for the sample at $Q^* = 0.23$ ($Q^* = Q/Q_{max}$) during the deionisation process with ionic exchanger. After 3 h a well-developed plateau, which manifests the slowing down of the density fluctuations, can be observed. After 7-h deionisation the α process has disappeared and a phase transition into a glassy phase has occurred. Additionally, a new feature in the correlation functions emerges. In the time range 0.01–0.1 s characteristic oscillations in the correlation functions can be seen. To investigate this in more detail we performed a Q-dependent measurement of the sample.

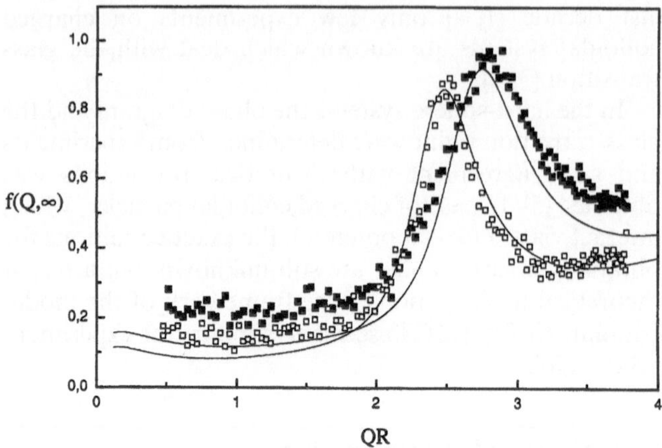

Fig. 1 Experimental Debye–Waller factors, $f(Q,\infty)$, at $\Phi = 0.27$ (*closed symbols*) and $\Phi = 0.21$ (*open symbols*). The *solid lines* are calculations according to the mode-coupling-theory approach

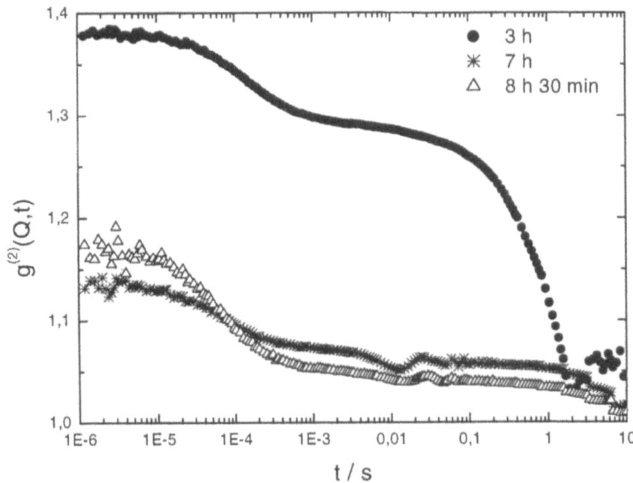

Fig. 2 Intensity correlation functions during the deionisation process for a highly charged sample at $\Phi = 0.25$ composed of perfluorinated colloidal particles at $Q^* = 0.23$

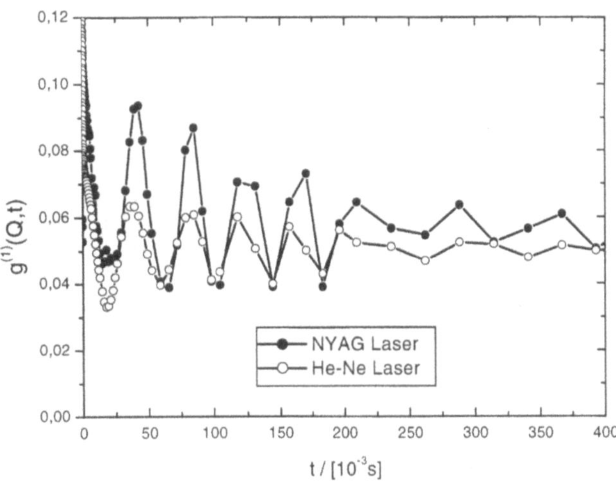

Fig. 4 Field autocorrelation functions of the sample in Fig. 2 for two different experimental configurations

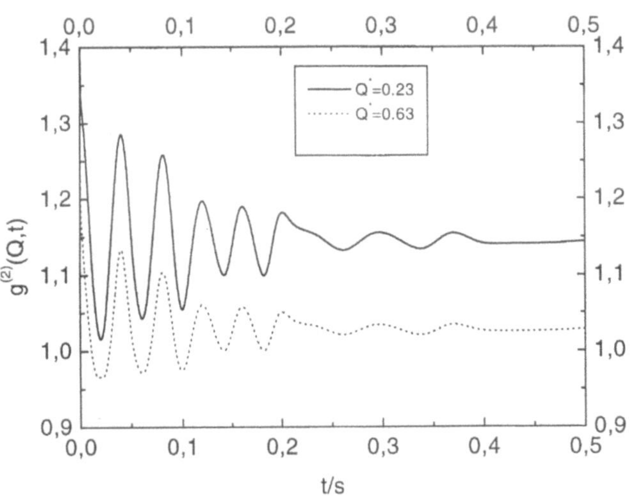

Fig. 3 Characteristic oscillations in the intensity correlation function of the sample in Fig. 2 for $Q^* = 0.23$ and $Q^* = 0.63$

As shown in Fig. 3 the oscillation frequency is not Q-dependent and only a slight decrease in the oscillation amplitude is detected with increasing Q values.

One explanation of the observed oscillations could be propagating shear waves, which are excited by vibrations of the whole experimental setup. The method of standing shear waves has been used to determine the elastic modulus of crystalline and glassy samples. Here the shear waves are excited mechanically by a loudspeaker membrane [13]. The second possibility to explain the experimental findings are intrinsic transverse phonons which are present in the glassy samples. The existence of long-wavelength transverse phonons have been discussed in the literature for crystalline samples but until now only overdamped modes could be detected [14]. To rule out the possibility of shear waves excited by external sources we performed a measurement of the sample with a different instrumental setup. As a light source a Nd:yttrium aluminium garnet laser was utilised and the whole setup was isolated from external vibrations by a marble table and a sand bed. The result of the measurement is compared with the data obtained from our setup. As can be seen in Fig. 4 the oscillations are still present and have the same frequency.

A second point which supports the picture of transverse phonons is given by the fact that the resonance frequency of the standing shear waves can be estimated to be between 200–600 Hz if one takes the elastic modulus values given in the literature for similar systems [15]. This frequency is at least 1 order of magnitude larger than the observed frequency in our measurements; therefore, we attribute the oscillations to propagating transverse phonons, which are characteristic for colloidal glasses. In a recent investigation of a hard-sphere colloidal silica system in a low-viscosity solvent we could confirm the present results.

Acknowledgements We want to thank H. Versmold for the use of his dynamic light scattering spectrometer and H. Löwen for helpful discussions.

References

1. van Megen W, Underwood SM (1993) Phys Rev Lett 70:2766
2. van Megen W, Underwood SM (1994) Phys Rev E 49:4206
3. Sirota EB, Ou Yang HD, Sinha SK, Chaikin PM (1989) Phys Rev Lett 62:1524
4. Härtl W, Versmold H, Zhang-Heider X (1995) J Chem Phys 102:6613
5. Götze W (1999) J Phys Condens Matter 11:A1
6. Wang GF, Lai SK (1999) Phys Rev Lett 82:3645
7. Stöber W, Fink A, Bohn E (1968) J Colloid Interface Sci 26:62
8. Beck C, Härtl W, Hempelmann R (1999) Angew Chem 111:1380
9. Härtl W, Zhang-Heider X (1996) J Colloid Interface Sci 185:398
10. van Megen W, Underwood SM, Pusey PN (1991) Phys Rev Lett 67: 1586
11. Fuchs M, Götze W, Hildebrand S, Latz A (1992) J Phys Condens Matter 87:43
12. Beck C, Härtl W, Hempelmann R (1999) J Chem Phys 111:8209
13. Dubois-Violette E, Pieranski P, Rothen F, Strzelecki L (1980) J Phys 41:369
14. Hurd AJ, Clark NA, Mockler RC, O'Sullivan WJ (1982) Phys Rev A 26:2869
15. Benzing DW, Russel WB (1981) J Colloid Interface Sci 83:178

Progr Colloid Polym Sci (2001) 117: 117–119

N. M. Nagy
J. Kónya
M. Beszeda
I. Beszeda
E. Kálmán
Z. Keresztes
K. Papp

Lead accumulation on montmorillonite

N. M. Nagy (✉) · J. Kónya · M. Beszeda
Isotope Laboratory, Faculty of Natural Sciences, University of Debrecen,
P.O. Box 8, 4010 Debrecen, Hungary
e-mail: noemi@tigris.klte.hu
Tel.: +36-52-310122
Fax: +36-52-310122

I. Beszeda
Department of Solid State Physics, Faculty of Natural Sciences, University of Debrecen,
4010 Debrecen, Hungary

E. Kálmán · Z. Keresztes · K. Papp
Institute of Chemistry, Chemical Research Center, Hungarian Academy of Sciences,
P.O. Box 17, 1525 Budapest, Hungary

Abstract The structure of lead and calcium–lead montmorillonites was studied by IR spectroscopy and scanning electron microscopy. The IR spectra showed no structural changes of the montmorillonite crystal lattice during calcium–lead cation exchange, only the intensities of hydrate water of the interlayer cation vary and the OH band on the edges disappears. Scanning electron microscopic studies showed that the distribution of lead is fairly even on the major part of the surface; however, there are places where the concentration of lead is higher than the mean surface concentration. The diameter of them is less than 1 μm. Similar lead enrichments were found on a natural clay sediment of a lake in Hungary. Lead ions adsorb on montmorillonite by cation exchange in the interlayer space and a heterogeneous nucleation on the particle surface followed by crystal growth.

Key words Lead ion · Calcium montmorillonite · Microparticles · IR spectroscopy · Scanning electron microscopy

Introduction

Nowadays the heavy-metal concentration of soils increases because of anthropogenic activity. Heavy-metal ions, including lead ions, can be present as different species, depending on the environmental conditions. The adsorption of lead ions on clay minerals has been studied by several authors; the important results have been summarized by Shen et al. [1]. They explained the adsorbed quantity of lead on montmorillonite by chemical speciation of lead.

By the X-ray absorption fine structure method Strawn and Sparks [2] indicated that lead could adsorb via two mechanisms, depending on the ionic strength. At low ionic strength, lead adsorption is pH-independent, consistent with an outer-sphere complexation. At high ionic strength, lead adsorption is pH-dependent, suggesting inner-sphere complexation, in which lead forms covalent bonds $[Pb(OH)_4^{2-}$ (aq)].

In this article, studies of the physical and chemical formations of sorbed lead ions on calcium montmorillonite and a Hungarian lake sediment by IR spectroscopy and scanning electron microscopy (SEM) are reported.

Experimental

Calcium montmorillonite was a product of the Central Laboratory of the National Mining Company, Hungary. X-ray diffraction and thermoanalysis showed 83% montmorillonite content, the other constituents were kaolinite (3%), calcite (5%), and quartz in trace and amorphous form (9%). The cation-exchange capacity was determined by the ammonium acetate method [3] and it was found to be 52 mmol/100 g for calcium ions (i.e. 104 mEq/100 g).

Air-dried calcium montmorillonite (40.0 mg) was measured into a beaker and 20.0 cm^3 bidistilled water and perchloric acid solution of different concentration was added and the mixture was stirred at constant speed for 3 min. Then, lead perchlorate solutions with different concentrations and labeled by radioactive ^{212}Pb isotope was added to the suspension and stirred again for 60 min. The concentration of the solution at the time of the addition of lead perchlorate was 5×10^{-4}, 1×10^{-3}, or 5×10^{-3} mol/dm^3 Pb^{2+} ion. The phases were separated with a membrane filter (0.45-μm pore size) and the pH of the liquid was measured. The solid was dried at room temperature (25 °C). The γ activity of the solid and the solution was measured using a NaI(Tl) scintillation detector. The adsorbed quantity of lead was calculated from the radioactivities. IR spectra, and SEM images were taken after the complete decay of ^{212}Pb (about 5 days).

The experiments were repeated in suspensions containing 5×10^{-4} mol/dm^3 lead ion and citric acid as a complex forming agent. The pH was adjusted to 3.2.

The experiments were made at room temperature. IR spectra of the dried solid samples were taken using a PerkinElmer Paragon 1000 PC Fourier transform IR spectrometer after the samples had been pressed with KBr. The dried solid samples were studied using an Amray-1830 and a Hitachi S-570 I scanning electron microscope and an Röntec EDR288 energy-dispersive X-ray spectrometer. Sediments from Lake Prod (next to road number 33 in eastern Hungary) were also studied. The lead content of the sediments was determined using a Spectroflame inductively coupled plasma optical emission spectrometer and was above 30 mg/kg.

Results and discussion

IR spectrometric studies of the lead montmorillonites showed no structural changes of the montmorillonite crystal lattice. The intensities of the hydrate water of the cation in the interlayer space vary during cation exchange. The band at 2,920 cm^{-1} disappeared as a result of lead adsorption. Since this band belongs to the OH vibrations at the edges [4], the disappearance of the band can be identified as proton–lead exchange on the edge charges of montmorillonite.

SEM studies showed that the distribution of lead is fairly even on the major part of the surface; however, there are places where the concentration of lead is higher than the mean lead concentration (Fig. 1).

The results show that lead ions are adsorbed on montmorillonite by two processes: by cation exchange in the interlayer space (outer-sphere complexation) and by adsorption on the edge sites (inner-sphere complexation). Cation exchange leads to the even distribution of the ions, while the adsorption on the edge site can act as the initial nucleus of precipitation of lead hydroxide observed by Strawn and Sparks [2]. The nanoparticles and microparticles (lead enrichments) seen by SEM can probably be formed on these nuclei.

The production of these particles is not expected from the thermodynamic properties under conditions of the bulk solution and cannot be observed in the absence of clay. On the basis of the solubility product of $Pb(OH)_2$ ($L = 6.8 \times 10^{-13}$) lead is present as Pb^{2+} at the pH values of the experiments (pH 3–5). The formation of lead enrichments requires the presence of the clay. In the presence of clay, lead enrichments are formed even in a solution containing a complex-forming agent (citric acid).

The chemical analysis by SEM shows that the elementary composition of lead enrichments (except

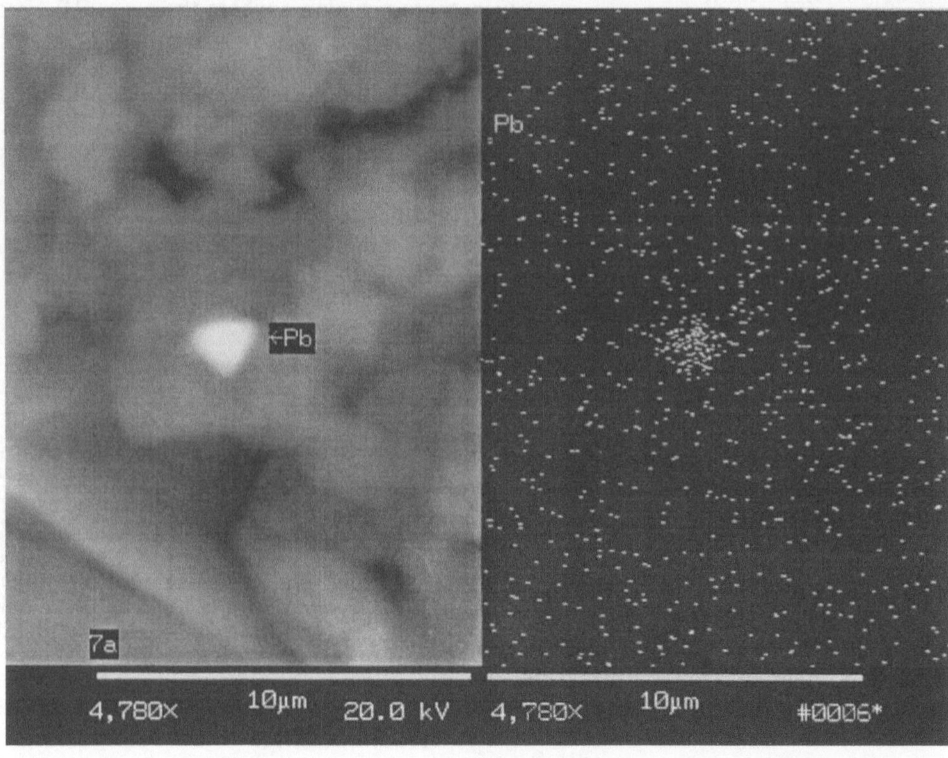

Fig. 1 Scanning electron microscope picture of montmorillonite treated with 5×10^{-4} moldm^{-3} lead perchlorate solution, pH 4.18. *Left*: morphology of the sample made by backscattered electrons. *Right*: lead map made by characteristic X-ray photons

Fig. 2 Scanning electron microscope picture of a natural clay sediment. The mean lead concentration is 36 mg/kg, lead map in the *window*

lead, of course) is usually the same as the mean composition of montmorillonite, so the enrichments are formed on montmorillonite, not on the impurities.

Similar lead enrichments were found on a natural clay sediment of a lake in Hungary (Fig. 2). The production of nanoparticles and microparticles containing lead ions under environmental conditions is especially interesting.

Conclusions

The results show that lead ions are adsorbed on montmorillonite by two processes: cation exchange in the interlayer space and a heterogeneous nucleation on the particle surface followed by crystal growth. Cation exchange leads to the even distribution of the ions, while the heterogeneous nucleation and crystal growth causes the formation of the enrichment of the lead ions. The production of these particles is not expected from thermodynamic aspects of the solution and cannot be observed in the absence of clay.

Acknowledgements The authors thank M. Földvári and P. Kovács-Pálffy for X-ray diffraction studies and P. Kovács for the clay sediment samples. The work was supported financially by the National Research Foundation (OTKA T23905 and CO278).

References

1. Shen S, Taylor WT, Bart H, Tu S-I (1999) Commun Soil Sci Plant Anal 30:2711
2. Strawn DG, Sparks DL (1999) J Colloid Interface Sci 216:257
3. Richards LA (1957) Diagnosis and improvement of saline and alkaline soils.US Department of Agriculture Handbook, p 60
4. Peker S, Yapar S, Besün N (1995) Colloids Surf 104:249

Progr Colloid Polym Sci (2001) 117: 120–125

A. Pozsgay
L. Papp
T. Fráter
B. Pukánszky

Polypropylene/montmorillonite nanocomposites prepared by the delamination of the filler

Abstract Polypropylene/montmorillonite (PP/MMT) nanocomposites were prepared by the exfoliation of the nanoclay during processing. The filler was used in various forms (sieved bentonite, sodium and organophillic MMT) in order to determine the effect of subsequent stages of organoclay preparation on the structure and properties of the composite. The filler content was varied over a relatively wide range to determine the extent of maximum reinforcement and the range of practically relevant compositions. Specimens were injection-molded with various weak sites (weld lines, gate section) in the test area to see the performance of the nanocomposite under practically relevant conditions. The results proved that extensive exfoliation of the organophilic MMT occurred during the homogenization and subsequent injection molding of the PP composites studied; however, organophilization of NaMMT was not complete, which hindered the perfect exfoliation of the MMT layers. Sieved bentonite and NaMMT behaved like traditional fillers, while the incorporation of the organoclay into PP yielded a true nanocomposite. Because of incomplete exfoliation and poor adhesion only a moderate improvement of mechanical properties was achieved in the study. Reinforcement depends on composition; above a certain filler content the nanoclay behaves like a particulate filler possibly owing to the stacking of exfoliated layers. The weld line strength of PP nanocomposites is very low; exfoliated clay particles orientate parallel to the weld line and strongly deteriorate the properties.

Key words Nanocomposite · Delamination · Organoclay · Strength · Weld line

A. Pozsgay · L. Papp · T. Fráter
B. Pukánszky (✉)
Budapest University of Technology and Economics, Department of Plastics and Rubber Technology, 1521 Budapest, P.O. Box 92, Hungary, and
Institute of Chemistry, Chemical Research Center, Hungarian Academy of Sciences 1525 Budapest, P.O. Box 17, Hungary
e-mail: pukanszky@muatex.mua.bme.hu
Tel.: +36-1-4632015
Fax: +36-1-4633474

Introduction

Recently, interest in nanocomposites has increased enormously both in industry [1] and academia [2–5]. This new class of materials can be prepared by various routes, resulting in composites with different structures. Molecular composites consist of rigid molecules embedded into a softer polymer matrix [6], colloidal composites are usually prepared by in situ sol–gel precipitation of the filler particles [7], while layered silicate nanocomposites contain exfoliated clay particles [8]. This third class of composites can be prepared by intercalation and in situ polymerization of a monomer in the galleries of the nanoclay or by the exfoliation of the layers during processing [2].

Layered silicate nanocomposites are claimed to have numerous advantages. The homogeneously dispersed thin clay layers increase stiffness and strength considerably even at very low filler contents. In contrast to traditional particulate-filled composites, in nanocomposites efficient reinforcement can be achieved at filler content as low as 3–5 wt%. Further advantages of these materials are their low flammability [1, 2], increased dimensional stability and heat deflection temperature as well as decreased permeability. Owing to the combination of the previously mentioned properties the potential application fields of these materials are mainly the automotive and the packaging industries [1]. However, in spite of their potential as well as the intensive research and development carried out on them, technical problems and their relatively high price hinder the extensive application of these materials; obviously further research is needed before a real breakthrough is achieved.

The goal of the present study was to prepare polypropylene/montmorillonite (PP/MMT) nanocomposites by the exfoliation of the nanoclay during processing. The filler was used in various forms in order to determine the effect of the various stages of organoclay preparation on the structure and properties of the composite. The filler content was varied over a relatively wide range to determine the extent of maximum reinforcement and the range of practically relevant compositions. Specimens were injection-molded with various weak sites (weld lines, gate section) in the test area to see the performance of the nanocomposite under practically relevant conditions.

Experimental

A PP homopolymer (Tipplen H 377, melt flow index of 9 g/10 min at 21.6 N and 230 °C) produced by TVK, Hungary, was used as a matrix material in the experiments. Bentonite mined in Mád, Hungary, was selected as the starting material for the preparation of organophillic MMT. The most important characteristics of the filler are compiled in Table 1. The filler consists of large aggregated particles with a layered structure. The raw mineral was sieved to obtain a filler of less then 90-μm maximum particle size (sieved sample). In the next step the bentonite containing mostly CaMMT was subjected to ion exchange in order to produce NaMMT. Ion exchange was carried out with NaCl solution of 1.0 mol/l concentration at 20 °C and was repeated three times. Subsequently, the product was treated with an amount of *N*-cetyl pyridinium chloride corresponding to the theoretical ion-exchange capacity of the filler in order to obtain the organoclay (organophillic MMT).

The three samples (sieved, Na and organophillic MMT) were prepared in sufficient quantities and homogenized with PP to produce nanocomposites of 1, 2, 5, 10 15 and 20 v% filler content. Homogenization was carried out in a Brabender DSK 42/7 twin screw compounder at set temperatures of 195, 200 and 205 °C. The granules produced in the compounder were injection-molded using a Battenfeld BA 200 CD machine at barrel temperatures of 205, 215 and 225 °C into dog-bone tensile bars of 3 × 2-mm dimensions in their test section. The temperature of the mold was set to 30 °C. The design of the mold made possible the simultaneous preparation of three kind of specimens representing various practical molding configurations. They differed in the position of the gate, which was located at one end, at both ends to produce a weld line in the test area, or in the mid section of the specimen, respectively.

Changes in the structure of the filler as an effect of treatment (ion exchange, organophilization) was followed by wide-angle X-ray scattering (WAXS) using a Phillips PW 1830/PW 1050 instrument with Cu Kα radiation at 40 kV and 35 mA. The same technique and conditions were used for the characterization of the PP/clay composites. The dispersion and orientation of the filler, as well as the failure behavior of the composites, were studied by scanning electron microscopy (SEM) using a JEOL JSM-5600 LV instrument. The melting and crystallization characteristics of the composites were determined using a PerkinElmer DSC 7 apparatus on 5-mg samples with heating and cooling rates of 10 °C. Tensile properties were measured using a Zwick 1445 machine at 10-mm gauge length and 50 mm/min cross-head speed.

Table 1 Characteristics of the filler used in the experiments (catalogue data)

Mineral composition (wt%)	
Montmorillonite	70
Quartz	30
Chemical composition (wt%)	
SiO_2	76.0
Al_2O_3	15.0
Fe_2O_3	2.6
CaO	2.0
MgO	1.4
Other	3.0
Properties	
Particle size (μm)	0–60
Bulk density (g/cm^3)	0.55
Moisture content (wt%)	7.5
Specific surface area (m^2/g)	375

Results

The results of the study are discussed in three sections. First, we present the changes in the MMT structure during the course of organoclay preparation. Subsequently, the structure and properties of the nanocomposites are discussed separately for the perfect part and for the specimen containing the weld line. Compared to the perfect specimen the presence of the gate in the test area did not change the properties of the composite significantly; thus we refrain from the discussion of these results in order to save space.

MMT structure

One of the filler samples introduced into the PP matrix was produced by the sieving of the raw mineral obtained from the mine. Sieving was expected to remove very large particles, which deteriorate mechanical properties. In the presence of large spherical particles the dominating deformation mechanism of composites is debonding [9–11] and both yield stress and strength decrease with increasing particle size [12, 13]. The maximum size of 90 μm used in the experiments is well over the usual particle size range of commercial fillers; however, the initial size of the particles does not play a role in nanocomposites, where complete exfoliation is assumed. The particle size of the sieved sample and NaMMT, where exfoliation does not take place, had to be the same to allow comparison. Moreover, the filler used in the experiments contains a significant amount of microcrystalline quartz, which may also influence the properties. The WAXS diffractogram of the three samples is presented in Fig. 1. Figure 1A shows the diffractogram taken from the sieved sample and indicates the presence of quartz, shown by the characteristic peak appearing at the 2θ angle of 26.7°.

The procedure to produce NaMMT from the sieved sample considerably changed the diffractogram (Fig. 1B); only small traces of quartz can be detected in this sample. The peaks of MMT remain practically unaltered. Quartz obviously settles in and can be removed from the vessel used for the ion exchange. As a consequence, this step of sample preparation has two results:exchange of Ca^{2+} by Na^{+} and removal of quartz. It is interesting to note that the gallery distance of MMT decreased in this step, probably owing to the intensive drying of the ion-exchanged filler (see WAXS peak at around 7.8°). The treatment of NaMMT with a solution of *N*-cetyl pyridinium chloride leads to further ion exchange. The success of the treatment is clearly proved by the increase in the gallery distance of the layers. After treatment, the peak detected at 7.3° shifts towards lower angles and appears at 4.2°, which corresponds to a gallery thickness of 21.3 Å. Unfortunately, the organophilization process was imperfect as shown by the appearance of the WAXS peak at 7.3 , which indicates the presence of nontreated bentonite with closed galleries.

Structure and properties of nanocomposites: perfect parts

Fillers and reinforcements often modify the morphology of crystalline polymers. Nucleation usually leads to a change in lamella thickness, spherulite size and crystallinity, but occasionally modifies the crystal form as well [14–16]. The effect of the three fillers on the temperature of crystallization of PP is presented in Fig. 2 as a function of composition. There may be several reasons for the changes observed in the figure. The sieved bentonite contains both MMT and quartz particles; both may nucleate PP. The layered structure of bentonite is somewhat similar to that of talc and may result in increased nucleation [16]. On the other hand, most of the quartz was removed during ion exchange and the nucleating effect of the filler decreased simultaneously, i.e. quartz might also be the nucleating component. We are convinced that the layered particle structure of bentonite is the source of strong nucleation. During ion exchange, MMT layers are separated, then they reform again during drying with a smaller gallery distance (Fig. 1). The filler particles formed in this process have a weaker nucleation effect than the original bentonite, probably owing to this change in the distance of the layers. As an effect of organophilization the layer thickness increases significantly and a large part of the layered structure is destroyed during processing. Both processes

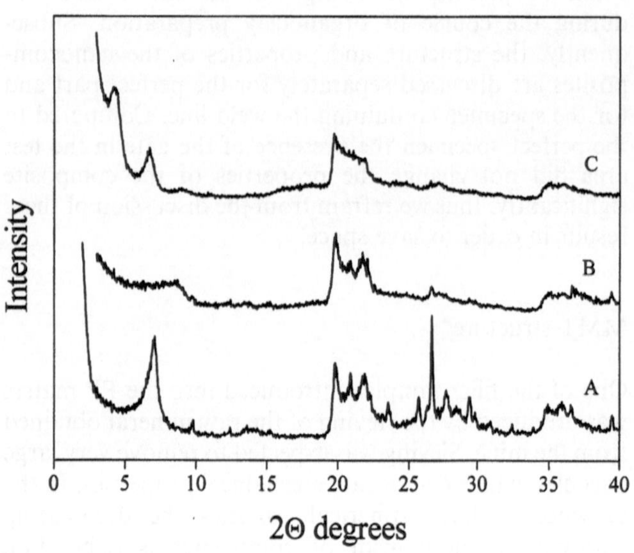

Fig. 1 Wide-angle X-ray scattering diffractogram of the filler samples used in the experiments:**A** sieved bentonite; **B** sodium montmorillonite (*NaMMT*); **C** organoclay

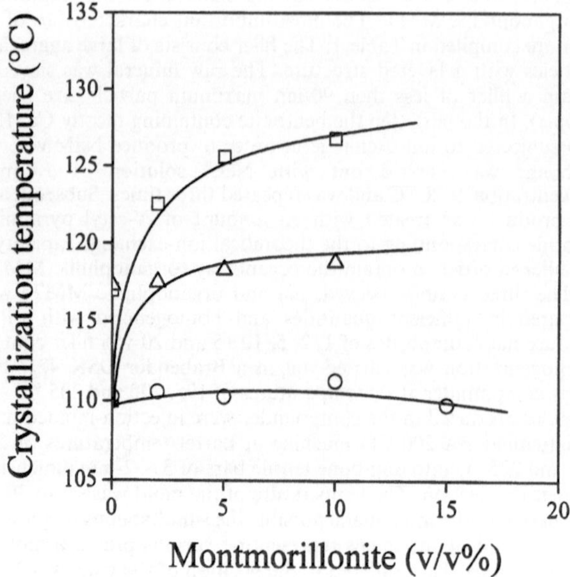

Fig. 2 Nucleation effect of the various filler samples in the polypropylene (*PP*) matrix used:sieved bentonite (□); NaMMT (△); organoclay (○)

change the distance of the layers and thus influence the nucleation efficiency of the filler. In this case nucleation is not observed at all.

The SEM micrograph taken from the fracture surface of the composite containing the sieved filler in 0.05 volume fraction is presented in Fig. 3. Very large particles can be observed on the surface and the dominating deformation mechanism is clearly debonding. The micrograph forecasts poor mechanical properties:the complete lack of a reinforcing effect. In other areas parallel orientation of plateletlike particles can be observed on the surface. This latter structure may result in reinforcement, but the local areas represented in Fig. 3 dominate the response of the material towards external loading.

The properties of the composites follow exactly the prediction made by the analysis of the SEM micrographs. The composition dependence of the yield stress is presented in Fig. 4. A strong decrease in this property occurs when the composite contains sieved bentonite and NaMMT, while the organoclay reinforces PP. The composites prepared with the first two fillers failed without yielding at very low filler contents, which proves again that debonding takes place during deformation and the voids formed merge rapidly into critical cracks. On the other hand, exfoliation changes the deformation mechanism, yielding dominates at low filler contents and both strength and deformation increase. At high filler loadings, however, the benefit is lost. Either exfoliation does not take place or the layers organize into large stacks and behave like commodity fillers. These results verify earlier observations which indicate that the advantages of nanocomposites appear only at low filler content in a limited composition range.

WAXS diffractograms were taken from the composites to check the extent of exfoliation and the structure of the composites. The diffractograms, not presented here to save space, show the characteristic peaks of PP, but at low angles two other peaks appear as well, which belong to the closed gallery structure of bentonite (6.7°) and to the opened layers of the organoclay (4.4°). Their presence indicates both incomplete organophilization and imperfect exfoliation. The composition dependence of the extent of exfoliation is expressed quantitatively as the ratio of the intensity of the two peaks (organophilic/bentonite) in Fig. 5. The intensity of the organoclay-related peak decreases with increasing filler content, which indicates increasing exfoliation, since delaminated MMT sheets do not give a WAXS pattern. The viscosity increases with filler content, which results in higher shear and a more complete exfoliation of the organophilized clay; however, the diffractograms reveal that a considerable amount of bentonite is also present in the composite, i.e. organophilization was not complete, resulting in the moderate improvement in properties.

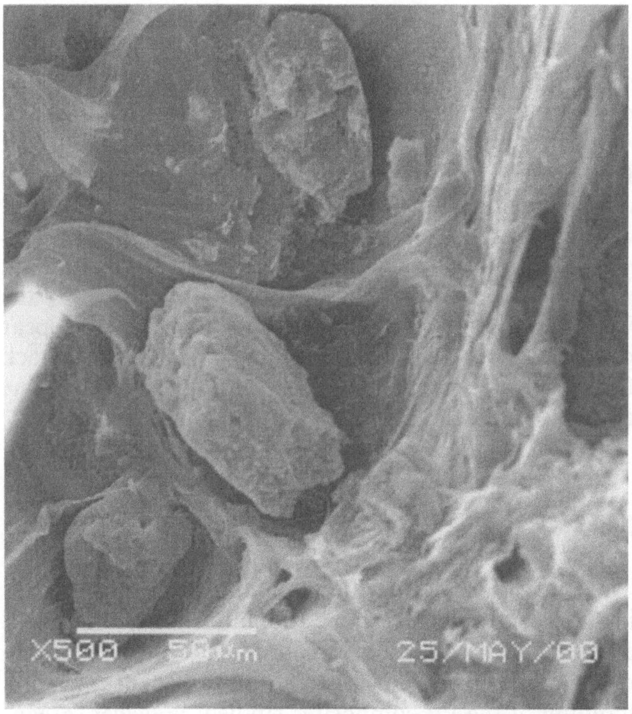

X500 50 μm 25/MAY/00

Fig. 3 Scanning electron microscopy (*SEM*) micrograph taken from the fracture surface of a PP nanocomposite containing 0.05 volume fraction sieved bentonite; large particles, debonding

Weld lines

The study of specimens containing weld lines might supply further information about exfoliation and the

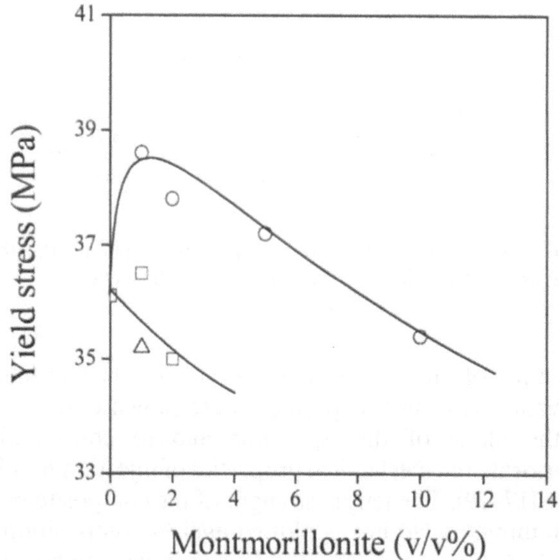

Fig. 4 Comparison of the yield stress of PP nanocomposites containing sieved bentonite (□), NaMMT (△) and organoclay (○)

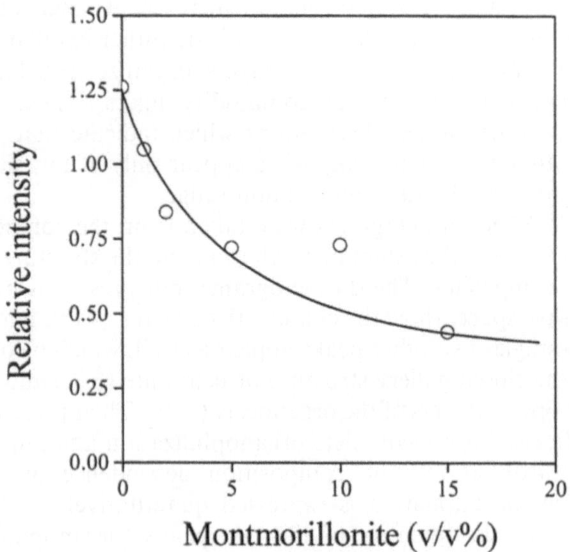

Fig. 5 Composition dependence of the relative intensity (organoclay/bentonite) of the two peaks assigned to closed and open gallery distance of the organoclay

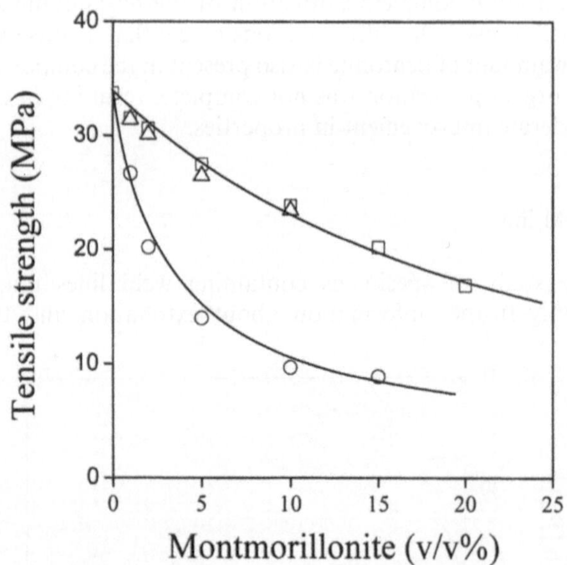

Fig. 6 Weld-line strength of nanocomposites containing the three fillers; sieved bentonite (□); NaMMT (△); organoclay (○)

structure of the composites, but it also has practical relevance. Anisotropic particles were shown to orientate in the plane of the weld line and to considerably deteriorate the mechanical properties of injection molded parts [17–19]. The tensile strength of the composite parts containing a weld line is plotted against composition in Fig. 6. A very significant difference can be seen between the organoclay and the other two fillers in their effect on weld-line strength. The sieved bentonite and NaMMT

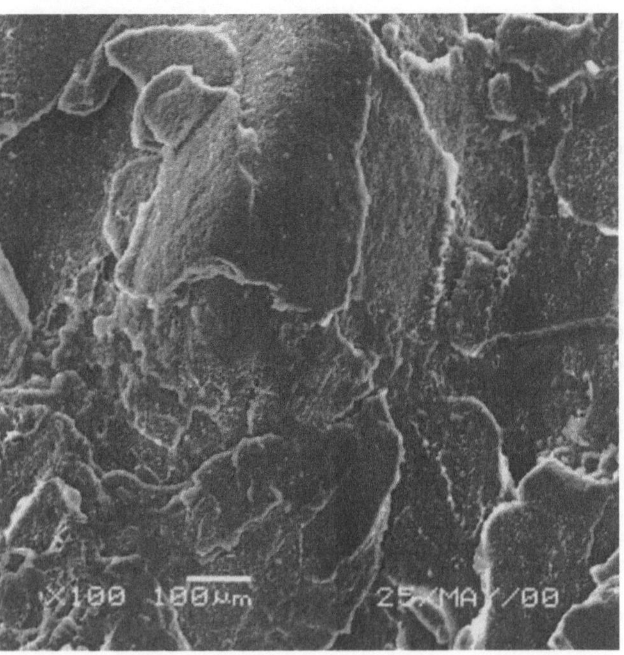

Fig. 7 SEM micrograph taken from the fractured weld line of an injection-molded specimen prepared from a composite containing 0.05 volume fraction organoclay

decrease the strength with increasing filler content, which is in line both with the structure and with the deformation mechanism (Fig. 3) of the composites as discussed earlier. The composite containing the organoclay possesses a strongly decreased strength. Obviously exfoliated clay sheets orientate in the plane of the weld line and inhibit interdiffusion of molecules in the merging melt fronts. This result proves again that a considerable amount of the organoclay exfoliates during homogenization and the processing of composite parts. This statement is further corroborated by Fig. 7, which shows a SEM micrograph taken from the fracture surface of the weld line. Very large, 200–400-μm smooth areas appear on the surface as an effect of a rapidly propagating crack front. The crack apparently proceeds along the weld line containing the clay sheets of parallel alignment.

Conclusions

The study of PP composites containing a bentonite filler in the various stages of organophilization proved that extensive exfoliation of organophilic MMT occurred during the homogenization and subsequent injection molding of the PP composites studied; however, organophilization of NaMMT was not complete in the study, which hindered the perfect exfoliation of the MMT layers. Sieved bentonite and NaMMT behaved like traditional fillers, while the incorporation of the organoclay into PP yielded a true nanocomposite. Because of incomplete exfoliation, and probably owing to poor adhesion, only a moderate improvement in the mechanical properties was achieved in the study. Reinforcement depends on composition; above a certain filler content the nanoclay behaves like a particulate filler, possibly owing to the stacking of exfoliated layers. As a consequence, the advantages of nanocomposites can be exploited only at low filler contents in a limited composition range. The weld line strength of the PP nanocomposites is very low, exfoliated clay particles orientate parallel to the weld line and they strongly deteriorate the properties. Both treatment and processing technology must be improved in order to utilize all the potential of nanoclays and achieve maximum reinforcement.

Acknowledgements The authors are truly indebted to Tamás Grósz and István Sajó for their assistance in completing the WAXS measurements. Péter Hargitai is acknowledged for the preparation of the SEM micrographs. The National Fund for Scientific Research (OTKA T 30579) and the Varga József Fund of the Faculty of Chemical Engineering at the Budapest University of Technology and Economics is greatly appreciated for the financial support of the research.

References

1. Sherman LM (1999) Plast Technol 45:52
2. Giannelis EP (1996) Adv Mater 8:29
3. Giannelis EP (1998) Appl Organometal Chem 12:675
4. Mülhaupt R, Stricker F (1997) Kuststoffe 87:482
5. Herron N, Thorn DL (1998) Adv Mater 10:1173
6. Ruckenstein E, Yuan Y (1997) Polymer 38:3855
7. Schmidt HK, Geiter E, Mennig M, Krug H, Becker C, Winkler RP (1998) J Sol-Gel Sci Technol 13:397
8. Liu P, Qi Z, Zhu Z (1999) J Appl Polym Sci 71:1133
9. Vollenberg P, Heikens D, Ladan HCB (1988) Polym Compos 9:382
10. Pukánszky B, Vörös G (1993) Compos Interfaces 1:411
11. Vörös G, Fekete E, Pukánszky B (1997) J Adhes 64:229
12. Pukánszky B, Turcsányi B, Tüdos F (1988) In:Isida H (ed) Interfaces in polymer, ceramic, and metal matrix composites. Elsevier, New York, pp 467–477
13. Pukánszky B (1990) Composites 21:255
14. Fujimama M, Wakino T (1991) J Appl Polym Sci 42:2739
15. Varga J (1989) J Thermal Anal 35:1891
16. Pukánszky B, Belina K, Rockenbauer A, Maurer FHJ (1994) Composites 25:205
17. Christie M (1986) Plast Eng 42:41
18. Fisa KB, Dufour J, Vu-Khanh T (1987) Polym Compos 8:408
19. Waxman A, Narkis M (1991) Polym Compos 12:161

Progr Colloid Polym Sci (2001) 117: 126–130

C. Beck
W. Härtl

Fullerenes as new colloidal model systems

C. Beck (✉)
Paul Scherrer Institute,
Laboratory for Neutron Scattering,
5232 Villigen, Switzerland
e-mail: christian.beck@psi.ch
Tel.: + 41-56-3104621
Fax: + 41-56-3102939

C. Beck · W. Härtl
University of Saarbrücken,
66123 Saarbrücken, Germany

Abstract Water-soluble $C_{60}(OH)_{22-27}$ (fullerol) particles with different concentrations in D_2O as solvents were used as colloidal spheres. By means of quasielastic neutron scattering (QENS) we determined the dynamic structure factor, $S(Q,\omega)$, for samples with concentrations ranging from 1 to 6 vol%. Recently, it has been shown that C_{60} in CS_2 is a new hard-sphere colloidal model system. In contrast to this lipophilic system with only low solubility in most solvents, hydrophilic $C_{60}(OH)_{22-27}$ can be treated as a new soft-sphere colloidal model system for weakly or even highly charged particles. We show that QENS is well suited for studying the dynamic and static behavior of these fullerol–D_2O systems with different concentrations. At 1 bar and 298 K, for scattering vectors$Q = 0.4$ Å^{-1} up to $Q = 2.6$ Å^{-1} $S(Q,\omega)$ can be well described by a sum of a narrow and a broad Lorentzian line. The first is attributed to the diffusion of $C_{60}(OH)_{22-27}$ colloidal particles in the solvents, whereas the broad Lorentzian is due to the motion of the solvents molecules themselves. From the width of the narrow line we find the translational diffusion coefficient of the fullerol particles to be about $2.74 \cdot 10^{-10}$ m^{2}s^{-1}. Taking into consideration the hydrodynamic shell and the resulting larger effective hydrodynamic diameter of the particles this value is comparable to those obtained from the Stokes–Einstein relation. Additionally, we show that the self-diffusion coefficient is significantly influenced by the volume fractions of the fullerol particles in D_2O. This is a direct hint for the beginning of particle–particle self-organization in the sense of a liquid-like-ordered system.

Key words Fullerenes · Scattering · Colloids · Diffusion · Self-organization

Introduction

The dynamics of dilute monodisperse suspensions of spherical colloidal particles has been extensively studied by means of dynamic light scattering [1] to prove their model system character for the atomic state, which is due to the similar interparticle interaction potential. A tremendous amount of work has been done in the field of soft-condensed-matter research to study and reveal the hydrodynamic behavior [2] or even the liquid–glass transition in systems of highly charged and hard-sphere particles [3].

In this article we report a study of the dynamic behavior of hydroxylated fullerene molecules, i.e. so-called fullerol $C_{60}(OH)_{22-27}$, which are intermediate between classical colloidal suspensions and binary mixtures of atomic liquids. We consider at a temperature of 298 K dispersions of fullerol, which are highly soluble in polar liquids, in D_2O with different volume fractions, ϕ, ranging from dilute up to 6%. Each of these systems is

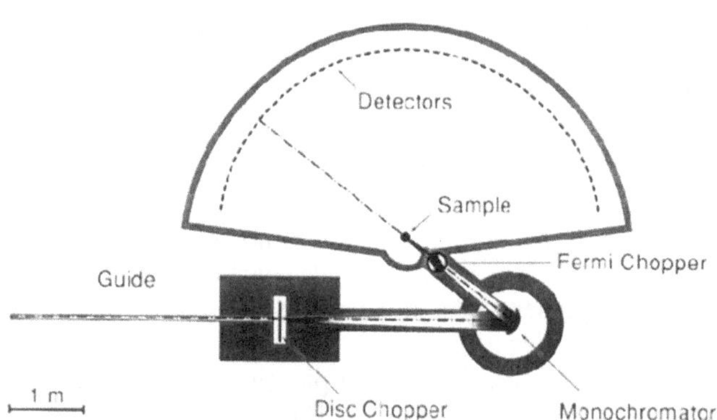

Fig. 1 Brief sketch of the spectrometer FOCUS

essentially a monodisperse solution of perfect $C_{60}(OH)_{22\text{–}27}$ with a calculated crystallographic (powder) diameter of about 0.9 nm.

We show that neutron time-of-flight (TOF) spectroscopy is a powerful tool to study these dynamic processes of fullerol particles in D_2O as a solvent by analyzing the quasielastic broadening of the elastic scattered intensity at Q values between Bragg reflections. The momentum transfer, Q, and the energy transfer, ω, of the neutron to the sample during the scattering process mainly characterize the method. In contrast to a triple-axis spectrometer, a TOF experiment simultaneously detects several points in the (Q,ω) phase space by time-resolved neutron detection within a large solid angle by the use of typically several hundreds of detectors. TOF spectroscopy uses the de Broglie relation between the wavelength of the neutron and its velocity:

$$\lambda = h/mv \Rightarrow \lambda = 3956/v \ .$$

Typical neutron velocities used in a scattering experiment range from 200 to 3000 ms^{-1}; therefore, the flight times for experimentally relevant distances of several meters are of the order of milliseconds. By measuring the time the neutrons need to pass a known distance one can calculate the wavelength that directly yields the energy, E.

Experimental

Spectrometer

The SINQ TOF spectrometer FOCUS, Fig. 1, is based on the hybrid-TOF principle and combines a double-focusing crystal monochromator with a Fermichopper. The disc chopper in front of the monochromator serves as an antioverlap device. At a distance of 2.5 m from the sample position, three banks of detectors equipped with 383 ^{3}He single-tube detectors cover a range of scattering angles between 30° and 130°. One main purpose of the concept is the high flexibility of the instrument. The pyrolythic graphite monochromator allows a continuous and large range of initial energies, E_i: 2.3 meV < E_i < 20 meV. A Fermichopper with a straight collimation of 2.0° and 1°, respectively, is located between the monochromator and the sample at a distance of 0.5 m in front of the sample. The elastic energy resolution ranges from 50 μeV (full width at half maximum, FWHM) at $\lambda_i = 6$ Å to 131 μeV at 4 Å and 1300 μeV at 2 Å [4]. An elastic scattering vector Q range from 0.2 to 5.6 Å^{-1} can be applied with the following relationship for Q:

$$Q = \frac{4\pi}{\lambda} \sin\Theta \ , \qquad (1)$$

with Θ the scattering angle and λ the wavelength. The optical parameters of FOCUS can be adapted to the requirements of the experiment by the variable curvatures of the monochromator in the horizontal and vertical directions and flexible distances between the main spectrometer components. A novel concept of FOCUS consists of the option to operate the machine either in time-focusing or in monochromatic-focusing conditions [5]. Whereas the time-focusing option allows a sharp energy resolution with enhanced intensity at a certain energy transfer window, the monochromatic-focusing option of the instrument provides an almost constant resolution within a broad range of energy transfers, especially on the neutron-energy-loss side of the spectrum.

We performed all measurements using a conventional closed-cycle refrigerator at 298 K with a standard aluminum hollow cylinder sample can. As a first instrumental setup we chose an incident wavelength of 6 Å, which corresponds to a primary energy, E, of 2.27 meV, with an instrumental resolution of $\Delta E = 50$ μeV covering a scattering vector window between $Q = 0.3$ and $Q = 2.0$ Å^{-1}. The second setting consisted of $\lambda = 4.2$ Å and $E = 4.6$ meV with $\Delta E = 120$ μeV and an elastic Q range of 0.8–2.6 Å^{-1}. Different detector efficiencies were corrected using the completely incoherent and elastic scatterer vanadium in the same sample geometry.

Sample

The hydroxylated fullerene, i.e. so-called fullerol, was purchased from MER, Tucson, USA, as a dry powder. The particles were further dissolved in analytical grade D_2O. All scattering curves from the sample were corrected using pure D_2O in an aluminum cuvette in order to gather the pure contribution of the fullerol. The determinations of the dynamic character of the more highly concentrated samples were carried out in the presence of a mixed-bed ion exchanger (Fluka) in the deuterated modification in order to remove all excess ions within the sample; only the counterions of the surface were present.

Theory

Static properties

The molecule has a total scattering cross-section of 2,485 barn (1 barn equals 10^{-24} cm^2), which is built up by an incoherent scattering cross-section of 2,005 barn mainly caused by the hydrogen atoms and a coherent scattering cross-section of 480 barn resulting from the carbon and the oxygen atoms. In a thought experiment, filling the volume of one particle completely with D_2O means that the particle scatters 1.5 times more strongly than the solvent particles themselves, which is not the case for pure C_{60}. For a spherical, monodisperse hollow particle the normalized integrated scattering intensity

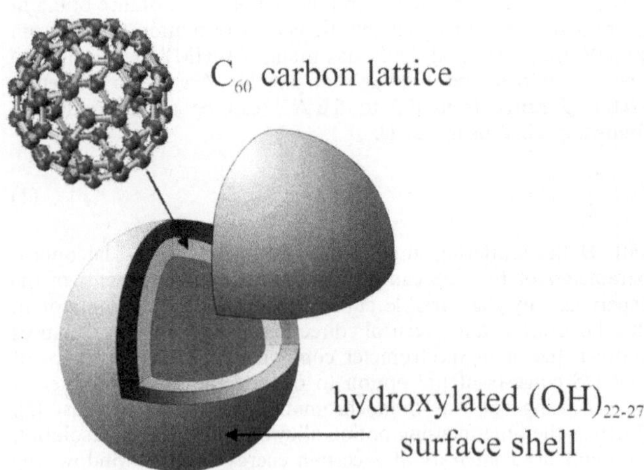

Fig. 2 The fullerol particle: $C_{60}(OH)_{22-27}$

due to intraparticle interactions, $P_{coh}(Q)$, the single-particle form factor, is:

$$P_{coh}(Q) = \left(\frac{A(Q)}{C}\right)^2 \qquad (2)$$

and

$$A(Q) = 3\left(\frac{\sin(QR_o) - QR_o\cos(QR_o)}{(QR_o)^3}\right) - 3\left(\frac{\sin(QR_i) - QR_i\cos(QR_i)}{(QR_i)^3}\right) \qquad (3)$$

$$C = 1/3(R_o^3 - R_i^3),$$

with R_i and R_o the inner radius and the outer radius of the hollow sphere, respectively. This relation only holds for the situation of coherently scattering surface elements in the sense that the scattered waves are able to interfere in a coherent manner and is directly related to the scattering properties of the carbon and the oxygen atoms of fullerol. By considering the scattering behavior of the hydrogen atoms a completely different picture is drawn. The incoherent scattered waves originating from the hydrogen atoms on the surface are not able to interfere coherently, with the result that $P_{inc}(Q) = 1$ over the complete Q range; therefore, the complete form factor should consist of 20% $P_{coh}(Q)$ and 80% $P_{inc}(Q)$.

For interacting particles a second static contribution, that is to say the static structure factor, $S(Q)$, also plays a prominent role. For a 20% coherent scatterer and the volume fractions we are dealing with this means that the static coherent scattering intensity is additionally superimposed by a 5–10% effect, which is essentially is negligible.

Dynamic properties

The neutron spectra of a diffusing particle are given by

$$I(Q,\omega) = IP(Q)\int_{-\infty}^{\infty} S(Q,\omega')R(Q,\omega-\omega')\mathrm{d}\omega' , \qquad (4)$$

where I is a Q- and ω-independent normalization factor, $P(Q)$ is the normalized single particle form factor, $S(Q,\omega)$ represents the dynamic structure factor of the diffusing species and $R(\omega)$ is the resolution function of the spectrometer. For isotropic processes $S(Q,\omega)$ is described by a Lorentzian line [6]:

$$S(Q,\omega) = \frac{1}{\pi}\frac{\Gamma(Q)}{\Gamma(Q)^2 + (\hbar\omega)^2} , \qquad (5)$$

with $\Gamma(Q)$ the Q-dependent half width at half maximum (HWHM) and $\hbar\omega$ the energy transfer with respect to the sample. The HWHM is directly connected to the nature of diffusion. The rotational diffusion is considered to be too fast and so with our instrumental setup not determinable [7]. For translational diffusive processes the well-known Q^2D law, with D the diffusive constant, describes the HWHM. For particles much larger then the solvent molecules, the Stokes–Einstein continuum relation for solvents with viscosity, η, holds for the free diffusion constant, D_{free}:

$$D_{free} = \frac{k_B T}{6\pi\eta R} . \qquad (6)$$

That is why we are on the cutting edge when applying this theory because the fullerol particles are only about 10 times larger then D_2O [8].

As mentioned earlier the incoherent scattering hydrogen atoms on the surface dominate the scattering

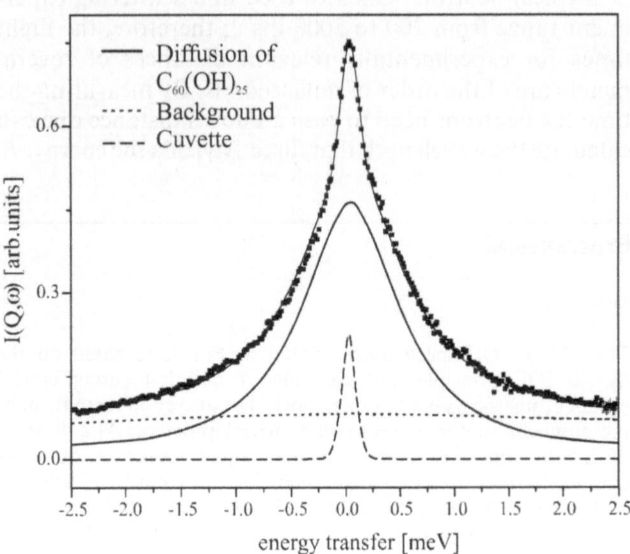

Fig. 3 Raw scattering function at highest applied Q (2.6 Å^{-1})

function of the fullerol particles. The direct consequence is the determination of the free diffusion constant for the dilute systems and a self-diffusion constant for the more highly concentrated samples.

Results and discussion

Dilute systems

The raw scattering curves of all the samples investigated are excellently describable by two Lorentzian curves due to the motion of water molecules and fullerol plus an elastic contribution of the aluminum container (Fig. 3). The data were further corrected with the neutron spectra of pure D_2O in the same cuvette to collect only the scattering curves of the colloidal fullerol particles.

For the dilute sample we determined from the Q dependence of the FWHM, $\Gamma(Q)$, at room temperature a diffusion constant of 2.74×10^{-10} m^2s^{-1} which, by applying the Stokes–Einstein relation leads to an effective particle radius of about 8.0 Å; for pure C_{60} the radius is about 3.5 Å [9] and for dilute C_{60} in CS_2 a value of 4.5 Å was recently obtained by Smorenburg et al. [8]. Taking into consideration the hydrodynamic solvent shell of one or even more molecular layer on the surface of all the diffusing particles in solution the discrepancy of the values is reasonable.

To gather information about the static properties of the dilute colloidal fullerol, i.e. the single particle form factor, we illustrate in Fig. 4 the intensity of the quasielastic peak with respect to Q. Using the function for a hollow sphere with nonnegligible thickness, Eq. (2), the Q dependence of the intensity is well describable with a fixed inner radius of 3.5 Å for pure C_{60} and an outer radius of 4.7 Å by taking into consideration the contribution ratio of the coherent and incoherent surface scattering of one particle, i.e. $P_{coh}(Q)$ and $P_{inc}(Q)$.

Concentrated systems

The high solubility of the fullerol particles in polar solvents makes a determination of the dynamic properties at high volume fractions possible; therefore, we were able to probe samples at a volume fraction of $\phi = 1$, 4 and 6%. The particle dispersions in D_2O showed an acidic pH of about 5, which can be interpreted as a direct hint for the partial deprotonation of the surface hydroxyl groups and leads to a categorization of the particles as at least weakly charged systems. After removing all the excess ions by means of a mixed-bed ion exchanger only the counterions are present and the surface charges are not shielded anymore. The direct consequence is a relatively long range repulsion, which at appropriate number densities can cause collective ordering phenomena, like a liquid-like-order phase, a crystalline phase or even a glassy phase [10].

The FWHM is displayed in Fig. 5 with respect to Q^2 for the three different volume fractions. The influence of the increasing concentration on the velocity of the particles is clearly visible. For the most concentrated system we determined a diffusion constant of 1.36×10^{-10} m^2s^{-1}, whereas the dilute system revealed a value of 2.74×10^{-10} m^2s^{-1} (Table 1). The slowing of the diffusion has to be interpreted in terms of the beginning of self-organization of the system, i.e. a liquid-like-ordered system. Thus the diffusion constant, D,

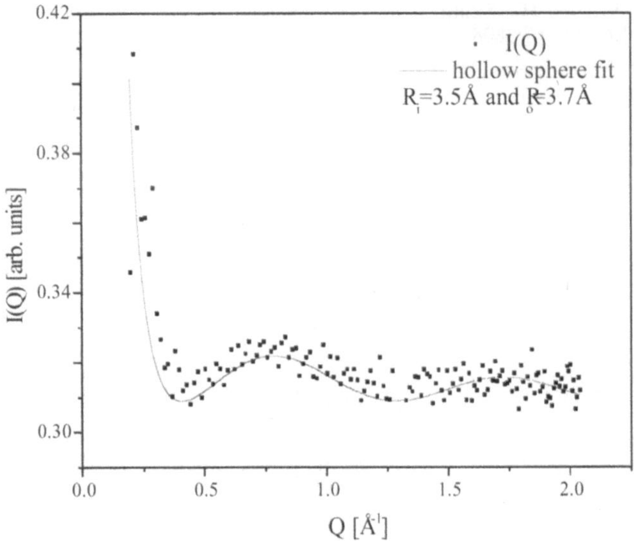

Fig. 4 Single-particle form factor, $P(Q)$, for a dilute fullerol system

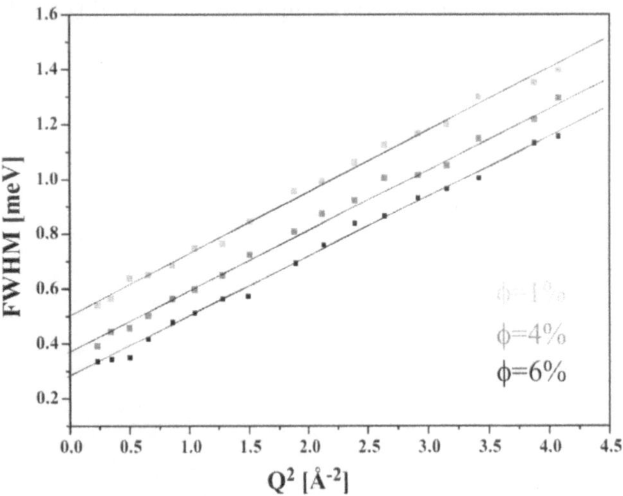

Fig. 5 Following the Q^2D law for fullerol samples with a volume fraction, ϕ, of 1, 4 and 6%

Table 1 Diffusion constants of the fullerol systems in comparison to dilute, pure C_{60} [8]

System	D (m^2s^{-1})	$D^* = D/D_{free}$
Fullerol: $\phi = 1\%$	2.74×10^{-10}	1
Fullerol: $\phi = 4\%$	2.02×10^{-10}	0.7
Fullerol: $\phi = 6\%$	1.36×10^{-10}	0.5
Pure C_{60}: $\phi < 1\%$	1.5×10^{-9}	1

mirrors the self-diffusional behavior of the colloidal fullerol particles: D shows no Q dependence in contrast to the collective diffusion constant [2]. The reduced self-diffusion constant, $D^* = D/D_{free}$, will slow to a value of 0.1 with respect to the concentration, where a phase transition will occur [11]. For our colloidal systems we measured a course of the value of D^* from 1 to 0.7 to 0.5, which indicates the regime of a liquid-like-ordered system.

Conclusion

We showed that water-soluble fullerol particles with different concentrations in D_2O as a solvent are suitable colloidal model systems. The inelastic neutron scattering facility FOCUS was used to determine the dynamic and static behavior for samples with concentrations ranging from 1 to 6 vol%. In contrast to the lipophilic hard-sphere C_{60} system in nonpolar solvents, hydrophilic fullerol can be treated as a new soft-sphere colloidal model system for weakly or even highly charged particles. After the correction of the raw scattering data the scattering function can be well described by a single Lorentzian line, which is attributed to the diffusion of colloidal particles in D_2O. For a dilute system we extracted from the FWHM the self-diffusion coefficient of the spheres to be about 2.74×10^{-10} m^2s^{-1}. By taking into consideration the hydrodynamic shell and the resulting larger effective hydrodynamic diameter of the particles this value is comparable to those obtained from the Stokes–Einstein relation. A direct hint for the beginning of particle–particle self-organization in the sense of a liquid-like-ordered system is revealed by the reduced self-diffusion constant with respect to the free counterpart.

Acknowledgements FOCUS was built and is being operated in close cooperation between the University of Saarbrücken, Germany, and the Paul Scherrer Institute, Switzerland. Generous financial support by the German BMBF (project nos. 03-HE4SA2 and 03-HE5SA2) is gratefully acknowledged.

References

1. Pusey PN (1991) In: Hansen JP, Levesque D, Zinn-Justin J (ed) Liquids, freezing and glass transition. Elsevier, Amsterdam, pp
2. Härtl W, Beck CH, Hempelmann R (1999) J Chem Phys 110:7074
3. Beck CH, Härtl W, Hempelmann (1999) J Chem Phys 11:3209
4. Janssen S, Mesot J, Holitzner L, Furrer A, Hempelmann R (1997) Physica B 1174:234
5. Janssen S, Rubio-Temprano D, Furrer A (2000) Physica B 283:355
6. Beé M (1988) Quasielastic neutron scattering. Institute of Physics, London
7. Neumann DA, Copley JRD, Capelletti RL, Kamitakahara WA, Lindstrom RM, Creegan KM, Cox DM, Romanow WJ, Coustel N, McCauley HP Jr, Maliszewskyi NC, Fischer JE, Smith AB II (1991) Phys Rev Let 67:3808
8. Smorenburg HE, Crevecoeur RM, de Schepper IM, de Graaf LA (1995) Phys Rev E 52:2742
9. Smorenburg HE, de Schepper IM, de Graaf LA (1994) Phys Lett A 187:204
10. Härtl W, Beck, CH, Hempelmann R (2001) Prog Colloid Polym Sci 117:113–116

Progr Colloid Polym Sci (2001) 117: 131–135

V. Socoliuc
D. Bica

Experimental investigation of magnetic-induced phase-separation kinetics in aqueous ferrofluids

V. Socoliuc (✉)
National Institute for Research and Development in Electrochemistry and Condensed Matter, Tirnava 1, 1900 Timisoara, Romania
e-mail: socoliuc@icmct.uvt.ro

D. Bica
Center for Fundamental and Advanced Technical Research, Romanian Academy, Bd. Mihai Viteazul 24, Timisoara, Romania

Abstract In this article we report microscopy and light scattering investigation of phase-separation phenomenon in ferrofluids. The influence of the temperature and an external magnetic field on the kinetics and on the quantitative extent of the phase separation was investigated.

Key words Ferrofluids · Colloidal stability · Phase transition · Phase separation

Introduction

Ferrofluids, or magnetic colloids, are suspensions of magnetic nanoparticles dispersed in a carrier liquid. In order to prevent agglomeration due to attractive van der Waals or magnetic dipole–dipole interactions, in addition to the Brownian motion, a repulsive force between the particles is created by means of steric hindrance or electrostatic repulsion. The ideal ferrofluid is a homogenous dispersion of isolated particles. Phase separation is one of the main phenomena that leads to radical changes in the ferrofluid colloidal stability and structure [1]. When the temperature or the external magnetic field strength exceed their critical values, the formation of highly concentrated phase droplets may occur in the ferrofluid.

In this article we present the results of research on the influence of the temperature and an external magnetic field on the kinetics of phase-separation phenomena in an aqueous ferrofluid as well as on the size, shape and space configuration of the condensed-phase drops.

Sample description

An aqueous ferrofluid was prepared for the investigations with a volume fraction of 3% magnetic particles stabilized with dodecyl benzene sulfonic acid produced by means of a chemical coprecipitation method [2, 3].

Experimental

Optical microscopy investigations of the ferrofluid condensation phenomena were carried out. A special electromagnet and a sample thermostat were designed and adapted to a metallographic microscope with a maximum magnification of 1500 ×. The magnetic field range was 0–1.5kOe, the field gradient in the sample region was less than 10Oe/mm at 0.5 kOe and about 50Oe/mm at 1.5kOe and the magnetic field transient time was less than 1 s.

Wide-angle forward scattering of light was investigated by recording the scattering pattern projected on a screen with a charge-coupled-device camera (Fig. 1). This technique allows the instantaneous sampling of the entire scattering pattern and the recording of its time evolution. A He–Ne laser beam polarized perpendicular to the magnetic field direction was used as incident light.

Detection of the light transmitted in the incident beam direction was performed (Fig. 2). The He–Ne laser incident beam was polarized at 54.7° with respect to the field direction in order to annul the dichroism induced by the magnetic particle orientation and agglomeration [4]. The scattered light was filtered out with an iris. In order to compensate the laser emission fluctuations the incident beam was split and both beams were filtered and focused on two identical large-area photodetectors (IPL10530DAW Integrated Photomatrix) with incorporated operational amplifiers. The signals from the photodetectors were fed into a National Instruments LabPC+ data acquisition board; data processing and signal

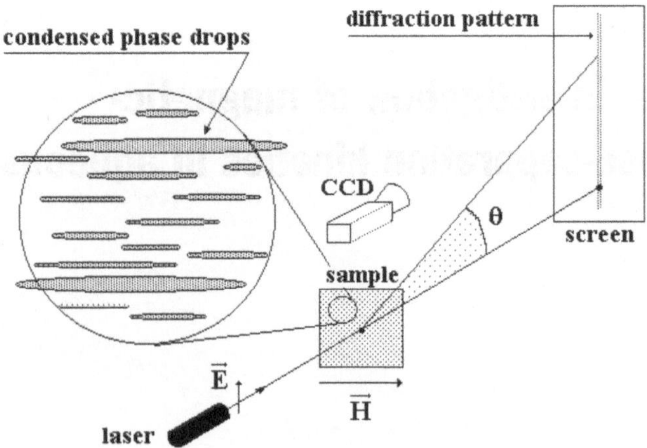

Fig. 1 Wide-angle light scattering experiment setup

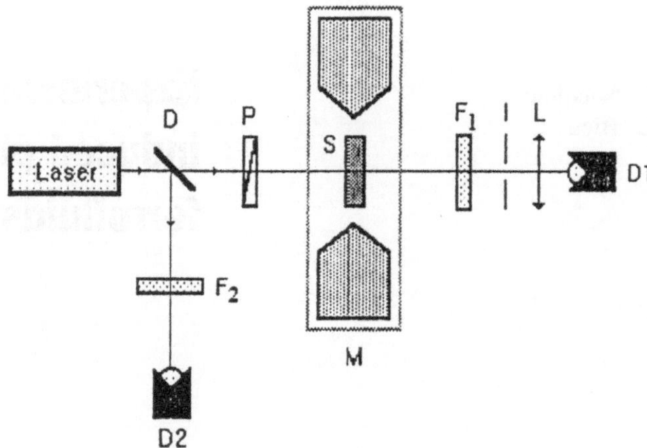

Fig. 2 Forward scattering experiment setup

referencing were made with a virtual instrument developed with LabView. The ferrofluid was contained in a 10-μm light-path cell with a detachable window. The magnetic field was perpendicular to the incident light beam and parallel to the cell plates. A Weiss electromagnet was used to produce a static magnetic field in the range 0–4kOe with less than 10Oe/cm gradient. The magnetic field strength was measured with 10Oe precision. The sample temperature was measured with 0.1 °C precision and was controlled with a water thermostat system and a specially designed cell mount.

Results and discussion

Optical microscopy investigation

After the sudden onset of the 1.5kOe magnetic field, a high density of small acicular drops (primary drops) of the condensed state form in the ferrofluid, aligned parallel to the magnetic field direction. Their length is about 5 μm and the thickness was estimated to be less than 1 μm. From the beginning of their formation the primary drops move chaotically inside the cell, while the formation of larger drops (secondary drops about 20 μm long and 1 μm thick) begins (Fig. 3). The primary drops vanish as the number of secondary drops increases. The secondary drops also drift within the uncondensed ferrofluid but their motion is much slower than that of the primary drops. Owing to their large magnetic moment aligned in the field direction and to the magnetic dipole attraction, the secondary drops stick together head to tail, thus leading to the formation of very elongated drops (some of them exceeding 100 μm). Similar evolution resulted from theoretical modeling of phase condensation in ferrofluids [5] and was also observed experimentally [6].

After about 1 min the drops no longer grew and they settled in a fairly stable space configuration owing to magnetic lateral repulsion. The temperature was found to have little influence on the dimensions of the drops but their density increases with a temperature decrease. At low temperatures the density of the drops increases significantly with increasing field intensity.

At temperatures above 30 °C no condensed-phase drops were observed even at the highest field value (1.5kOe), while at temperatures below 20 °C condensed-phase drops were observed for field values as low as 10Oe.

Light scattering experiment

The pasted (negative) images of the scattering pattern at several moments of time after the field settling subsequent to its sudden onset are shown in Fig. 4. The spots at the bottom of the picture are the images of the unscattered light beam projection. The sample temperature was 27 °C, the applied field was 3kOe and the field transient time was less than 1 s. The pronounced lack of circular symmetry of the scattering pattern is characteristic for very elongated scatterers [7]. One can observe that immediately after the onset of the field the diffraction pattern presents a minimum in the vicinity of the central spot. Over time the minimum moves toward the central spot and the scattered light intensity increases over the entire angular range. The angular dependence of the scattered light intensity is plotted in Fig. 5 for several moments after the onset of the field. The data were obtained from vertical gray-level distribution sampling of the patterns plotted in Fig. 4. Once again one can observe the displacement of the minimum toward the central spot and also the increase in the scattered intensity at the minimum corresponding angle.

An important question arises regarding the origin of the local minim in the scattered light pattern (Fig. 5). It is well known that light scattering in ferrofluids with droplike aggregates originates from the light diffraction on the drops of the condensed phase which, since their

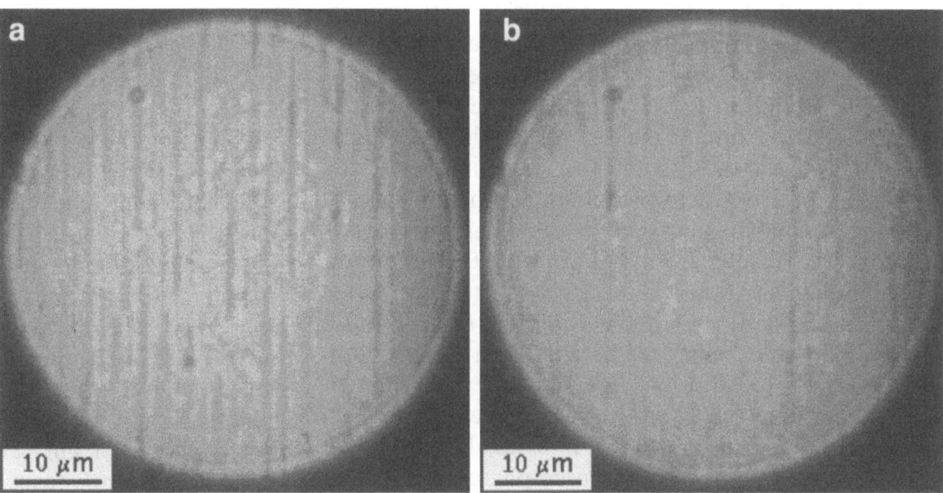

Fig. 3 Optical microscopy image of the condensed phase drops elongated in the external magnetic field direction for $H = 1.5$kOe at **a** 21 °C and **b** 28 °C

particle density is much greater than the uncondensed state, are opaque to visible light. Following Ref. [7], in the case of very elongated scatterers, the scattered light angular dependence can be well approximated as

$$I(\theta) \approx \left(\frac{\sin u}{u}\right)^2 , \qquad (1)$$

where $u = 2\pi n s/\lambda \cdot \sin(\theta/2)$, where s is the drop thickness, n is the refractive index of the uncondensed phase, λ is the wavelength of light in a vacuum and θ is the scattering angle. If one assumes that the light scattered by the ferrofluid sample is the result of the incoherent superposition of light diffracted by individual drops, the diffraction function $(\sin u/u)^2$ should fit the experimental

Fig. 4 Time evolution of the scattered light patterns after the field setup as recorded with a charge-coupled-device camera (27 °C, $H = 3$kOe)

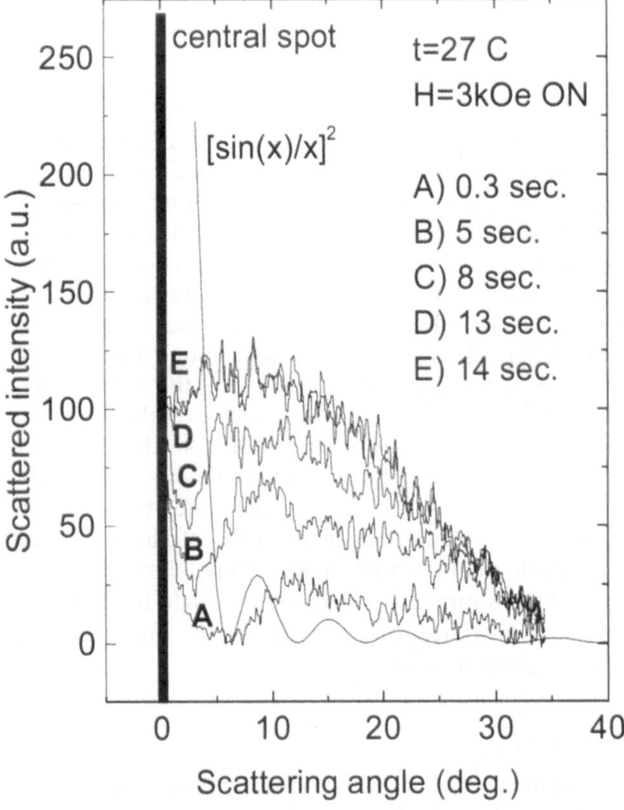

Fig. 5 Angular distribution of the scattered light at several moments of time after the onset of the field

data quite well. The diffraction function with a first-order minimum angle and a first-order maximum amplitude corresponding to the experimental data at $t = 0.3$ s is plotted in Fig. 4. The experimental angle of the first-order maximum is nearly twice the theoretical value and both the width and magnitude of the experimental zero-order maximum are much smaller then the theoretical values. On the other hand, the drops thickness that results from Eq. (1) with the minimum corresponding angle and the refractive index of water ($n = 1.5$) is about 10 nm, which is an order of magnitude greater than the value estimated from microscopy investigations; therefore, we infer that the minimum in the scattered light originates from the coherent interference of the light diffracted by individual drops. The space configuration of the condensed-phase drops in the ferrofluid could be modeled as the superposition of the light scattered by a series of ideal diffraction gratings with the space between neighboring drops distributed over a certain range of values. Thus, one can model the angular dependence of the scattered light as the incoherent superposition of the light coherently scattered by pairs of neighboring drops as

$$I(\theta) \approx \cos^4\theta \left(\frac{\sin\left(\frac{2\pi ns \sin(\theta/2)}{\lambda}\right)}{\frac{2\pi ns \sin(\theta/2)}{\lambda}} \right)^2 \times \int_D \left(\frac{\sin\left(\frac{\pi NnD \sin\theta}{\lambda}\right)}{\sin\left(\frac{\pi nD \sin\theta}{\lambda}\right)} \right)^2 g(D, d_0, w)\mathrm{d}D \ . \tag{2}$$

The term $\cos^4(\theta)$ describes the perpendicular polarization of light relative to the elongation direction of the drops, the second term describes the angular dependence of the light diffracted by individual drops and the integral describes the incoherent summation of the interference term of pairs of drops ($N = 2$) over the distribution $g(D,d_0,w)$ of the distance between neighboring drops, where d_0 and w are the distribution parameters. In Fig. 6 the angular dependence of the interference integral is plotted for $d_0 = 2.5$ μm and several values of w assuming a Gaussian distribution of the distance between drops and for $d_0 = 5$ μm and $w = 2$ μm. As the distribution width increases, the integral becomes smother and constant at high values of the scattering angle, while the first-order minimum remains dependent on d_0. The minimum corresponding angle is related to the mean distance between drops by the following approximate equation

$$n\bar{D} = \frac{\lambda}{2\sin\theta_{\min}} \ , \tag{3}$$

while at high scattering angles the experimental data are approximately well described by the polarization term times the diffraction term, since the integral of the interference term is independent of the scattering angle:

$$I(\theta) \approx \cos^4\theta \left(\frac{\sin\left(\frac{2\pi(n\bar{s})\sin(\theta/2)}{\lambda}\right)}{\frac{2\pi(n\bar{s})\sin(\theta/2)}{\lambda}} \right)^2 \ . \tag{4}$$

Following the previous discussion one can use Eq. (3) to compute the mean distance between drops and Eq. (4) to determine the mean thickness of the drops times the uncondensed phase refractive index by fitting the experimental data at high scattering angles. The time dependence of the mean thickness and spacing of the drops times the matrix refractive index is plotted in Fig. 7. At the beginning of the condensation process the drops thickness is constant at about 0.8 μm for about 5 s and afterwards it slowly and asymptotically increases toward about 1.1 μm. The spacing between the drops increases constantly from 3 to 8 μm, mainly owing to the process of coalescence of primary drops as observed in the microscopy investigations. Although it was not possible to determine the refractive index of the condensed phase, it is reasonable to assume that it is independent of the drop size.

Light extinction experiment

The time dependence of the scattered light intensity (normalized to the transmitted intensity in the absence of the field) was measured between 15 and 35 °C, after the sudden onset of the magnetic field for several values of its magnitude (the data measured at 23 °C are plotted in Fig. 8). The magnetic field transient time was less than 1 s. It was found, as an example, that for field values below 0.5kOe no light scattering occurs at 27 °C, while at 23 °C light scattering occurs at field values as low as 0.15kOe. The lower the temperature and the higher the magnetic field value, the more pronounced the light scattering.

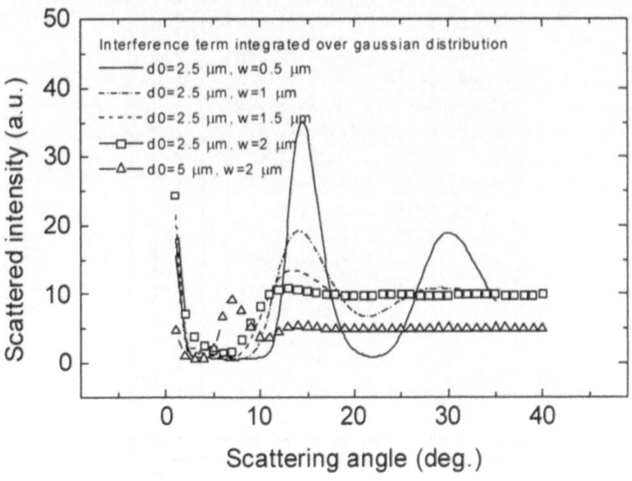

Fig. 6 The interference term for a pair of drops integrated over the Gaussian distribution of the spacing between drops for several values of the parameter w

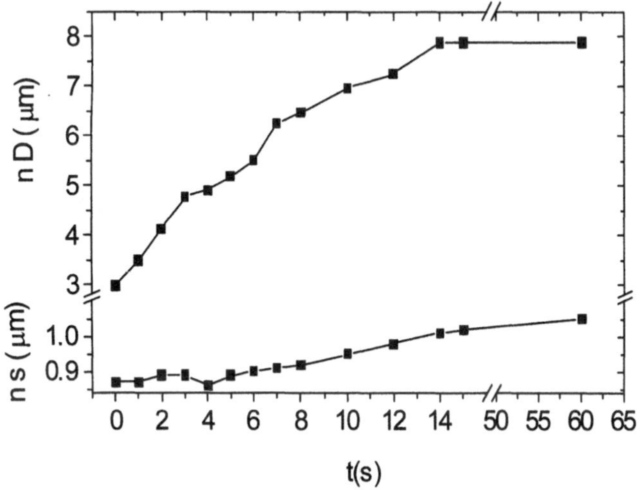

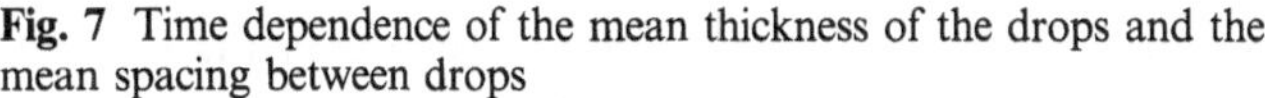

Fig. 7 Time dependence of the mean thickness of the drops and the mean spacing between drops

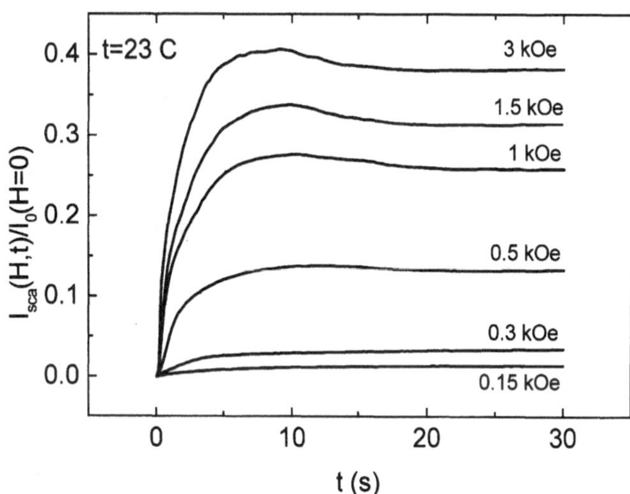

Fig. 8 Time dependence of the scattered light intensity normalized to the transmitted intensity in the absence of the field at 23 °C for several values of the magnetic field intensity

Depending on temperature, above a certain value of the magnetic field the scattered light intensity reaches a local maximum. The time until the scattered light intensity reaches the local maximum increases with temperature and decreases with magnetic field value. One can divide the time evolution of the scattered light into two stages: prior and subsequent to the local maximum corresponding moment. On basis of the microscopy observations one can infer that at the beginning of the first stage light scattering is mainly due to the formation of primary drops. The transient time of the scattered light was found to be independent of field intensity but it increases with decreasing temperature.

At temperatures below 20 °C light scattering occurs for magnetic field values as low as 10Oe and above 30 °C no scattering was observed even at the highest magnetic field value (3kOe), in good agreement with microscopy observations. For several temperature values the critical field value (H_c) was determined at which the drops begin to scatter the light. The data was fitted with the equation derived by Cebers [8] for the dependence on temperature of the critical field at which phase condensation occurs:

$$H_c(t) \approx \frac{1}{t - t_\infty} , \qquad (5)$$

where t_∞ is the temperature above which phase condensation does not occur no matter how strong the magnetic field is. As a result of the fitting of the experimental data with Eq. (5), a value of 31.7 °C was found for t_∞, which is in good agreement with the observations from microscopy and scattering investigations.

Conclusions

The effect of magnetic-induced phase condensation on the aqueous ferrofluid investigated is that the ferrofluid becomes a biphasic system i.e. condensed-phase droplets form in equilibrium with the uncondensed phase matrix. The density of the condensed-phase droplets increases with decreasing temperature and increasing magnetic field intensity. While the length of the droplets at equilibrium increases with field intensity, their thickness is independent both of temperature and field intensity. Above 32 °C, magnetic-induced phase condensation does not occur.

The kinetics of phase condensation was found to be influenced mainly by the temperature. The growing process of the condensed-phase drops was found to evolve in two stages: small primary drops grow and stick together into large secondary drops, similar to the theoretical predictions reported in Ref. [4].

References

1. Berkovski B, Bashtovoy V (1996) Magnetic fluids and applications handbook. Begell House, New York
2. Bica D (1985) Romanian Patent 90078
3. Bica D (1995) Rom Rep Phys 47:265
4. Kopcansky P, Koneracka M, Tomasovicova N, Tomco L (1999) J Magn Magn Mater 201:204
5. Yu Zubarev A, Ivanov AO (1997) Phys Rev E 55:7192
6. Jayedevan B, Nakatani I (1999) J Magn Magn Mater 201:62
7. van de Hulst HC (1957) Light scattering by small particles. Wiley, New York
8. Cebers A (1991) Physical properties and models of magnetic fluids. European Advanced Course of UNESCO, Minsk, April 1991

Progr Colloid Polym Sci (2001) 117: 136–140

I. Varga
T. Gilányi
R. Mészáros

Characterisation of ionic surfactant aggregates by means of activity measurements of a trace probe electrolyte

I. Varga (✉) · T. Gilányi · R. Mészáros
Department of Colloid Chemistry,
Loránd Eötvös University,
P.O. Box 32, 1518 Budapest 112, Hungary
e-mail: imo@para.chem.elte.hu
Tel.: 36-1-2090555
Fax.: 36-1-2090602

Abstract On the basis of the Poisson–Boltzmann cell model a new trace probe electrolyte method was applied to determine the surfactant aggregation number in Poly(ethylene oxide)– Sodium dodecyl sulphate systems. It is demonstrated that activity measurements on a probe electrolyte added to a polymer/surfactant solution in a trace amount can be used for the determination of the mean aggregation number of the surfactant aggregates. An important new finding of the experimental part of this work is that the aggregation number depends only on the equilibrium surfactant concentration and is independent of the polymer concentration.

Key words Polymer-surfactant complex · Aggregation number · Poisson–Boltzmann theory · Poly(ethylene oxide) · Sodium dodecyl sulphate

Introduction

The interaction of surfactants with macromolecules has been the subject of investigations since the early 1950s. First, the protein–surfactant systems were investigated owing to their biological importance. Later, with the appearance of well-defined synthetic polymers the research was extended to the polymer–surfactant interaction. Beyond their scientific interest, polymer–surfactant systems have received considerable attention because of their numerous industrial applications (e.g. in pharmaceutical and biomedical applications, oil recovery, food and mineral processing) [1–3].

The binding of ionic surfactants to nonionic polymers is a cooperative process which starts at a well-defined concentration that is often called the critical aggregation concentration (cac). Below this concentration there is no interaction between the surfactant and the polymer. Above the cac the surfactant starts to bind to the polymer. With increasing amount of bound surfactant the equilibrium surfactant concentration increases. When it reaches the critical micelle formation concentration (cmc) free micelles form and the equilibrium surfactant concentration becomes practically constant [3].

The polymer–surfactant complex has been described as a "string of beads" in which the polymer chain connects micellelike surfactant aggregates by wrapping around them [4]. This illustrative picture probably well characterises the separated individual complex molecules, for example, at low polymer concentration of long-chain polymers with high degree of surfactant binding.

For the description of the polymer–surfactant interaction several thermodynamic and molecular interaction models have been proposed [5–9]. A common feature of these models is that one of their central parameters is usually the surfactant aggregation number. The number of methods available for the determination of the surfactant aggregation number in the polymer–surfactant complex is rather limited. In a few cases small-angle neutron scattering measurements were used [10], while recently mainly fluorescence probe methods have been applied for this purpose [11]; however, these techniques require quite special and expensive instrumentation, which makes their use rather limited.

Recently, we have described a new trace probe electrolyte method that is suitable for the determination of the surfactant aggregation numbers [12]. In this contribution we demonstrate the usefulness of this method by its application to the poly(ethylene oxide) (PEO)/sodium dodecyl sulphate (NaDS) system.

Experimental

Materials

The polymer investigated was PEO (Aldrich $M_w = 1.0 \times 10^5$ and $M_w = 1.0 \times 10^6$). The PEO solutions were purified by mixed-bed anion-cation exchange, by a similar method to that used in case of latex dispersions [13] for the elimination of ionic contaminants. This purification was necessary in order to obtain well-reproducible results.

The surfactant was NaDS (Merck) recrystallised twice from a 1:1 hot benzene–alcohol mixture. The cmc was found to be 8.2 mM from conductometric measurements.

Potentiometric measurements

The electromotive force (emf). values of the following two galvanic cells were determined by means of a Radelkis research pH meter at 25.0 ± 0.1 °C:

1. Cell 1: Na–glass| c NaDS, c_p PEO |Au|Hg|HgIDS
2. Cell 2: Na–glass| 10^{-4} M NaI, c NaDS, c_p PEO|Ag|AgI

The emf values were converted into mean activities (a_{NaDS} and a_{NaI}) using the Nernst equation. NaI was chosen as a probe electrolyte, but this choice is not exclusive. The trace electrolyte should meet the following requirements: it does not bind specifically (chemically) to the polymer or to the surfactant and reversible electrodes are available to measure its mean activity. The binding on PEO was checked by emf measurements against the polymer concentration at constant 10^{-4} M NaI concentration. The emf was constant within 0.5 mV, i.e. binding of the iodide ions to PEO in the concentration range investigated cannot be detected. It was checked experimentally that the iodide ion electrode was not sensitive to the presence of the surfactant ions.

Theory: trace probe electrolyte method

The distribution of small ions in a solution is strongly influenced by the presence of macroions. Consequently, if we can find an appropriate method for monitoring this influence we can get information about the characteristic features of the macroions. The addition of a probe electrolyte in a trace amount to a polymer–surfactant solution can be such a method if there is no specific interaction between the probe electrolyte and the polymer molecules. In this case the local distribution of the small ions can be given by the following equation:

$$c_B(x) = c_{B,e} \exp(y) \ , \tag{1}$$

where $y = e\Psi/kT$ is the reduced electric potential with a zero point chosen in a polyme-free equilibrium solution. The analytical concentration of the small ions can be given by the integration of their local concentration for the volume of the solution:

$$c_B^0 = c_{B,e} \frac{1}{V} \int_V \exp(y) \mathrm{d}V \ . \tag{2}$$

The main lesson from Eq. (2) is that if a (polymer–surfactant) solution contains macroions ($y \neq 0$) then the analytical concentration of the small ions cannot be equal to their concentration in the equilibrium solution. The traditional interpretation of this experience is based on the counterion dissociation of the macroions:

$$c_B^0 \left(c_B^0 + \alpha Z c_{mac} \right) = c_e^2 \ , \tag{3}$$

where c_e refers to the concentration of the monomeric surfactant, α is the degree of counterion dissociation and Z and c_{mac} are the total charge and the concentration of the macroions, respectively. However, it can be shown that the experimentally determinable apparent degree of dissociation does not have the physical meaning that is usually attributed to it [14].

The effect of a macroion on the distribution of an electrolyte is demonstrated in Fig. 1. In the absence of macroions the electrolyte is distributed uniformly in the volume of the solution. When macroions are added to the solution then the ions having similar charge are expelled from the close vicinity of the macroions. Owing to this exclusion effect the equilibrium concentration of these ions is increased. This concentration increase can also be viewed as if the free volume available for the electrolyte decreased. This is demonstrated by the square concentration profile in the figure. Using the latter approach we can define the excluded volume of the system (V^*):

$$V^* = \left(1 - \frac{c_B^0}{c_{B,e}} \right) V \ . \tag{4}$$

In order to evaluate the integral in Eq. (2) the "chemical" structure of the system under consideration must be specified. For the sake of simplicity we assumed that the aggregate distribution is statistically uniform in the solution. Certainly, this assumption is fulfilled only in special cases (e.g. when the polymer concentration is high enough). Our model and measurements were restricted to these cases.

Dividing the volume of the solution into electrically neutral equivalent spherical cells in such a manner that every cell contains a single aggregate and the appropriate amount of electrolyte [15, 16], the excluded volume of the system can be expressed in the following form:

$$\frac{V^*}{V} = 1 - \frac{3}{R^3 - a^3} \int_a^R \exp(y) r^2 \mathrm{d}r \ , \tag{5}$$

where R is the radius of the cell,

$$R = \left(\frac{3m}{4\pi N_A c_{mac}} \right)^{1/3} \ , \tag{6}$$

a is the core radius of a surfactant aggregate,

$$a = \left(\frac{3mV_0}{4\pi N_A} \right)^{1/3} \ , \tag{7}$$

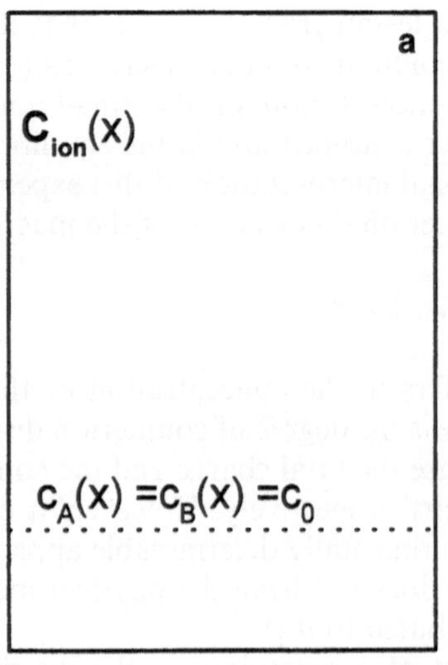

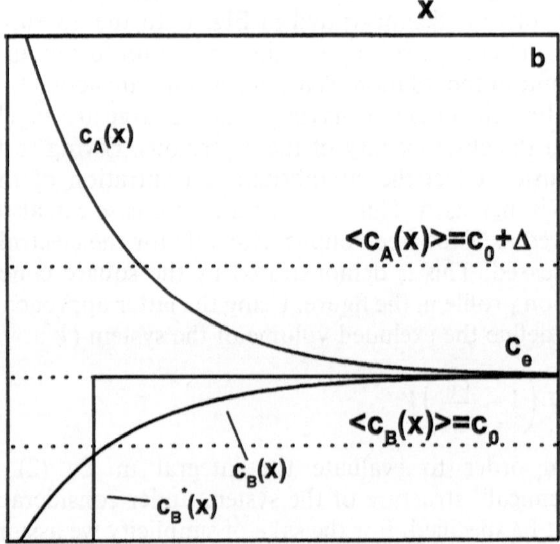

Fig. 1 The local distribution of small ions **a** in the absence and **b** in the presence of a macroion

$V_0 = 212.4\ \mathrm{cm^3 mol^{-1}}$ [17] is the molar volume of the surfactant aggregates and m is the surfactant aggregation number. The reduced electric potential, $y(r)$, can be computed from the Poisson–Boltzmann equation for spherical symmetry and 1:1 electrolytes by means of the following equation

$$r^{-2}\frac{\mathrm{d}}{\mathrm{d}r}\left(r^2\frac{\mathrm{d}y}{\mathrm{d}r}\right) = \frac{2e^2 c_{\mathrm{B,e}}}{\varepsilon kT}\sinh y \tag{8}$$

using the boundary conditions $y = y_0$ at $r = a$ and $(\mathrm{d}y/\mathrm{d}r)_{\mathrm{R}} = 0$. This means that the surfactant aggregation number can be calculated if we determine the excluded volume of the system experimentally.

In the previous calculations the electrical structure of the aggregate surface is approximated by the simplest Gouy–Chapman model. The closest distance of approach of the mobile ions to the surface is a, the core radius of the surfactant aggregate. The polymer segment density is assumed to be small and its effect on the ion distribution is neglected. In the close vicinity of the core this model is probably not correct, but fortunately the integral in Eq. (5) is not sensitive to the inner electrical structure of the aggregate.

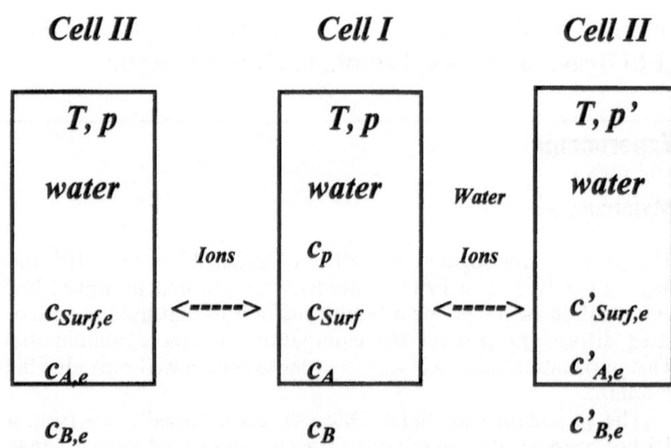

Fig. 2 Scheme of an osmotic and a nonosmotic membrane equilibrium to study the polymer–ionic surfactant solutions

The scheme of different experimental arrangements for the determination of the excluded volume of a polymer–surfactant system is depicted in Fig. 1. One of the cells (cell I) contains the polymer–surfactant solution and a 1:1 probe electrolyte in a trace amount. Let us suppose that cell I is in a nonosmotic membrane equilibrium with cell II. This means that the temperature and the pressure are equal in the two cells and the membrane is permeable only for ions. The condition of the thermodynamic equilibrium between the two cells is the constancy of the electrochemical potential of the mobile components. Since the pressure and the temperature are equal in the two cells this condition reduces to the equality of the mean activities of the surfactant (ASurf) and the probe electrolyte (AB) in the cells:

$$a^{\mathrm{I}}_{\mathrm{ASurf}} = a_{\mathrm{ASurf,e}} = \gamma_{\mathrm{ASurf,e}}\left(c_{\mathrm{A,e}}c_{\mathrm{Surf,e}}\right)^{1/2}, \tag{9}$$

$$a^{\mathrm{I}}_{\mathrm{AB}} = a_{\mathrm{AB,e}} = \gamma_{\mathrm{AB,e}}\left(c_{\mathrm{A,e}}c_{\mathrm{B,e}}\right)^{1/2}, \tag{10}$$

where γ is the corresponding mean activity coefficient and the indexes I and e refer to the polymer–surfactant solution in cell I and the equilibrium solution (cell II), respectively. Consequently, the emf measured either in the polymer–surfactant solution (cell I) or in its equilibrium solution (cell II) is the same:

$$E = E_0 - \frac{kT}{e}\ln\left(\gamma^2_{\mathrm{AB,e}}c_{\mathrm{A,e}}c_{\mathrm{B,e}}\right). \tag{11}$$

Taking into account that $c_{\mathrm{A,e}} = c_{\mathrm{Surf,e}} + c_{\mathrm{B,e}}$ and $c_{\mathrm{Surf.e}} \gg c_{\mathrm{B,e}}$

$$E = E_0 - \frac{kT}{e}\ln\left(\gamma^2_{\mathrm{AB,e}}c_{\mathrm{Surf,e}}c_{\mathrm{B,e}}\right). \tag{12}$$

If the surfactant concentration is below the cac the ions are in single dispersed form and both the surfactant and the probe ion concentration are the same in the polymer–surfactant solution and in its equilibrium solution, i.e. $c_{B,e} = c^{I}_{B}$ and $c_{Surf,e} = c^{I}_{Surf}$. Furthermore, the equilibrium solution is a dilute electrolyte and according to the Debye–Hückel theory at constant temperature and P_o its activity coefficient depends only on the ionic strength, which is $c_{ASurf,e}$ in cell II since $c_{B,e}$ is negligible. The change in $\gamma_{AB,e}$ in the concentration range between c_{cac} and the cmc is usually negligible, so the difference between the emf values measured at c_{ASurf} and c_{cac} can be given as

$$E(c_{ASurf}) - E(c_{cac}) = \frac{kT}{e} \ln \frac{c_{B,e} c_{Surf,e}}{c^0_B c_{cac}} \quad . \tag{13}$$

Following the determination of $c_{Surf,e}$ (e.g. by means of galvanic cell 1), Eq. (13) can be used for the calculation of the quotient $c_{B,e}/c^0_B$ and consequently for the determination of the excluded volume of the system.

It is noted that the potentiometric method is only a tool for the rapid determination of $c_{I,e}$. The measurements can be performed in a real Donnan equilibrium (between cells I and III in Fig. 1) by analysing the composition ($c'_{Surf,e}$ and $c'_{B,e}$) of the equilibrium solution and correcting the activities because of the pressure difference between the two cells. The pressure correction on the activity is usually negligible and $c'_{B,e} \cong c_{B,e}$.

Results and discussion

We made excluded-volume measurements in PEO–NaDS solutions at several polymer concentration in the case of two different molecular mass PEOs (1×10^5, 1×10^6).

The relative excluded volume, V^*/V, is plotted against the concentration of the surfactant in complex form ($c_{aggr} = c_{NaDS} - c_{cac}$) in Fig. 3 at different constant polymer concentrations. V^*/V depends on the number and the size of the surfactant aggregates, the ionic strength of the solution and the interaction between the aggregates.

In the case of ordinary micelle formation ($c_p = 0$) the ionic strength is approximately constant and for noninteracting micelles with constant size V^*/V should increase linearly with the complex concentration. In fact, the V^*/V function increases linearly at low micelle concentrations but declines from linearity at higher micelle concentrations, which can be interpreted by an increase in the aggregation number and/or micelle–micelle interactions because both lead to smaller excluded volumes at the same micelle concentration.

In the presence of polymer the V^*/V versus c_{aggr} function is a saturation-type curve. The initial slope is significantly higher than in the case of micelle formation.

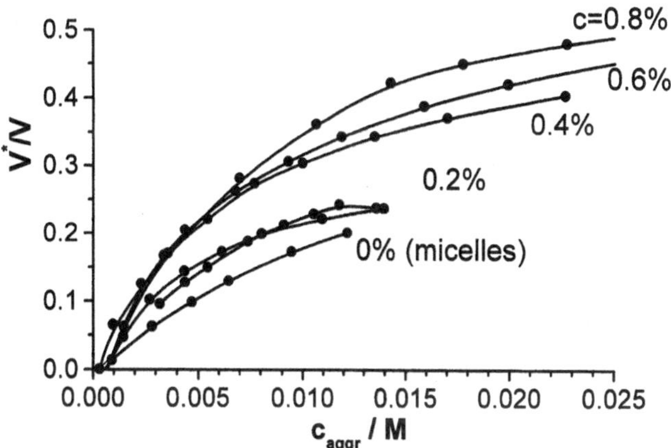

Fig. 3 The relative excluded volume against the concentration of the aggregated surfactant for the poly(ethylene oxide) (*PEO*)–sodium dodecyl sulphate (*NaDS*) system

This qualitatively indicates that the surfactant aggregates in the polymer complex are smaller than the free micelles. With increasing complex concentration all the parameters affecting V^*/V tend to lower the excluded volume: the ionic strength increases from 5 mM to the cmc and the aggregation number and the interaction between the aggregates also increase.

For the free micelle formation the trace probe electrolyte method yields $m = 57$, which is in good agreement with the aggregation number measured by static light scattering [18].

The aggregation numbers calculated from the trace probe electrolyte measurements for the PEO–NaDS solutions are plotted against the equilibrium surfactant concentration in Fig. 4. The reproducibility is good: the m values fall into a narrow range ($\sigma = \pm 3$), although they were measured at several polymer concentrations of two different molecular weight polymers. From these measurements the following statements can be made:

- The aggregation numbers are significantly smaller than those of ordinary micelles.
- The aggregation number increases monotonously with the equilibrium surfactant concentration.
- Within experimental error the aggregation number is invariant to the polymer concentration and the polymer molecular weight and depends only on the equilibrium surfactant concentration (which is the ionic strength in these experiments performed without addition of supporting electrolyte).

The dependence of the aggregation number on the equilibrium surfactant concentration can be partly explained by the well-known effect of the ionic strength on the electrical free energy of the micelle formation [19]. The equilibrium surfactant concentration determines the ionic strength of the solution. $c_{NaDS,e}$ increases from the critical interaction concentration up to the cmc, while

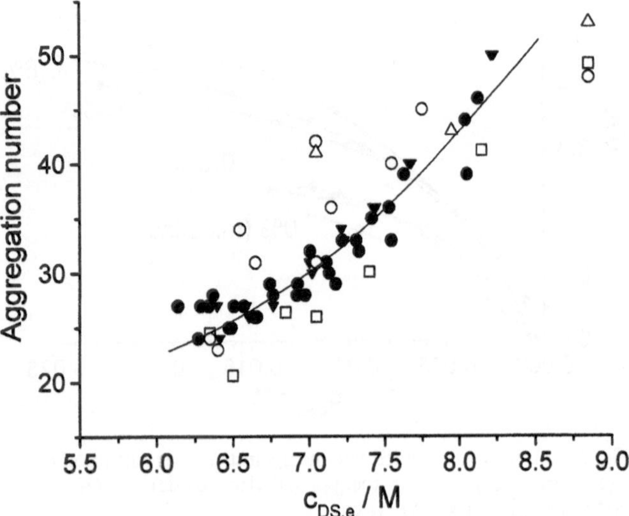

Fig. 4 The aggregation number of NaDS in a PEO complex against the equilibrium surfactant concentration. *Solid symbols*: trace probe electrolyte method; $M = 1.0 \times 10^5$ (○), $M = 1.0 \times 10^6$ (●). Open symbols: Zana et al. [20] (○), van Stam et al. [22] (□), Francois et al. [21] (△)

the amount of the polymer-bound surfactant tends to saturation. The electrical free energy of micelle formation increases with increasing aggregation number and counteracts with the growth of the micelles. At higher ionic strength the electrical free energy is smaller, leading to the formation of larger micelles. On the other hand, for entropy reasons the increase in the equilibrium concentration in itself leads to the shift of the distribution curve of the aggregation number to higher values, resulting in larger mean aggregates. The third effect that may affect the aggregation number is the interaction between the aggregates. The equilibrium aggregate size is determined by the free-energy minimum of the system and not that of a single aggregate. This effect has not been analysed even for the more simple ordinary micelle formation because the existing models are restricted for noninteracting dilute micellar systems.

In the literature the aggregation numbers are usually plotted against the total surfactant concentration, which results in different curves in the case of different polymer concentrations, implying the dependence of *m* on the polymer concentration. In order to demonstrate that this dependence is only an apparent one we collected the NaDS aggregation numbers from the literature measured in PEO–NaDS solutions [20, 21, 22]. In Fig. 4 we also plotted these aggregation numbers against the equilibrium surfactant concentration. The equilibrium surfactant concentrations were calculated from the total concentrations using the binding isotherms of the PEO–NaDS system [23]. Although, these aggregation numbers were measured over a very wide polymer and surfactant concentration range they transformed practically to the same curve that we measured by the trace probe electrolyte method.

Conclusion

Summarising our results it can be concluded that the activity measurements on a probe electrolyte added to a surfactant solution in a trace amount give a simple experimental method to estimate the mean aggregation number of the surfactant aggregates either in the form of a polymer–surfactant complex or in the form of free micelles. An important new finding of the experimental part of this work is that the aggregation number depends only on the equilibrium surfactant concentration and is independent of the polymer concentration.

Acknowledgement This work was supported by the Hungarian Scientific Research Fund (no. T029780).

References

1. Li Y, Dubin PL (1994) In: Herb CA, Prud'homme RK (eds) Structure and flow in surfactant solutions. ACS symposium series 578. American Chemical Society, Washington, DC, pp 320–336
2. Goddard ED (1986) Colloids Surf 19:255
3. Goddard ED, Ananthapadmanabhan KP (1992) Interaction of surfactants with polymers and proteins. CRC, Boca Raton,
4. Nagarajan R, Kalpakci K (1982) Polym Prep Am Chem Soc Div Polym Chem 23:41
5. Nagarajan R (1985) Colloid Surf 13:1
6. Nagarajan R (1986) Adv Colloid Interface Sci 26:205
7. Nagarajan R (1989) J Chem Phys 90:1980
8. Ruckenstein E, Huber G, Hoffmann H (1987) Langmuir 3:382
9. Nikas YJ, Blankschtein D (1994) Langmuir 10:3512
10. Cabane B, Duplessix R (1982) J Phys 43:1529
11. Zana R (1986) In: Zana R (ed) Surfactant solutions. Surfactant science series, vol 22. Dekker, New York, pp 241–294
12. Gilanyi T, Varga I (1998) Langmuir 14:7397
13. Vanderhoff JW (1980) Pure Appl Chem 52:1263
14. Gilányi T (1988) J Colloid Interface Sci 125:641
15. Gunnarson G, Jönsson B, Wennerström H (1980) J Phys Chem 84:3114
16. Marcus RA (1955) J Chem Phys 23:1057
17. Vass S, Török T, Jákli G, Berecz E (1989) J Phys Chem 93:1758
18. Huisman HF (1964) Proc K Ned Akad Wet Ser B 57:407
19. Stigter D (1975) In: van Olphen H, Mysels KJ (eds) Physical chemistry: enriching topics from colloid and surface science. Theorex, La Yolla, p 181
20. Zana R, Lang J, Lianos P (1985) In: Dubin PL (ed) Microdomains in polymer solutions. Plenum, New York, p 357
21. Francois J, Dayantis J, Sabbadin J (1985) Eur Polym J 21:165
22. van Stam J, Almgren M, Lindblad C (1991) Prog Colloid Polym Sci 84:13
23. Gilányi T, Wolfram E (1985) In: Dubin PL (ed) Microdomains in polymer solutions. Plenum, New York, p 383

Progr Colloid Polym Sci (2001) 117: 141–144

T. Gilányi
R. Mészáros
I. Varga

Determination of binding isotherms of ionic surfactants in polymer gels

T. Gilányi (✉) · R. Mészáros · I. Varga
Department of Colloid Chemistry,
Loránd Eötvös University, P.O. Box 32,
1518 Budapest 112, Hungary
e-mail: tobe@ludens.elte.hu
Tel.: +36-1-2090555
Fax: +36-1-2090602

Abstract A thermodynamic analysis of the equilibrium of ionic surfactant solutions with polymer solutions, polymer microgels and macrogels is given. It is concluded that the amount of surfactant bound to the polymer cannot be calculated exactly from the total and equilibrium surfactant concentration measurements. A trace probe electrolyte method is suggested for the determination of the binding isotherm of the surfactant.

Key words Binding isotherm · Surfactant · Microgel · Macrogel · Thermodynamics

Introduction

Surfactants associate into micelles above a critical micelle concentration (cmc) in aqueous solution. In the presence of water-soluble polymers polymer–surfactant complexes form in which surfactant aggregates are distributed along the polymer chains [1–3]. The main driving force of the surfactant aggregation, known as the "hydrophobic effect", is the transfer of the surfactant hydrocarbon chain from water into the micelle-like aggregate core. One of the manifestations of the polymer–surfactant interaction is the interaction of surfactants with polymer gels. In principle, conceptual differences cannot be expected in the polymer–surfactant interaction; either the polymer is free or it is in a loosely cross-linked gel.

Several studies have been reported in the last decade on stimuli-responsive hydrogels, which can change their swelling and shrinking in response to external stimuli. The discovery of a discontinuous volume-phase transition in gels, which is often called a collapse transition, has rendered such soft materials technologically useful [4]. The stimuli that have been investigated to induce changes in polymer gels are diverse and they include temperature, pH, solvent and ionic composition, electric field, light intensity and specific molecules such as surfactants.

Concerning the interaction between surfactants and polymer gels one of the most important pieces of thermodynamic information is the binding isotherm of the surfactant on the gel. On one hand, this function is necessary to the correct evaluation of several physicochemical and scattering measurements. On the other hand, the isotherm itself provides basic information about the way of binding (e.g. monomer binding or micelle-like collective interaction) and the standard free-energy change of the interaction.

Surfactant binding measurements on gels are rare and the results are contradictory [5]. Abuin et al. [6] studied the interaction between sodium dodecyl sulphate (NaDS) and poly(*N*-isopropylacrylamide) (PNIPAM) microgel particles. Surface tension measurements indicated that the interaction was similar to that of the surfactant with free polymer. The interaction starts from a critical surfactant concentration ($cac_1 \approx 1$ mM) and above a second critical concentration (cac_2) there is no further surfactant binding, as is the general experience in the case of surfactant–(free) polymer interaction [1, 2]. Mylonas at al. [7] determined the binding of NaDS on PNIPAM polymer by equilibrium dialysis method. They estimated a cac_1 value that was similar to the one found previously in case of PNIPAM microgel. Mears at al. [5] determined the binding of NaDS in PNIPAM microgel particles by separating the equilibrium surfactant solution by centrifugation. They did not observe a well-defined cac; however, the bound amount increased sharply above 3 mM equilibrium NaDS concentration. A further increase in binding was experienced when the equilibrium surfactant concentration exceeded the cmc.

The discrepancies in these investigations have different reasons. There may be a difference in the interaction if the polymer is in the form of free coils, macrogel or microgel latex particles. The differences in the sample preparation, for example, the degree of cross-linking and the type of the cross-linking monomer may play a role as well. A sound basis for the evaluation of the different experimental methods used to calculate the surfactant binding isotherm is also lacking. The general route of calculating the amount of bound surfactant as a difference of the total and equilibrium surfactant concentration seems to be trivial but, as will be shown in this work, is erroneous.

In order to investigate the role of the previously mentioned parameters in the interaction systematically, the first step is the definition of the binding isotherm and the determination of an exact way to calculate it from experiments. In this work an analysis is given to explore the relation of the amount of bound surfactant to the measurable quantities.

Binding isotherm

The mass balance of the surfactant distributed in different molecular forms (as free monomers, polymer-bound surfactants and "free" micelles) in the microgel latex can be given as

$$c_o V = (\langle S \rangle + c_{\mathrm{mic}})C + Bc_p V \tag{1}$$

or by treating the latex as a two-phase system as

$$c_o V = (\langle S \rangle_g + c_{\mathrm{mic,g}})V_g + (\langle S \rangle_s + c_{\mathrm{mic,s}})V_s + Bc_p V \ , \tag{2}$$

where the subscripts g and s refer to the gel and solution phase, respectively, c_o is the total (analytical) surfactant concentration, $\langle S \rangle$ is the volume-average free surfactant monomer concentration, V is the volume of the system, c_{mic} is the concentration of the surfactant micelles, c_p is the polymer concentration (mass/volume) and B is the number of moles of bound surfactant per unit mass of polymer.

Equation (2) can also be applied for a macrogel ($V_s = 0$, $V = V_g$) and for a polymer solution ($V_g = 0$, $V = V_s$). Independently of the model investigated (macrogel, microgel latex or polymer solution) we are facing the same problem: Eq. (1) or Eq. (2) defines the amount of bound surfactant, B, but the quantities in brackets, $\langle S \rangle$, cannot be measured directly. The experimentally available quantities are the mean surfactant activity in the microgel latex ($a_{\pm}$) and either the mean surfactant activity ($a_{\pm e}$) or the surfactant concentration (c) in the gel (polymer)-free solution which is in equilibrium with the latex. The scheme of the system investigated is given in Fig. 1a. In the following we discuss the relation between B and the measurable quantities.

The thermodynamic condition of the equilibrium in the system concerned is the constancy of the

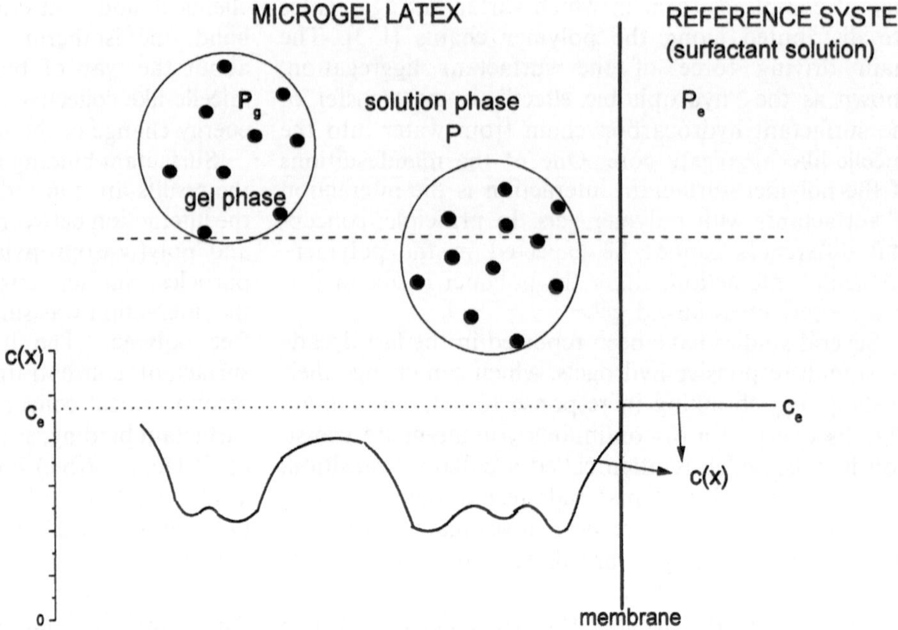

Fig. 1 Scheme of the microgel latex–ionic surfactant system in equilibrium with a polymer-free reference solution. The *solid spheres* represent the surfactant aggregates bound in the microgel particles and the $c(x)$ function denotes the local concentration of the free surfactant ions

electrochemical potential of the surfactant ion, ($\hat{\mu}_-$), throughout the system:

$$\hat{\mu}_- = \mu_- + ez_-\psi \ , \tag{3}$$

where μ_- is the chemical potential of the surfactant ion with valency z_- and Ψ is the electrical potential. For the sake of simplicity, systems containing a 1:1 surfactant electrolyte without added foreign salt will be discussed here. In order to eliminate the electrical potential from Eq. (3) the mean chemical potential of the surfactant electrolyte can be given as $\mu_\pm = (\hat{\mu}_+ + \hat{\mu}_-)/2$. The equilibrium condition of the system is the invariance of the mean chemical potential in the gel phase ($\mu_{\pm g}$), in the solution phase ($\mu_{\pm s}$) and in the gel-free equilibrium solution ($\mu_{\pm e}$). As an alternative view we can consider the microgel solution as a one-phase system. In this case the equilibrium is determined by the constancy of the mean chemical potential in the latex phase ($\mu_\pm$) and in the gel-free equilibrium solution ($\mu_{\pm e}$). The mean chemical potential of the surfactant electrolyte in the latex can be expressed as

$$\mu_\pm = \mu_\pm^{\mathrm{o}} + KT \ln a_\pm(\text{latex}) \ , \tag{4}$$

where $\mu_\pm^{\mathrm{o}}$ is the standard-state chemical potential defined at temperature T and pressure p. If we assume that owing to the very low segment density in the polymer-containing phases (latex, gel) the segment density dependence of the standard potential can be neglected and we suppose that the partial molar volume of the surfactant electrolyte ($\bar{V}_\pm$) does not depend on the pressure then the mean chemical potential of the solution, gel and equilibrium phases can be given in the following forms:

$$\mu_{\pm s} = \mu_\pm^{\mathrm{o}} + kT \ln a_{\pm s} \quad (\text{solution phase}) \ , \tag{5}$$

$$\mu_{\pm g} = \mu_\pm^{\mathrm{o}} + (p_g - p)\bar{V}_\pm + kT \ln a_{\pm g} \quad (\text{gel phase}) \ , \tag{6}$$

$$\mu_{\pm e} = \mu_\pm^{\mathrm{o}} + (p_e - p)\bar{V}_\pm + kT \ln a_{\pm e} \quad (\text{equilibrium solution}) \ , \tag{7}$$

where $a_\pm$ is the mean activity of the surfactant electrolyte and p is the pressure in the appropriate phases. It can be shown that the pressure term is negligible compared to the activity term in Eq. (7) if the pressure difference between the latex and its equilibrium solution is small enough (e.g. less than 1 atm). The contribution of the pressure difference to the chemical potential in Eq. (6) is also neglected although it is questionable that the pressure difference between the gel phase and its environmental solution is sufficiently small in the case of strongly interacting polyelectrolyte type gels.

If it is justified to neglect the terms just discussed the equilibrium condition reduces to the constancy of the mean surfactant activity in the different phases: $a_\pm = a_{\pm s} \cong a_{\pm g} \cong a_{\pm e}$. This means that in an equilibrium we may express the mean surfactant activity by means of the polymer-free reference solution

$$a_{\pm e} = \gamma_{\pm e}(c_+ c_-)^{1/2} \approx c_e \ , \tag{8}$$

where the symbols refer to the polymer-free surfactant solution in equilibrium with the gel–surfactant–solvent system. c_+ and c_- are the concentration of the surfactant monomer and its counterion, respectively. By neglecting the small ion interactions for dilute surfactant solutions ($\gamma_{\pm e} \approx 1$) and by taking into account that $c_+ = c_- = c_e$ the mean activity of the surfactant electrolyte can be approximated by c_e, called the equilibrium monomer concentration ($a_{\pm e} \approx c_e$). Note that c_e is not the total surfactant concentration of the equilibrium solution; it is the surfactant monomer concentration. This is important if the equilibrium concentration (c) exceeds the cmc and micelles form. In this concentration range $c = c_e + c_{\mathrm{mic}} \approx \mathrm{cmc} + c_{\mathrm{mic}}$.

By taking into account only the electrostatic interactions the relation between $\langle S \rangle$ and c_e can be formally expressed in case of a 1:1 surfactant electrolyte as

$$\langle S \rangle = \frac{c_e}{V'} \int_V e^y \mathrm{d}V' = c_e \langle e^y \rangle \ , \tag{9}$$

where $y = e\psi/kT$ is the reduced electrical potential with a reference potential chosen as $\psi = 0$ at the polymer-free reference system and V' is the volume of the system concerned. y and the surfactant monomer concentration are local functions of the space coordinates as depicted schematically in Fig. 1. The concentration of the surfactant anions is lower than c_e around the negatively charged micelles and polymer-bound surfactant aggregates; consequently, $\langle e^y \rangle$ may be significantly smaller than unity. By rewriting Eqs. (1) and (2) by means of Eq. (9) one obtains

$$c_o = \{\langle e^y \rangle_g \Phi + \langle e^y \rangle_s (1 - \Phi)\} c_e + c_{\mathrm{mic,g}} \Phi + c_{\mathrm{mic,s}}(1 - \Phi) + Bc_p \ , \tag{10}$$

where Φ is the volume fraction of the gel phase.

The binding isotherm is defined by the $B(c_e)$ function. As a route B is generally calculated from the relation [5–8]

$$B = \frac{c_0 - c}{c_p} \ , \tag{11}$$

and plotted against c, the total equilibrium surfactant concentration.

By comparing Eq. (11) to Eq. (10) it can be stated that

1. If $c > \mathrm{cmc}$, the calculated B is not the amount of the surfactant bound.
2. Plotting B against c has no thermodynamic significance above the cmc because $c \neq c_e$.
3. The calculated B values can be accepted as an approximation if $c < \mathrm{cmc}$ when $c \approx c_e$ (the error in B changes with the volume fraction of the gel and with

the amount of bound surfactant via the value of the $\langle e^y \rangle$ term).

The concentration of the equilibrium solution can be determined by analysing the supernatant solution separated from the microgel particles by ultrafiltration and ultracentrifugation [5], assuming that the separation does not influence the equilibrium concentration, or by the equilibrium dialysis method [7]. The separation of the microgel particles from a polymer-free equilibrium solution is not necessary. Potentiometric measurements with two reversible electrodes for the surfactant ion and its counterion yield directly the mean surfactant activity or c_e independently of the concentration range investigated. However, it is important to note that a surfactant-sensitive electrode [8] cannot be applied against a reference electrode connected via a salt bridge because of the development of an anomalous diffusion potential (the so-called suspension potential [9]) in the presence of colloid electrolytes, which is the case concerned here.

In order to determine the correct binding isotherm the $\langle e^y \rangle$ term in Eq. (8) must be determined from independent measurements. Recently, we suggested a method for the determination of $\langle e^y \rangle$. This method is based on the activity measurements of a foreign electrolyte added to the system in a trace amount. The probe electrolyte should meet the following requirements: it does not bind specifically to the polymer or to the surfactant and its concentration is negligible compared to that of the surfactant. If we write an equation analogous to Eq. (9) for the anion of an added 1:1 trace electrolyte $[\langle \mathrm{Tr} \rangle = c_{e,\mathrm{Tr}} \langle e^y \rangle$, where $\langle \mathrm{Tr} \rangle$ and $c_{e,\mathrm{Tr}}$ are the total (analytical) and the equilibrium concentrations of the trace anion, respectively] then the $\langle e^y \rangle$ term can be eliminated from Eq. (9):

$$\langle S \rangle = \frac{c_e - c}{c_{e,\mathrm{Tr}}} \langle \mathrm{Tr} \rangle \ . \tag{12}$$

In the right-hand side of Eq. (12) all the quantities are determinable experimentally. Further details of the application of the trace probe electrolyte method are given in Refs. [10, 11]. Another possibility to calculate B from Eq. (10) is to perform the measurements in the presence of a large amount of inert electrolyte when $\langle e^y \rangle \approx 1$; however, in this case the system investigated is different from the salt-free one because the salt may influence the interaction between the polymer and surfactant.

The previous discussion considers the interaction of an ionic surfactant with electrically neutral gels (polymers). In case of polyelectrolyte solutions or polyelectrolyte gels an ion-exchange process must also be taken into account. In these cases the determination of the binding isotherms requires further analysis. It should also be stressed that in Eq. (9) only electrostatic interactions were taken into account; however, the surfactant ions may be excluded from the vicinity of the polymer chain owing to nonelectrostatic interactions (e.g. because of the preferential sorption of water). This exclusion theoretically leads to negative B values (negative surfactant adsorption) at low surfactant concentrations if the surfactant does not bind specifically to the polymer. Negative surfactant binding (i.e. exclusion of the surfactant) was measured below the critical interaction concentration for poly(vinyl alcohol) macrogel and alkyl sulfate systems [12].

Acknowledgement This work was supported by the Hungarian Scientific Research Fund (no. T029780).

References

1. Goddard ED (1986) Colloids Surf 19:255
2. Goddard ED, Ananthapadmanabham KP (1992) Interaction of surfactants with polymers and proteins. CRC, Boca Raton
3. Nagarajan R, Kalpakci (1982) Polym Prep Am Chem Soc Div Polym Chem 23:41
4. Gandhi MV, Thomson BS (1992) Smart materials and structures. Chapman and Hall, London
5. Mears SJ, Deng Y, Cosgrove T, Pelton R (1997) Langmuir 13:1901
6. Abuin E, Leon A, Lissi E, Varas JM (1999) Colloids Surf 147:55
7. Mylonas Y, Staikos G, Lianos P (1999) Langmuir 15:7172
8. Okuzaki H, Osada Y (1991) Macromolecules 28:4554
9. Overbeek JTG (1956) Prog Biophys Biophys Chem 6:58
10. Gilanyi T, Varga I (1998) Langmuir 14:7397
11. Varga I, Gilanyi T, Meszaros R (2000) Prog Colloid Interface Sci 117
12. Kralova K (1970) Thesis. Loránd Eötvös University, Budapest

Progr Colloid Polym Sci (2001) 117: 145–152

Á. Csiszár
A. Bóta
C. Novák
E. Klumpp
G. Subklew

Changes in the thermotropic and the structural behaviour of 1,2-dipalmitoyl-*sn*-glycero-3-phosphatidylcholine/water liposomes effected by 2,4-dichlorphenol

Á. Csiszár · A. Bóta (✉) · C. Novák
Department of Physical Chemistry, Department of General and Analytical Chemistry, Budapest University of Technology and Economics, Müegyetem rkp 3. 1111 Budapest, Hungary
e-mail: bota.fkt@chem.bme.hu

E. Klumpp · G. Subklew
Institute of Applied Physical Chemistry, Research Centre Jülich, 52428 Jülich, Germany

Abstract Toxic 2,4-dichlorophenol (DCP) molecules, originating from the environment, drastically affect the main functions of organic cell membranes. To yield some information on the destroyed membrane structures, 1,2-dipalmitoyl-*sn*-glycero-3-phosphatidylcholine (DPPC)/water liposome systems containing different amounts of DCP were studied by differential scanning calorimetry and small-angle X-ray scattering (SAXS) methods. The transition points of the main transition and its enthalpy are changed only in a relatively narrow range of about 0–2.5%, while the enthalpy of the pretransition is affected by DCP more strongly. The disappearance of the pretransition occurs in a range of about 3 orders of magnitude of the DCP/DPPC ratio (from 10^{-5} to 10^{-2}). The changes in the first Bragg profiles of the SAXS curves depend not only on the temperature but also very strongly on the DCP/DPPC molar ratios. Especially, at higher concentrations DCP causes a strong broadening in the Bragg profiles and a difference from the profiles of the nonrippled phase. The general destruction in the layer arrangements and the reductions in the transition enthalpies between them indicate defect structures consisting of domains with nonlayer structure which are formed simultaneously by DCP molecules.

Key words Liposome · Multilamellar structure · Phase transition · Differential scanning calorimetry · Small-angle X-ray diffraction

Introduction

1,2-Dipalmitoyl-*sn*-glycero-3-phosphatidylcholine (DPPC) is one of the most common phospholipid components in biological membranes [1]. On dispersing this molecule in water, concentric onion-like shells of alternating water and lipid–bilayer regions form spontaneously as unilamellar or multilamellar vesicles (liposomes); therefore, DPPC is widely applied for the preparation of vesicles which are considered as model systems of membranes in living cells. DPPC liposomes exhibit at least four different multilamellar structures in excess water, depending on the temperature [2–5]. These structures are characterized by specific lattice parameters of periodical shells. The lamellar arrangement is the main structural feature of liposomes. Moreover, within the hydrocarbon region of the shells the chain packing is also varied and different subcells are formed. On increasing the temperature, different multilamellar phases are formed: a crystalline phase, L_c; a nonrippled gel phase, $L_{\beta'}$; a rippled gel phase, $P_{\beta'}$; and a liquid-crystalline phase, L_α. These four parent phases consist of regular multilamellar structures with periodicities of 59, 64, 70, and 67 Å, for the L_c, $L_{\beta'}$, $P_{\beta'}$, and L_α phases, respectively as measured by small-angle X-ray scattering (SAXS). These structures are depicted schematically in Fig. 1. Accordingly, there are three phase transitions: the subtransition (L_c–$L_{\beta'}$) at T_{m1}(about 18 °C), the pretransition ($L_{\beta'}$–$P_{\beta'}$) at T_{m2}

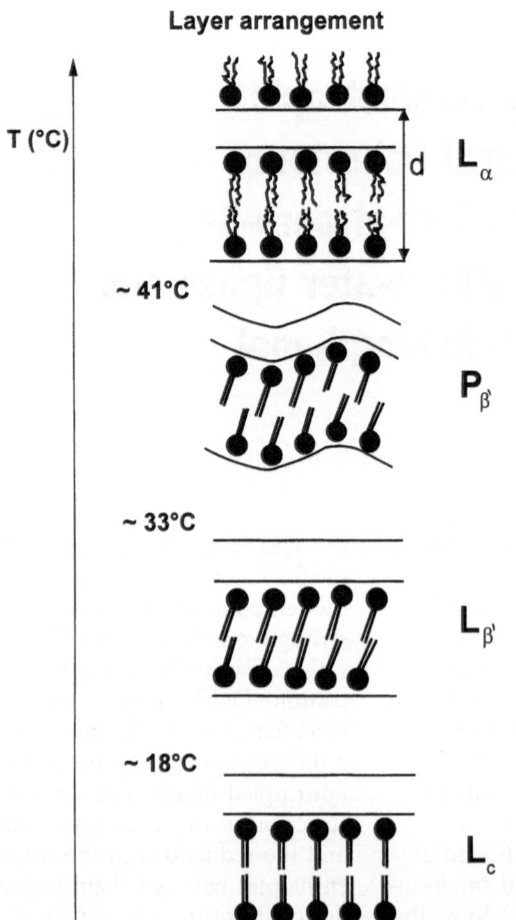

Fig. 1 Layer structures of the fully hydrated 1,2-dipalmitoyl-*sn*-glycero-3-phosphatidylcholine (*DPPC*)/water liposomes as a function of temperature

(about 33 °C), and the main transition ($P_{\beta'}$–L_α) at T_{m3} (about 41 °C). The enthalpy changes accompanying the three transitions are about 14.2, 5.6, and 32.2 kJ/mol lipid, respectively. The value of the pretransition enthalpy is significantly smaller than the other two effects observed for the subtranstion and the main transition. During the pretransition, fluctuations and correlation distances increase significantly in the whole system and defect structures can be formed [6, 7]; therefore the effect of impurities becomes very strong. The effect of each molecule, the molecules being known to be toxic, can be especially drastic. The toxic molecules can induce nonlamellar local structures [8–10]. If the temperature of the organism falls into the transitional region of the lipid domains of the membrane, the transport properties can be drastically changed; therefore, studies on the transitional states of the model system are expected to have considerable relevance in biology. Toxic 2,4-dichlorophenol (DCP) molecules, originating from the environment, drastically affect the main functions of organic cell membranes [11]. To yield some information on the destroyed membrane structures we performed differential scanning calorimetry (DSC) and SAXS measurements in DPPC/water liposome systems containing different amounts of DCP. Calorimetric data (e.g. the change in the transition enthalpy, the temperature range of the transition) are fundamental to characterize the system and to distinguish between the different states related to the different phases [12, 13]. The changes in the calorimetric data compared to those of the pure system also indicate changes in the structure. As the periodicity of the lamellar arrangement of liposomes falls between 60 and 70 Å, SAXS is very powerful for the study of this structure [14, 15]. The periodic, unoriented liposomes show radially symmetric SAXS patterns, which are detected in one-dimensional forms and are presented as the scattering intensity as a function of the scattering variable. The incorporation of DCP into the double layers can be proved by means of fluorescence spectroscopy. For steady-state measurements pyrene is widely used as a fluorescence label in studies of vesicle systems [16, 17]. Using these methods, we have concentrated on the changes in the thermotropic behaviour and on the destruction of the layer arrangement of the gel and liquid phases of the liposome systems.

Experimental

Materials and preparation

Synthetic L-α-dipalmitoylphosphatydilcholine (DPPC, purity higher than 99%) was purchased from Avanti Polar Lipids (Alabaster, Ala, USA) and was used without further purification. Deionized, triple-quartz-distilled water was added to the dry lipid powder under a nitrogen gas atmosphere to yield a lipid concentration of 30 w/w%. The mixture was kept at 50 °C for about 10 h and vortexed frequently. After incubation the sample was quenched to 4 °C, then reheated to 50 °C again and vortexed intensively. The process was repeated ten times to achieve homogeneous hydration. The sample was stored at 40 °C. The preparation of the systems consisting of DCP was the same, only DCP solutions with different concentrations were used instead of pure water. The half value of the solubility of DCP was chosen as the maximal DCP concentration, which yielded a molar ratio of 0.04 (DCP/DPPC). The DCP concentration was varied in a range of 3 orders of magnitude of the molar ratio down to 4×10^{-5}. For fluorescence spectroscopy, the samples were diluted to about 0.1 w/w% by water. The incorporation of DCP into the vesicle bilayers was verified by fluorescence spectroscopy, but the distribution ratios of DCP between the liposomes and the excess water were not measured. Therefore the molar ratios of DCP referred to DPPC are expressed for the whole aqueous systems and not only for the bilayer regions.

Methods

The DSC measurements were made with a TA Instruments DSC 2920 instrument. The scan rate was 1 °C/min and the reproducibility of the measurements of the temperature values was within ±0.005 °C. The DSC curves were recorded in the heating direction in all cases and they are presented in w/g unit.

The SAXS measurements were performed using a Kratky camera and a proportional counter (Anton Paar, Graz, Austria). The scattering of Ni-filtered Cu Kα radiation (λ = 1.542 Å) was recorded in the 6×10^{-3}–0.6 1/nm range of the scattering variable, defined as $h = (4\pi \sin \theta)/\lambda$, where 2θ is the scattering. The primary beam was line-focused. The intensity curves were corrected considering the geometry of the beam profile in order to obtain point-focused curves. For X-ray measurements the lipid dispersion was transferred into thin-walled quartz capillaries (Hilgenberg, Germany) with a diameter of 1 mm. In order to remove air bubbles the capillaries were centrifuged for 5 min at 500*g* at room temperature. The capillaries were sealed with a two-component synthetic resin and transferred into metal capillary holders placed into an aluminium block. This block was positioned directly into the beamline and was used as a thermal gradient incubator for controlled annealing at different temperatures around the pretransition.

The fluorescence spectra were recorded using a luminescence spectrometer (PerkinElmer, LS 50, UK). The fluorescence activity was measured on pyrene molecules added to the systems using 10^{-3} mol/mol pyrene/DPPC ratios. The pyrene was weighed into the empty fluorescence capillary as a solution in chloroform, then it was dried and finally the liposome system was added. After 1-h incubation, the fluorescence activity was maximal and nearly constant for 6 h at the reference temperature (26 °C). The pyrene molecules are solubilised in the chain region of the double layers and are sensitive to all changes which are in progress in the lipid region.

Results

Thermal behaviour of the systems

The pure, fully hydrated DPPC/water system shows three transitions in the DSC curves, which are the subtranstion, pretransition, and main transition, as can be seen in Fig. 2. The explanation of the characteristic data (transition enthalpy, transition point, T_m, beginning point of the transition, i.e. onset temperature, T_{os}) is presented in Fig. 2b. It is noteworthy that the subtransition can be detected in the first scan and only when the sample has been kept for several hours below 4 °C. The two other transitions were reversible when the DSC measurements were carried out in the temperature range from 4 to 50 °C and were repeated in this temperature range. DCP causes changes in the thermograms over a wide concentration range. The dimensions of the changes of the transition signals are very different; therefore, the DSC curves related to the main transition, the pretransition and the subtransition are presented separately in Figs. 3, 4, and 5. The main transition can be observed for the entire concentration range over 3 orders of magnitude of the DCP and are found not to be affected strongly by the actual concentration. The transition point is shifted to lower temperature values for higher DCP concentrations only, as can be seen in Fig. 3. The transition enthalpy decreases slightly as the DCP ratio increases. The characteristic parameters of the main transition are summarized in Table 1. The pretransition

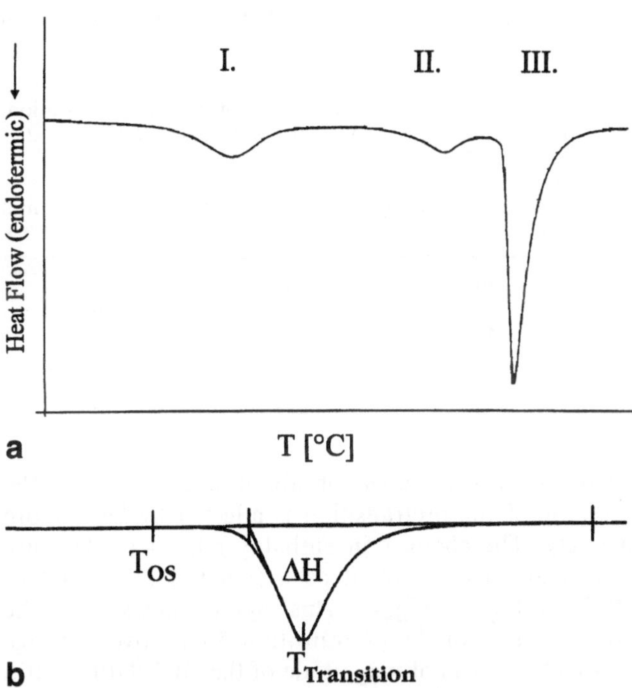

Fig. 2 a Typical differential scanning calorimetry (*DSC*) curves of fully hydrated DPPC/water liposomes in the temperature range from 4 to 50 °C. **b** The characteristic temperature data (transition point, T_m, beginning point of the transition, i.e., onset temperature, T_{os})

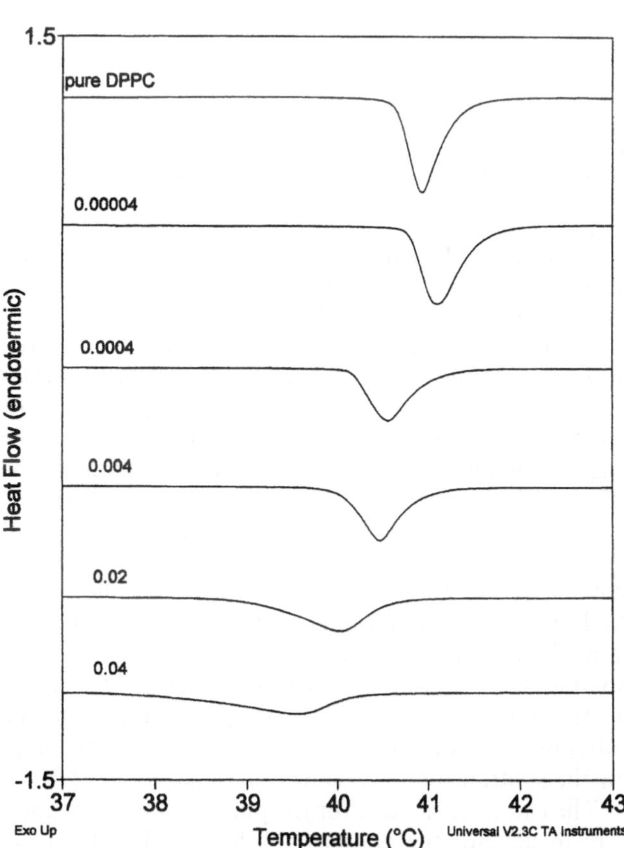

Fig. 3 DSC curves of the main transition in the pure and in the 2,4-dichlorophenol (*DCP*) loaded DPPC/water systems. The concentration of DCP is expressed as the molar ratios of DCP related to DPPC and is marked on each curve

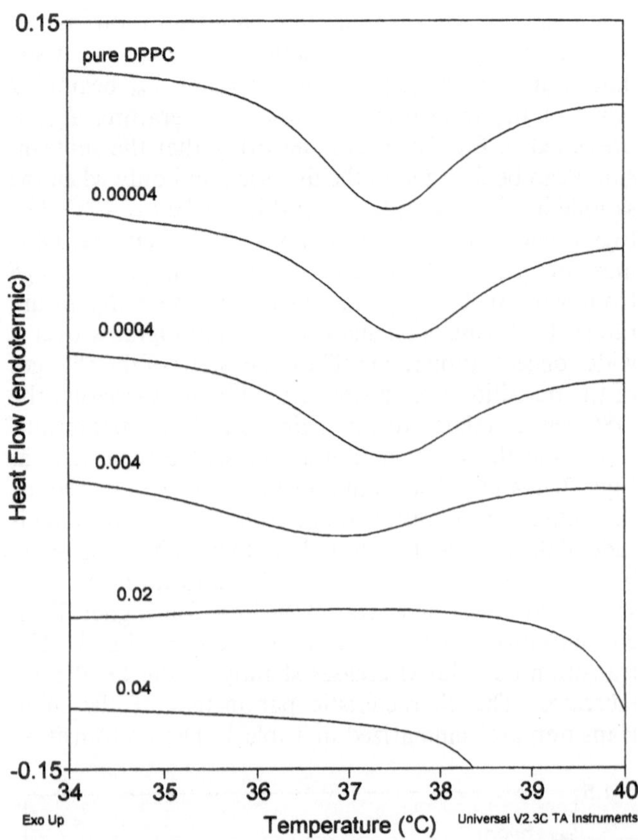

Fig. 4 DSC curves of the pretransition in the pure and in the DCP-loaded DPPC/water systems. The concentration of DCP is expressed as the molar ratios of DCP related to DPPC and is marked on each curve

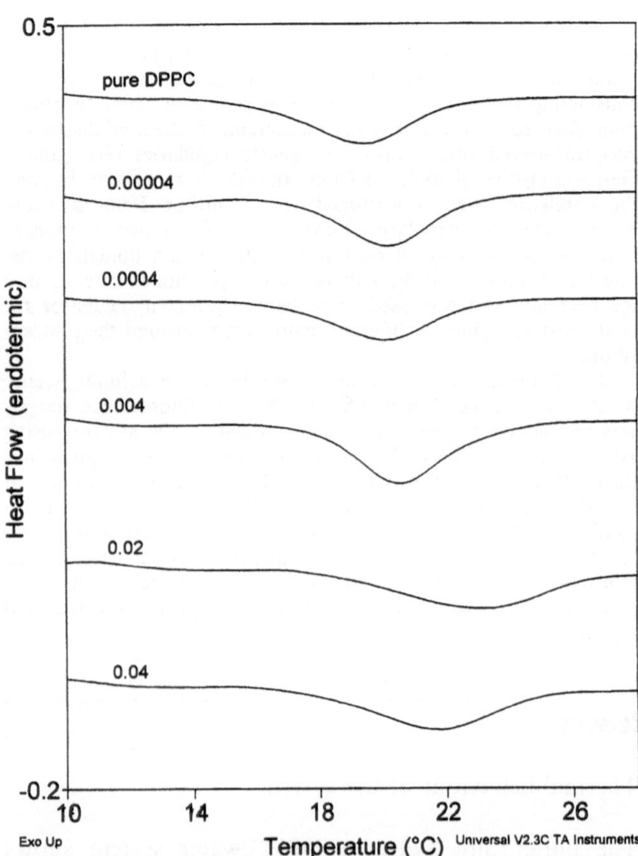

Fig. 5 DSC curves of the subtransition in the pure and in the DCP-loaded DPPC/water systems. The concentration of DCP is expressed as the molar ratios of DCP related to DPPC and is marked to each curve

is much more strongly affected by DCP than the main transition, especially in the higher DCP molar ratio regime. The pretransition point is higher in the presence of the smallest DCP quantity applied than that measured in the pure DPPC/water system. At a DCP/DPPC molar ratio of 4×10^{-3} the same pretransition point can be observed as in the pure system. Above this ratio, the higher the DCP/DPPC molar ratio, the stronger the effect of DCP on the calorimetric data. At a DCP/DPPC molar ratio of 0.04 the pretransition cannot be identified at all. The calorimetric data for the pretransition are summarized in Table 2. The subtransition is affected by DCP, but not strongly, similarly to the case of the main transition, as can be observed in Fig. 5. The transition point is shifted to lower temperature values as a function of the DCP/DPPC ratio, while the changes in the enthalpy are not monotonous. The calorimetric data for the subtransition are shown in Table 3.

The calorimetric data for the pure system are in good agreement with the values published in the literature. Having the biological relevance in sight, we focused on the changes in the pretransition and the main transition. It may be concluded that the transition points of the main transition and its enthalpy are changed only in a relatively narrow range of about 0–2.5%, while the enthalpy of the pretransition is affected by DCP more strongly. The changes in enthalpy related to the pure system are demonstrated as a function of the DCP/DPPC ratio in Fig. 6. This figure shows that the disappearance of the pretransition occurs over a range of about 3 orders of magnitude of the DCP/DPPC ratio (from 10^{-5} to 10^{-2}). The relative changes in the transition points extend to about 10% in the case of the pretransition and are more drastic than those observed for the main transition as can be seen in Fig. 7. It must be

Table 1 The characteristic parameters of the main transition. 1,2-Dipalmitoyl-*sn*-glycero-3-phosphatidylcholine (*DPPC*), 2,4-dichlorophenol (*DCP*)

DCP/DPPC (mol/mol)	T_{os} (°C) ±0.1	T_m (°C) ±0.1	ΔH (kJ/mol lipid) ±0.5
Pure DPPC	40.7	41.3	32.2
4×10^{-5}	40.6	41.6	30.5
4×10^{-4}	40.6	41.3	30.1
4×10^{-3}	40.5	41.2	30.9
2×10^{-2}	39.6	40.9	31.8
4×10^{-2}	38.7	40.3	30.5

Table 2 The characteristic parameters of the pretransition

DCP/DPPC (mol/mol)	T_{os} (°C) ±0.2	T_p (°C) ±0.1	ΔH (kJ/mol lipid) ±0.5
Pure DPPC	33.0	34.8	5.6
4×10^{-5}	33.7	35.5	5.3
4×10^{-4}	33.3	35.1	4.0
4×10^{-3}	33.2	34.7	2.8
2×10^{-2}	29.6	32.0	0.8
4×10^{-2}	–	–	–

Table 3 The characteristic parameters of the subtransition

DCP/DPPC (mol/mol)	T_{os} (°C) ±0.2	T_s (°C) ±0.2	ΔH (kJ/mol lipid) ±0.5
Pure DPPC	15.8	19.10	14.2
4×10^{-5}	17.0	19.78	13.8
4×10^{-4}	17.3	19.66	11.7
4×10^{-3}	17.6	20.30	15.5
2×10^{-2}	17.7	22.60	14.6
4×10^{-2}	17.2	21.25	14.6

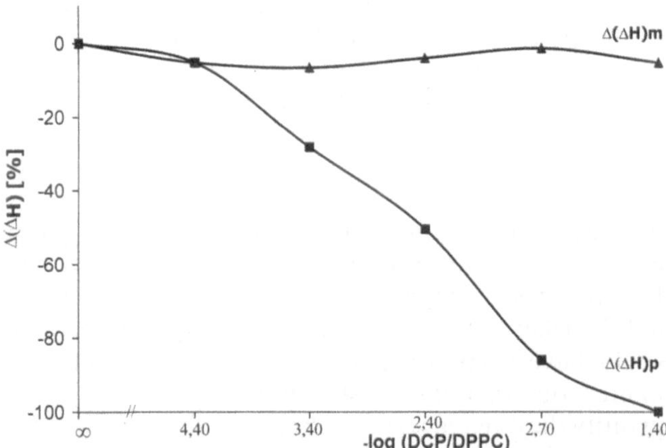

Fig. 6 Relative changes of the transition enthalpies in the pretransition and the main transition

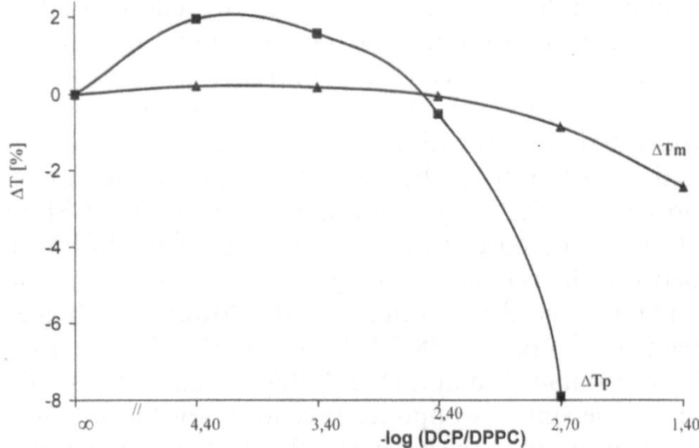

Fig. 7 Relative shifts of the transition points in the pretransition and the main transition

mentioned that the measurement of the calorimetric data of the pretransition is questionable at the highest DCP concentration, as the base line of the curves is uncertain.

SAXS studies on the layer structure

Study of the parent phases

The scattering curves (the intensity as a function of the absolute value of the scattering variable) exhibit more-or-less sharp small-angle diffraction peaks, indicating long-range correlation between the lamellae. The reference parent phases, for example, the gel and liquid phases show characteristic small-angle X-ray diffraction patterns corresponding to the regular lamellar arrangements of the DPPC/water liposomes investigated. The effect of the DCP concentration was studied at 26, 38, and 44 °C, i.e., in the temperature domains of the nonrippled gel ($L_{\beta'}$), the rippled gel ($P_{\beta'}$), and the liquid-crystalline (L_α) phases, respectively. The Bragg profiles of the reference and perturbed states measured at 26 °C can be observed in Fig. 8.

Samples which contain DCP in small quantities (DCP/DPPC molar ratios of 4×10^{-5} and 4×10^{-4}) exhibit higher and sharper Bragg maxima than the pure system. In the higher molar-ratio range (between 4×10^{-3} and 4×10^{-2}) drastic changes in the Bragg profiles can be observed. In this range the Bragg maximum diminishes significantly and the positions of the Bragg maxima shift from $s=0.0159$ to $s=0.0141$ 1/Å The lamellar structure related to the 2×10^{-2} molar ratio has a higher periodic distance than the regular $L_{\beta'}$ phase; moreover, the other lamellar structure related to the 4×10^{-2} molar ratio exhibits a slightly larger periodic

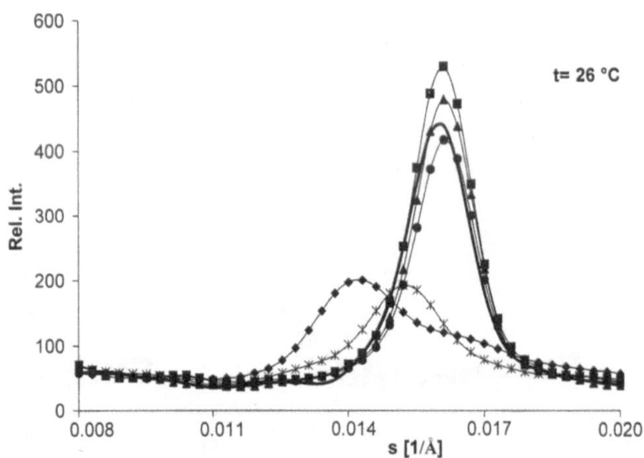

Fig. 8 First Bragg profile of the small-angle X-ray scattering (*SAXS*) curve detected in the pure and in the DCP-loaded DPPC/water systems at 26 °C. (pure: *solid line*; 4×10^{-5}: *squares*; 4×10^{-4}: *triangles*; 4×10^{-3}: *circles*; 2×10^{-2}: *crosses*; 4×10^{-2}: *rhomboids*)

distance than the $P_{\beta'}$ phase. Considering the calorimetric data, as no transition occurs between the subtransition and the main transition regions in the systems loaded with the highest DCP concentration, these shifts in Bragg reflections cannot be interpreted as the appearance of the $P_{\beta'}$ phase, but rather as the direct effect of the high DCP concentration on the layer structure. The broadening of these Bragg profiles is enormous, indicating a drastic loss in the layer regularity in the systems.

The Bragg profiles observed at 38 and 44 °C exhibit the same shape and their maxima are located at $s=0.0144$ and $s=0.0150$ 1/Å, corresponding to the characteristic periodic distances of the rippled gel and liquid-crystalline phases, respectively, as can be seen in Figs. 9 and 10. In the SAXS patterns detected at 38 and 44 °C the effect of the DCP concentration is similar; this means that only the maximum of the Bragg peaks is reduced, except for the system having a DCP molar ratio of 4×10^{-2}. The latter system exhibits an average periodic distance of about 66.8 Å instead of 63.4 Å in the temperature domains of the $L_{\beta'}$ and $P_{\beta'}$ phases and shifts to the value of 61.8 Å instead of 66.7 Å in the temperature domain of the $L_{\beta'}$ phase.

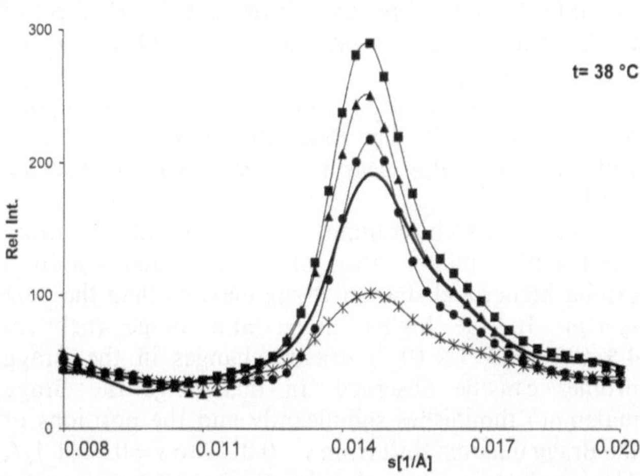

Fig. 9 First Bragg profile of the SAXS curve detected in the pure and in the DCP-loaded DPPC/water systems at 38 °C. (pure: *solid line*; 4×10^{-5}: *squares*; 4×10^{-4}: *triangles*; 4×10^{-3}: *circles*; 2×10^{-2}: *crosses*)

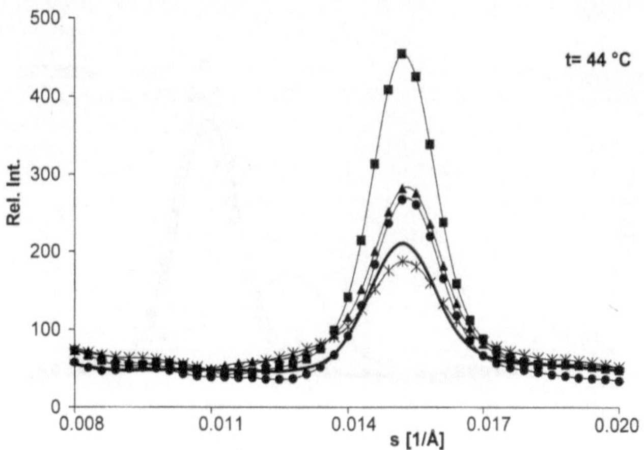

Fig. 10 First Bragg profile of the SAXS curve detected in the pure and in the DCP-loaded DPPC/water systems at 46 °C. (pure: *solid line*; 4×10^{-5}: *squares*; 4×10^{-4}: *triangles*; 4×10^{-3}: *circles*; 2×10^{-2}: *crosses*)

Studies in the pretransition range

The pretransition was strongly affected by DCP; therefore, the states of this transition were studied closely. The SAXS curves of the transitional states were recorded over a wide temperature range from 28 to 38 °C with a step width of 2 °C. The changes in the first Bragg profiles depend not only on the temperature but also very strongly on the DCP/DPPC molar ratios. Especially, higher DCP concentrations caused a strong broadening in the Bragg profiles and a difference from the profiles of the parent (nonrippled and rippled) phases. The characteristic changes induced by DCP can be expressed by a control parameter. The reciprocal value of the full width at half maximum (FWHM) was defined as a control parameter. The reciprocal value of FWHM (n) recorded for the first Bragg profile in the $L_{\beta'}$ phase of the pure system at 28 °C was used for the normalization of the other cases. This parameter indicates that the $L_{\beta'}$ phase has a higher degree of order than the $P_{\beta'}$ phase because the latter has a significantly wider first Bragg profile. A multilamellar system with no regular layer correlation would have a control parameter of $n=0$. It must be pointed out that this parameterisation serves only to quantify the differences of the Bragg profiles of the SAXS curves and should not be taken as a physical definition of the order parameter. The n versus T functions are shown in Fig. 11. The changes in the control parameter demonstrate the different regularities of the lamellar packing in the parent phases and thereby reflect the effect of DCP concentration. The local minimum of the $n(T)$ functions indicates temperature values which are close to the transition points in the three systems investigated (pure and systems with DCP at molar ratios of 4×10^{-3} and 2×10^{-2}). It can be seen that the pretransition is in progress in the temperature domain from about 31 to 36 °C in the pure system. The characteristic difference between the gel and rippled gel phases expressed by the difference in the n values at the lower and higher temperature ranges (28–30 °C and 36–38 °C, respectively) was diminished at a DCP/DPPC molar ratio of 10^{-3} and turned into the opposite direction at the highest DCP concentration. In the latter case the change in the control parameter also indicates that the pretransition ceases to exist in this DPC concentration domain.

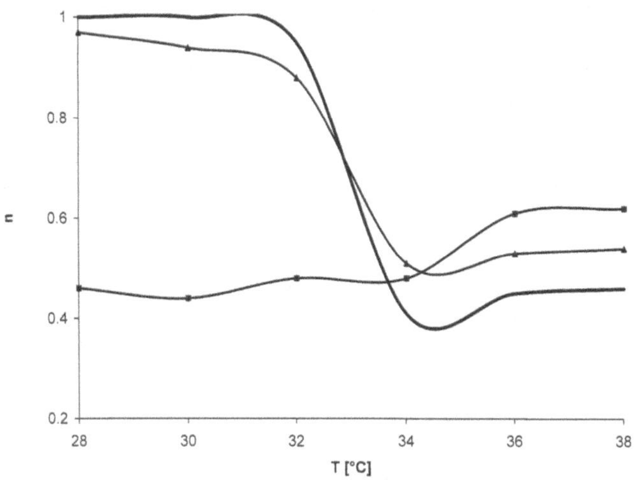

Fig. 11 Change of the control parameter in the pretransition range (pure: *solid line*; 4×10^{-3}: *triangles*; 2×10^{-2}: *squares*)

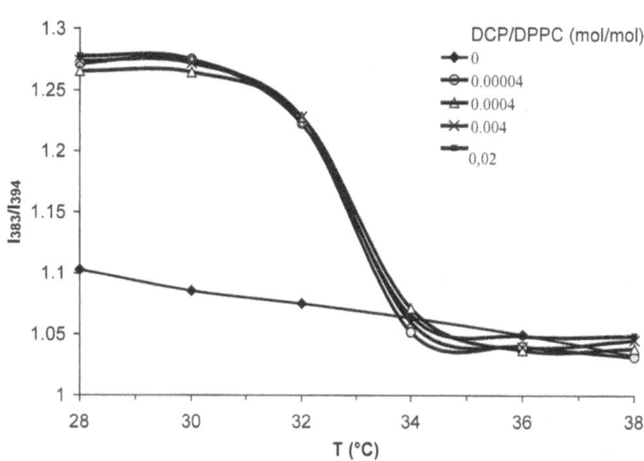

Fig. 12 Change in the intensity ratio of the first and the third peak, I_1 (at 383 nm)/I_3 (at 394 nm) during the pretransition observed in the fluorescence emission spectra of pyrene

Fluorescence spectroscopy study on the location of DCP molecules

The effect of DCP on the vesicles can be directly observed in the fluorescence emission spectra of pyrene. The higher the DCP/DPPC molar ratio, the more drastic the decrease in the fluorescence intensity. This effect can be observed in the entire temperature domains of rippled and nonrippled gel phases. The tendency of the decrease is the same in both phases. Presumably, the interaction between the label and the DCP molecules yields a decrease in the fluorescence intensity. However, the shape of the spectra remains characteristic of the parent phases, indicating the thermotropic character of the double-layer matrix. The ratio of the intensity values of the first and the third peaks, I_1 (at 383 nm)/I_3 (at 394 nm), is characteristic of the parent phases [18], namely the I_1/I_3 ratio quantifies the hydrophobicity of the surroundings of the pyrene molecules. The fluorescence spectra were recorded in the wide temperature range of the pretransition and the I_1/I_3 ratios are plotted as a function of the temperature in Fig. 12. In the pure DPPC/water system the changes in the I_1/I_3 ratio indicate that the hydrophobicity of the surrounding of pyrene is less in the rippled gel phase than in the nonrippled gel phase. The change in this character is continuous during the pretransition. The change is more pronounced in the presence of DCP molecules, but no dependence on the DCP concentration was observed. It is noteworthy that in the nonrippled gel phase the I_1/I_3 ratio reflects a higher hydrophobicity in the systems which are loaded with DCP molecules than in the pure system. We can conclude that the chemical character of the surroundings of the pyrene molecules is formed already at the lowest DCP concentration and this character is connected to the nonrippled phase being more rigid than the rippled phase, the latter having fluid crystalline-phase behaviour.

Conclusion

The changes in the pretransition were expected to be the strongest of the three transitions. This prediction, mentioned in the Introduction, was supported by DSC and SAXS measurements. DCP causes drastic changes in both the calorimetric and structural behaviour at higher concentrations. The diffraction peaks in the SAXS patterns show very similar shapes at 26 and 38 °C in the samples containing the highest quantity of DCP; moreover, the pretransition disappears between these states. We conclude that the changes occur in the temperature domain of the gel phase. However, it still remains unclear whether the same Bragg profile indicates identical or different structures.

DCP affects the transition temperatures, especially for the pretransition. A small quantity of DCP causes a slightly higher transition point than that of the pure system for all three transitions. Presumably, the defect structures present in the pure system can be changed to more regular forms by addition of small quantities of DCP.

The shifts of the transition points revealed by the DSC curves can be interpreted as a direct effect of DCP on the melting points of the self-organized phospholipid bilayers, but the transition points themselves are higher than those connected to the equilibrium states. Namely, the slow scan rates applied in the experiments proved not to be slow enough to attain the thermodynamic equilibrium states during the transition. In these cases, like in our calorimetric experiments, the transition point appears at a higher temperature than that observed by means of SAXS measurements.

Knowing the chemical character of chlorinated phenol, we can suppose that DCP molecules are located close to the polar groups in the bilayers. This assumption was supported by fluorescence measurements. The existence of the rippled gel phase is connected with the hydrophobic effect of the polar headgroup, especially to that of the choline group; therefore, the hydrophobic interaction is strongly perturbed by DCP molecules. However, destruction was observed not only in the rippled gel phase, but in all thermally adjacent phases. The results, i.e., the general destruction in the layer arrangements and the reductions in the transition enthalpies between them, do not exclude the possibility of the existence of defect structures consisting of domains with nonlayer structure which are formed simultaneously by DCP molecules.

Acknowledgements This work was supported by the Hungarian Scientific Funds OTKA (T 014396, T 21781) and a bilateral German–Hungarian Program TÉT (D-42/1998).

References

1. Cevc G., Marsh D (1987) Phospholipid bilayers. Physical principles and models. Wiley, New York
2. Tardieu A, Luzatti V, Reman FC (1973) J Mol Biol 75:711–733
3. Ruocco MJ, Shipley GG (1982) Biochim Biophys Acta 684:59–66
4. Ruocco MJ, Shipley GG (1992) Biochim et Biophys Acta 691:309–320
5. Maulik PR, Ruocco MJ, Shipley GG (1990) ChemPhys Lipids 56:123–133
6. Bota A, Kriechbaurm M (1998) Colloids Surf A 141:441–448
7. Bota A, Drucker T, Kriechbaurm M, Palfia Z, Rez G (1998) Langmuir 15:3101–3108
8. Lohner K (1991) Chem Phys Lipids 57:341–362
9. Lohner K, Degovics G, Laggner P, Gnamusch E, Paltauf F (1993) Biochim Biophys Acta 1152:69–77
10. Weber FJ, de Bout JAM (1996) Biochim Biophys Acta 1286:225–245
11. Escher BI, Schwarzenbach RP (1996) Environ Sci Technol 30:260–270
12. Biltonen RL, Freire E (1978) CRC Crit Rev Biochem 5:85–124
13. Grabielle-Madelmont C, Perron R (1983) J Colloid Interface Sci 95:471–482
14. Luzatti V (1968) In: Chapman D (ed) Biological membranes, vol 1. Academic, New York, pp
15. Laggner P (1988) Top Curr Chem 145:173–200
16. Lehtonen JYA, Holopainen JM, Kinnunen PKJ (1996) Biophys J 70:1753–1760
17. Socaciu C, Lausch C, Diehl HA (1999) Spectrochim Acta A 55:2289–2297
18. Somasundaran P, Krishnakumar S (1997) Colloids Surf A 123/124:491–513

Progr Colloid Polym Sci (2001) 117: 153–158

A. Marton
Y. Miyazaki

Correlation between equilibrium and NMR spectroscopic data in the study of the selectivity of cross-linked ionic polymers

A. Marton (✉)
Atmospheric Chemistry Research Group of the Hungarian Academy of Sciences, University of Veszprém, 8201 Veszprém, P.O. Box 158, Hungary
e-mail: marton@anal.venus.vein.hu
Tel.: +36-88-422022
Fax: +36-88-423203

Y. Miyazaki
Department of Chemistry, Fukuoka University of Education, Fukuoka 811-4192, Japan

Abstract Ion-exchange equilibrium and ^{31}P NMR measurements were used for the study of the ion-exchange distribution of phosphonic acid anions in various types of ion-exchange materials. An equation was derived for the calculation of ion-exchange selectivity coefficients of the monovalent and bivalent anions from the pH dependence of the measured distribution coefficient of the acid. The equation derived was also used to calculate the hydrogen ion concentration inside the polymer phase. ^{31}P NMR spectra of the equilibrium resin phase were studied as a function of pH. On the basis of the relationship derived for the pH dependence of the observed ^{31}P NMR shift of phosphonic acid the resin phase the chemical shifts of the monovalent and bivalent anions were calculated. The linear relationship observed between the calculated selectivity coefficients and the resin-phase ^{31}P chemical shifts proves that interactions responsible for ion-exchange selectivity (thermodynamic data) and chemical shifts (spectroscopic data) are strongly correlated in the systems studied.

Key words Ion exchange · Selectivity · ^{31}P NMR spectroscopy · Phosphonic acid · Probing interactions in swollen gels · Ions in macromolecular systems

Introduction

The selectivity of ion-exchange reactions is usually characterised by the thermodynamic equilibrium constant calculated either by the Gaines and Thomas equation [1] or by the relationship recently suggested by Högfeldt [2] or by using the equation derived on the basis of the concentrated electrolyte solution model of ion-exchange polymers [3]. Enthalpy and entropy data of the ion-exchange reactions may further contribute to our understanding of selectivity [4] but the interpretation of these data in terms of the physicochemical properties of the exchanging counterions still remains to be studied. Although a direct thermodynamic study of the interactions in the resin itself is hardly possible NMR spectroscopy provides a unique method to probe the chemical environment inside the polymer phase. Ion pairing [5], diffusion coefficients [6, 7] and hydration properties of the counterions [8] have been measured in various synthetic ion-exchange resins and membranes. The chemical shift [9] and the relaxation time [10] of the water protons in the hydration sphere of the counterion have also been studied with the intention to predict ion-exchange selectivity. Our current article is concerned with the observation of the NMR spectra of the counterion itself and with the study of the expected correlation between its resin-phase chemical shift and its selectivity coefficient. In our studies phosphonic acid was selected as a particular model compound because of the similarity of its protonation sites to the proton binding groups of biological energy-transfer molecules [11]. The intracellular ionisation of these groups was successfully followed by the sensitive pH dependence of the ^{31}P NMR signal [12].

Experimental

Materials

Analytical grade disodium phosphonate (Na_2PHO_3) was obtained from Wako Pure Chemicals, Japan. Strongly basic polystyrene-based anion exchange resins (100–200 mesh) with tetramethyl ammonium functionalities were obtained in various cross-linkings (1-X2, 1-X4 and 1-X8) from Muromac, Japan. Polysaccharide gel-type exchanger QAE Sephadex A-25 was received from Pharmacia Biotech, Sweden. The ion exchangers were converted into chloride form, washed free from all invaded electrolyte and were stored in an air-dried form. The ion-exchange capacities (Q), determined by the usual standard method [13], were as follows 1-X2: 3.39, 1-X4: 3.44, 1-X8: 3.00, QAE A-25: 2.69.

Equilibrium measurements

Exactly known amounts of each resin sample (1-X2: 0.65 g, 1-X4: 1.2 g, 1-X8: 1.6 g, QAE A-15: 0.5 g) were weighed into 25-cm^3 flasks into which 20 cm^3 disodium phosphonate solution (1.5×10^{-3} $moldm^{-3}$) was added. The ionic strength of the systems was set to 0.1 $moldm^{-3}$ by sodium chloride. The pH of the solution was varied in the range 1–12 by addition of concentrated hydrochloric acid or sodium hydroxide solutions. The pH was measured using a glass electrode (Horiba 6069-10C) calibrated for three buffer solutions. The systems were placed in a thermostatted water bath set to 25 °C. After equilibration overnight the resin and solution phases were separated by filtration and the pH of the solution was measured. The concentration of the phosphonate was determined by a spectrophotometric method [14] both in the starting and in the equilibrium solutions (c_o and c_e). The overall distribution coefficient, D_m, was calculated using the following equation:

$$D_m = \frac{V(c_0 - c_e)}{mc_e}, \quad (1)$$

where V is the volume of the solution (20 cm^3) and m is the mass of the weighed-chloride-form air-dried resin. The change in the distribution coefficient as a function of the pH of the equilibrium solution is shown for the systems studied in Fig. 1.

Measurement of ^{31}P NMR spectra

^{31}P NMR spectra of the equilibrium system were recorded using a JEOL JNM-GSX 500 spectrometer with a 10-mm multinuclear probe at a resonance frequency of 202.5 MHz. Typical instrumental parameters used to obtain spectra of the equilibrium systems were as follows: flip angle 45° (10 μs), pulse repetition time 3 s, spectral width 36 kHz, lock D_2O containing 0.0125% H_3PO_4 in a 2-mm coaxial capillary. The chloride concentration of the resin phase was also determined by an NMR spectroscopic technique according to the procedure described in our earlier publication [15]. The variation of the measured ^{31}P NMR chemical shifts is shown in Fig. 2 as a function of the pH.

Fig. 1 Distribution coefficient of phosphonic acid (D_m) as a function of pH

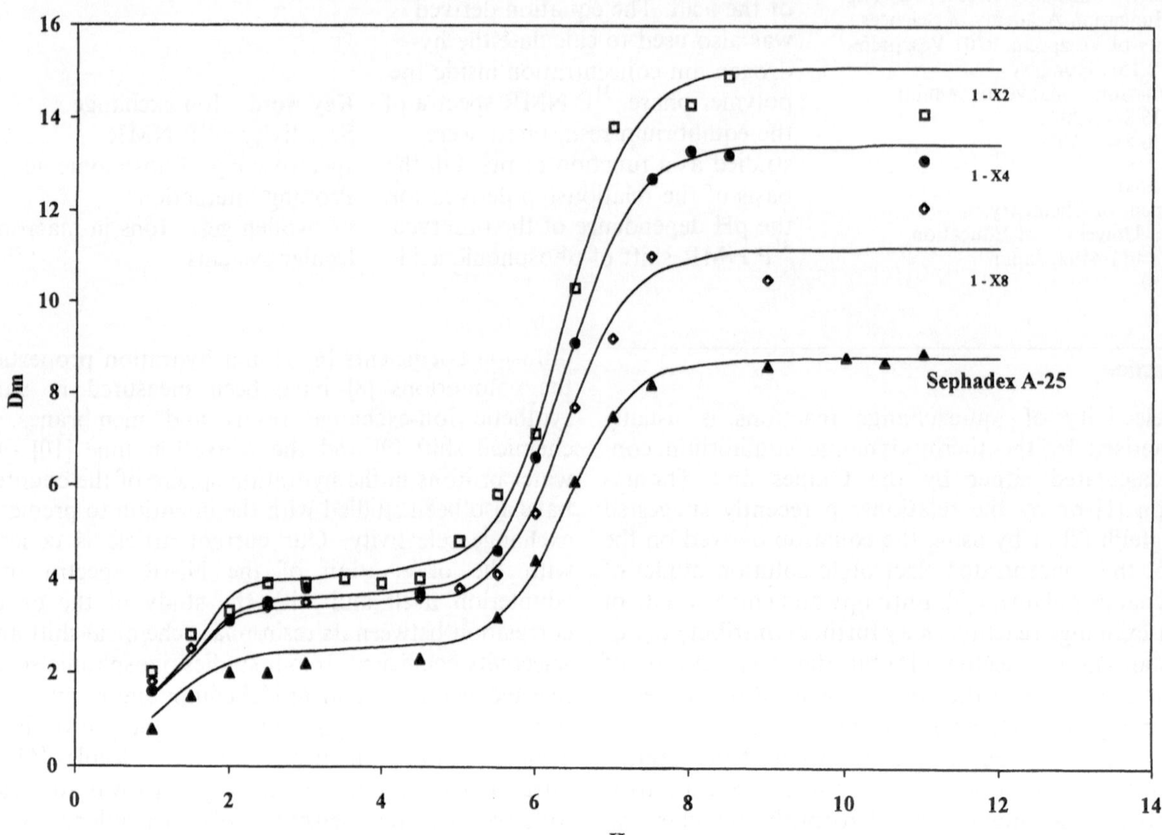

Results and discussion

In order to simplify notation, the various phosphonic acid species (PHO_3^{2-}, $HPHO_3^-$ and H_2PHO_3) will be

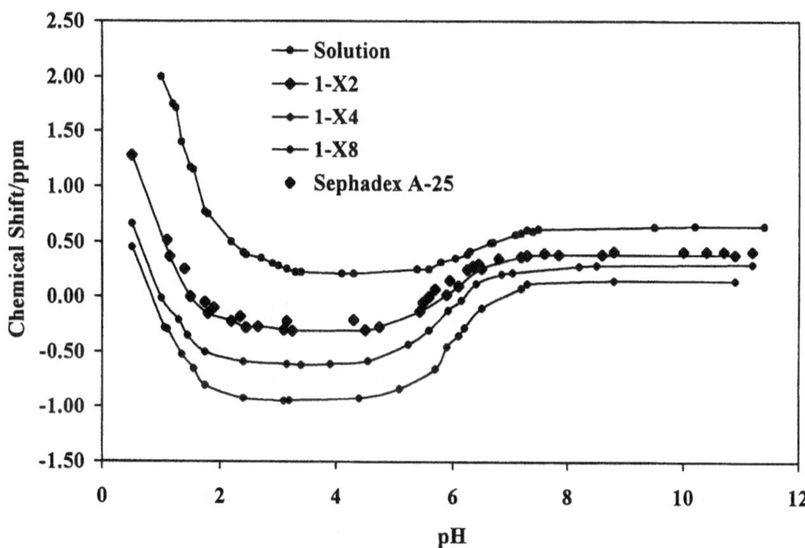

Fig. 2 ^{31}P chemical shift as a function of pH of the solution. For the curve "solution" chemical shifts are solution-phase data which were used to calculate the protonation constants. For the rest of the curves the chemical shifts were measured in the resin phase

denoted as P^{2-}, HP^- and H_2P. Using these symbols the ion-exchange reaction, the selectivity coefficient (K_{HP}) and the distribution coefficient (d_{HP}) of the monovalent ions can be described by the following equations:

$$RCl + HP^- = RHP + Cl^- \ , \tag{2}$$

$$K_{HP} = \frac{(HP^-)[Cl^-]}{[HP^-](Cl^-)} = d_{HP}[Cl^-]/(Cl^-) \ , \tag{3}$$

$$d_{HP} = \frac{(HP^-)}{[HP^-]} \ . \tag{4}$$

The exchange equilibria of the bivalent ions can be represented by a similar set of equations:

$$2RCl + P^{2-} = R_2P + 2Cl^- \ , \tag{5}$$

$$K_P = \frac{(P^{2-})[Cl^-]^2}{[P^{2-}](Cl^-)^2} = d_P\{[Cl^-]/(Cl^-)\}^2 \ , \tag{6}$$

$$d_P = \frac{(P^{2-})}{[P^{2-}]} \ . \tag{7}$$

In these equations the parentheses and the square brackets indicate resin- and solution-phase equilibrium concentrations respectively. The resin invasion of the neutral species will be neglected. Since the overall distribution coefficient (D_c) is defined as the ratio of the analytical concentration of the phosphonic acid in the resin and in the solution phases it can be shown that D_c is related to the ionic distribution coefficients (d_{HP} and d_P) by the following equation:

$$D_c = \phi_0 d_P + \phi_1 d_{HP} \ , \tag{8}$$

where ϕ is the mole fraction of the phosphonic acid species,

$$\phi_i = \frac{\beta_i[H^+]^i}{\sum_{i=0}^{2} \beta_i[H^+]^i} \ , \tag{9}$$

defined in terms of the cumulative protonation constant β_i:

$$\beta_i = K_0K_1K_2(i = 0, 1, 2 \text{ and } K_0 = 1) \ . \tag{10}$$

It is noted here that the values of the K_1 and K_2 constants were calculated from the pH dependence of the ^{31}P NMR chemical shift of the 1.5×10^{-3} mol/dm^3 disodium phosphonate solution (Fig. 2). By performing the least-squares calculation according to Eq. (16) the following values were obtained: $K_1 = 10^{6.34}$, $K_2 = 10^{1.12}$.

The set of d_P and d_{HP} distribution coefficients giving the minimum sum of the squares of the deviations,

$$\sum (D_m - D_c)^2 \ , \tag{11}$$

was obtained by a nonlinear regression. The calculated distribution coefficients are shown in the second and third columns of Table 1.

Ion-exchange selectivity coefficients were calculated for the resins studied using Eqs. (3) and (6) and are also presented in Table 1. (columns four and five). As a consequence of the electroselectivity of the resin the distribution coefficients are higher for the bivalent ions than for the monovalent phosphonate ions.

Knowing the pH dependence of the distribution coefficient (Fig. 1.) Eq. (8) offers a convenient way to estimate the internal pH of the resin phase. For this the protonation equilibria of the phosphonate anion are expressed in terms of resin-phase components and, in the absence of a better approximation, the values of the protonation constants are assumed to be the same as in the solution phase. This consideration leads to a

Table 1 Ion-exchange equilibrium data (d, K) and ^{31}P NMR chemical shifts ($\bar{\delta}$) calculated for the monovalent and bivalent phosphonate ions (HP^- and P^{2-})

	d_{HP}	D_P	K_{HP}	K_P	$\bar{\delta}_{HP}$	$\bar{\delta}_P$
AG 1-X2	3.62	13.1	0.25	0.066	2.42	3.15
AG 1-X4	4.19	15.3	0.20	0.035	2.15	3.05
AG 1-X8	3.53	11.33	0.12	0.012	1.81	2.91
QAE A25	2.08	7.29	0.366	0.25	2.55	3.20
Solution					3.01	3.43

relationship similar to Eq. (8) except that $[H^+]$ is now replaced by $[\bar{H}^+]$ (the bar refers to the resin phase). The quadratic equation obtained can be solved for $[\bar{H}^+]$ and the pH inside the resin phase can be estimated. It should be born in mind, however, that the protonation constants of the base studied in the resin and in the solution phases are certainly different owing to the high density of the positive charges of the immobilised cationic functional groups. If Ψ designates the average electrostatic potential generated by these charges then it can be shown that the protonation constants in the resin and solution phases are interrelated by the following general equation:

$$\log \bar{K} = \log K + 0.434\frac{ze\Psi}{kT} \ , \tag{12}$$

where z is the valence of the ion (including sign), e is the electronic charge of the proton, k is Boltzmann's constant and T is the temperature [16]. Since the K values in the two phases are equal only when $\Psi = 0$ (which obviously does not apply here), the calculated resin phase pH values should be considered as estimates.

When the calculated pH values were plotted as a function of the pH of the solution then a linear relationship was obtained:

$$-\log\left[\bar{H}^+\right] = -a\log[H^+] + b \ . \tag{13}$$

The values of the parameters a and b are given for the systems studied in Table 2.

As can be seen, the slope of the straight lines is very close to unity in all the cases, while the intercept varies for the different types of resins and indicates a higher pH value inside the resin than in the outside equilibrium solution phase.

In order to correlate the calculated ion-exchange selectivity data (K_{HP} and K_P) to the appropriate resin-phase NMR parameters the measured chemical shifts have to be decomposed into the contribution characteristics for the individual phosphonic acid species. These data are not directly accessible from the NMR spectra but they can be calculated by the method shown later. In the evaluation of the NMR spectra the following two features of the system have to be kept in mind. The rate of exchange of the phosphonic acid species between the solution and resin phases is very slow on the time scale of the NMR experiment. For this reason, the ^{31}P NMR spectra of the equilibrium system exhibit two well-separated resonance signals (the high-field signal is due to the resin-phase species, while the lower-field signal is due to the solution-phase phosphonic acid species). This feature of the system offers a convenient way to study the pH dependence of the chemical shifts in both the solution and the resin phases.

As opposed to the slow exchange of the phosphonic acid species between the equilibrium phases, the rate of proton exchange among the phosphonic acid species is fast on the time scale of the NMR spectroscopy. As a consequence, the species are indistinguishable and in each phase only one ^{31}P signal can be detected at a so-called population-averaged chemical shift value [17]. In the case of fast-exchange conditions the resin-phase chemical shift ($\bar{\delta}_c$) is defined as

$$\bar{\delta}_c = \bar{\phi}_0\bar{\delta}_P + \bar{\phi}_1\bar{\delta}_{HP} + \bar{\phi}_2\bar{\delta}_{H_2P} \ , \tag{14}$$

where $\bar{\delta}$ represents the individual chemical shift of the species shown in the subscript and the bars above the symbols indicate resin-phase parameters. A similar equation (without the bar) can, of course, also be written for the solution-phase ^{31}P chemical shifts. If the previous hypothesis for the protonation constants is accepted (i.e. $\bar{K}_1 = K_1$ and $\bar{K}_2 = K_2$) then by using Eqs. (9) and (14) the following relationship can be derived for the interpretation of the pH dependence of the experimentally measured population-averaged chemical shift:

$$\bar{\delta}_c = \frac{\bar{\delta}_P + \bar{\delta}_{HP}\bar{K}_1[\bar{H}^+] + \bar{\delta}_{H_2P}\bar{K}_1\bar{K}_2[\bar{H}^+]^2}{1 + \bar{K}_1[\bar{H}^+] + \bar{K}_1\bar{K}_2[\bar{H}^+]^2} \ . \tag{15}$$

Knowing the measured resin-phase chemical shifts ($\bar{\delta}_m$) and the chemical shifts calculated by Eq. (15) at different pH values ($\bar{\delta}_c$) the resin-phase chemical shifts of the monovalent and bivalent phosphonic acid anions ($\bar{\delta}_{HP}$ and $\bar{\delta}_P$) can be calculated by finding the minima of the function

Table 2 Values of the parameters a and b of Eq. (13) calculated from pH data (values in *parentheses* were estimated from ^{31}P NMR shifts)

	a	b
AG 1-X2	0.994	0.598 (0.544)
AG 1-X4	1.010	0.567 (0.614)
AG 1-X8	0.999	0.524 (0.456)
QAE A25	0.995	0.614 (0.565)

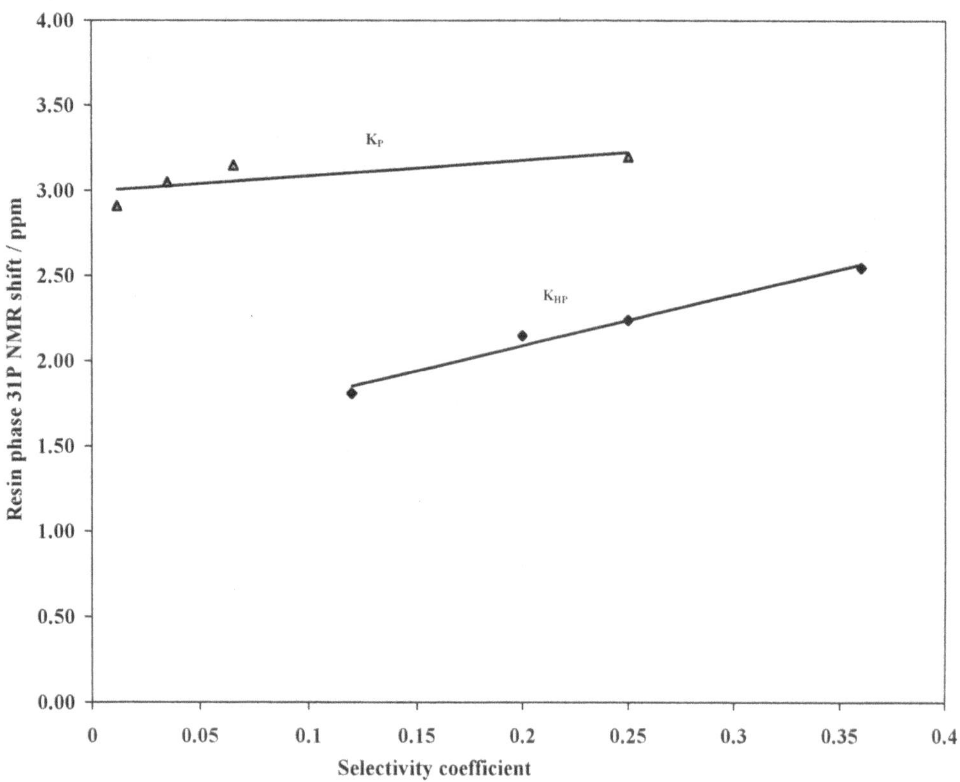

Fig. 3 Calculated resin-phase ^{31}P chemical shifts of the monovalent and bivalent phosphonic acid species as a function of their ion-exchange selectivity coefficients calculated for the resins studied (K_{HP} and K_P)

$$\sum (\bar{\delta}_m - \bar{\delta}_c)^2 \tag{16}$$

by using the least-squares approximation. The calculated chemical shift characteristics for the resin-phase environment of the anionic species studied are shown in Table 1. The correlation between the ion-exchange selectivity coefficients and the ^{31}P chemical shifts calculated for the various resins are presented for the monovalent and bivalent ions in Fig. 3. The figure indicates a linear correlation between the interactions governing the selective uptake of ions and the chemical environment controlling the ^{31}P chemical shift of these species in the polymer phase. A further support to this correlation is obtained if the value of the b coefficient is calculated from the available NMR data. For this purpose, the resin-phase hydrogen ion concentration in Eq. (15) is substituted by the expression $[\bar{H}^+] = 10^{-(pH+b)}$. In this way the b coefficient defined by Eq. (13) can be estimated from the spectroscopic data by applying the least-squares calculation procedure according to Eq. (16). Values of the b parameter calculated from pH and from ^{31}P chemical shift data are compared in Table 2. The good agreement between the two sets of data is a further indication of the existence of a strong correlation between selectivity and NMR spectroscopic data.

The observed correlation can be interpreted theoretically by considering the origin of the electronic screening of the ^{31}P nuclei and their relation to the thermodynamic parameters controlling ion-exchange selectivity.

Conclusion

The correlation between the ion-exchange selectivity coefficient and ^{31}P NMR chemical shifts has been proved and discussed for the ion-exchange distribution of monovalent and bivalent phosphonic acid anions on resins with various cross-linkages. A similar relationship may be expected to exist in systems where a selective uptake of ions takes place in phases like polyelectrolyte gels, synthetic or natural membranes or living cells.

Acknowledgements Thanks are due to Hirofumi Sakashita for obtaining NMR data at the Centre of Advanced Instrumental Analysis, Kyushu University, Japan. The JSPS grant (no. L97541) to A.M. is hereby acknowledged.

References

1. Gaines GL, Thomas HC (1953) J Phys Chem 21:714
2. Högfeldt E (1988) React Polym 7:81
3. Marton A (1997) Pure Appl Chem 69:1481
4. Marton A, Inczédy J (1995) In: Ohya H (ed) Proceedings of the International Conference on Ion Exchange. Japan Association of Ion Exchange, Takamatsu, pp 87–93
5. Komoroski RA, Mauritz KA (1978) J Am Chem Soc 100:7487
6. Zawodzinski T, Neeman M, Sillerud L O, Gottesfeld S (1991) J Phys Chem 95:6040
7. Ohuchi M, Horiuchi H, Sakai Y (1995) Kobunshi Ronbunshu 52:512
8. Busch M, Boldammer EV (1982) J Solution Chem 11:777
9. Reichenberg D, Lawrenson I J (1963) Trans Faraday Soc 59:141
10. Marton A, Elvidge J A Inczédy J (1980) J Chromatogr 201:79
11. Brindle KM, Fulton AM, Williams S (1995) In: Brown GC, Cooper CE (eds) Bioenergetics. Oxford University Press, Oxford, pp 159–187
12. Cantor CL, Schimmel PR (1980) Biophysical Chemistry. Freeman, New York, pp 481–538
13. Inczédy J (1966) Analytical applications of ion exchangers. Pergamon, New York, pp 116–130
14. Murphy J, Riley JP (1962) Anal Chim Acta 27:31
15. Kura G, Miyazaki Y, Marton A (1998) React Polym 38:197
16. Daune M (1999) Molecular biophysics. Oxford University Press, Oxford, pp 332–339
17. Popov AI, Hallenga K (1991) Modern NMR techniques and their application in chemistry. Dekker, New York, pp 485–520

Progr Colloid Polym Sci (2001) 117:159–166

L. Halász
Z. Németh
J. P. T. Horányi
A. Bóta

Structural and viscoelastic properties of lamellar systems formed from concentrated nonionic surfactant solutions

L. Halász (✉)
Technikon Pretoria, P.O. Box X 680, Pretoria 0001, South Africa

Z. Németh · J. P. T. Horányi
A. Bóta
Technical University of Budapest, 1111 Budapest, Budafokiút 8, Hungary

Abstract The structure and rheological properties of a concentrated lamellar liquid-crystalline nonionic surfactant system were investigated. The morphology of the samples was analyzed by polarization light microscopy, electron microscopy and small-angle X-ray scattering measurements. The rheological properties, the frequency-dependent storage and loss modulus, the creep compliance and the steady-state viscosity function were measured at different temperatures with systems containing surfactant in different concentrations. The rheological properties were described by the modified slip-plane theory. Discrepancies between the fitted curves of the model and the experimental curves were estimated as the result of the Frank stress.

Key words Rehology · Lamellar structure · Surfactant self-assembly

Introduction

Surfactant-based liquid crystals have a broad range of application in the cosmetic and food industries [1–6]. The most important systems are those in which bilayers are formed from the surfactants: these are the lamellar mesophase, the bicontinous cubic phase, the liposome phase and the so-called onion phase. The rheological properties of such systems were reported in some articles [7–17]. In spite of its importance, the microscopic origin of the rheological behavior is not yet well understood.

In explaining the flow curves of a three-component lamellar mesophase Bohlin and Fontell [7] modeled the lamellar structure as flexible layers of water in a liquid hydrocarbon chain environment. Flow was associated with cooperative changes of conformation of flexible layers subject to frictional forces in the liquid layers. The flow coefficients were found to be dependent on the thickness of the water layer, the bilayer interaction and the mobility in the hydrocarbon chain layer.

In investigations of the temperature dependence of flow Oswald and Allain [8] took into consideration the importance of the undulation of the bilayers and in the higher temperature range the number of the dislocation loops.

The viscoelastic behavior of a lamellar anionic surfactant, Aerosol OT–water system was investigated by Robles-Vasquez and coworkers [13, 14]. They considered the lamellar mesophase as a weak gel, but quantitative descriptions were not used for the explanation of the experimental results.

Other authors [9–11] believe in the importance of the orientation of lamellae under shear. The orientation of domains in the case of an unsheared sample is dependent on the domain–domain interaction. If stress is applied to the sample the orientation changes and the competition between the interactions and the external stress define the viscosity of the system. Penfold et al. [12] investigated the shear-induced ordering of the lamellar phase of hexaethylene glycol monohexadecyl ether in a Couette shear cell by small-angle neutron scattering. Two distinct lamellae orientations were identified. At low shear rates the lamellae are ordered parallel to the flow-vorticity plane, whereas at higher shear rates the lamellae order parallel to the flow-shear gradient plane.

Viscoelastic behaviour of liposomes was investigated intensively by Hoffmann and coworkers [15, 16] and they found a frequency-independent storage and loss modulus.

Onion phases was investigated by Panizza et al. [17]. The creep compliance curve was described well by a Burger model, consisting of a Maxwell and a Voigt element in series. In the analysis of the results of the dynamic tests the same model was used with a Laplace transformation of the creep equation. The calculated and measured results were in good agreement in the case of the storage modulus but a significant difference could be observed in the case of the loss modulus, especially in the high-frequency region.

Jones and McLeish [18] and Radiman et al. [19] used a slip-plane theory to describe the frequency dependence of the storage and loss moduli of cubic phases. The theory used the results of Doi et al. [20] and Harden and Doi [21] which were given for the ordered phases of block copolymers. For describing the rheological properties of liquid-crystalline phases formed from polymer melts Larson and Doi [22] introduced the so-called Frank stress, which was used first by Marucci and Maffettone [23] for other systems. The role of the Frank stress is to change the orientation of the domains by the shear.

In our previous work [24, 25] we investigated the rheological and structural properties of nonionic, lamellar surfactant system. We tried to describe the rheological properties of these system by an empirical equation and the original form of the slip-plane theory. In the present article we summarize the results of the rheological investigation done both in and out of the linear viscoelastic region of a nonionic lamellar system as a function of the surfactant concentration and the temperature. We use the modified version of the slip-plane theory to describe the rheological properties of the system.

Experimental

Materials

We studied a lamellar liquid-crystal system consisting of a nonionic surfactant (Synperonic A7, ICI) and water. The surfactants were used as supplied. Synperonics were made by ethoxylation of synperonic alcohol, which consists of 66% C13 and 34% C15 alkyl chains. The ethoxylation process gives rise to a wide distribution of poly(ethylene oxide) chains; therefore, only the average ethoxylation number can be declared. Though the properties of the samples should not change from batch to batch, the preparation of the lamellar systems investigated here was always done from surfactants from the same batch.

The phase diagram of the synperonic A7–water system was established by Halász and Macskási [26] and Dimitrova et al. [27] by rheological, calorimetric and NMR measurements and by polarization microscopic observations. They found normal micellar (L_1), hexagonal liquid-crystalline (H_1), lamellar liquid-crystalline (L_α) and inverse micellar (L_2) phases with increasing surfactant concentration at 23 °C. At this temperature the lamellar phase exists in a concentration range between 55 and 80% (w/w). The coexistence of the hexagonal and lamellar phases was observed at 50–55% (w/w). The boundaries of the lamellar phase in the phase diagram do not change markedly with temperature, but the lamallae melt at about 50 °C and at a higher temperature an optical isotropic liquid forms with low viscosity.

The lamellar samples were prepared by heating the aqueous mixtures to about 50–55 °C, where they were easily homogenized and, after homogenization, the mixtures were left to cool to room temperature. Then, the samples were stored for a 1-week period before the measurements. Special care was taken to prevent the evaporation of the water content of the samples during both their preparation and their storage. Mixtures were prepared with distilled water having a surface tension of 72–73 mN/m. The samples prepared in this way were checked by observing the texture of the samples in a polarization microscope before the rheological measurements.

Methods

Rheological measurements were done with a Haake RS 100 apparatus. A cone–plate sensor was used, with diameter of 20 mm, and cone angle of 4°. The sample thickness in the middle of the sensor was 0.134 mm, so in each case about 0.15 ml sample was tested rheologically. During the measurements the maximum allowed deviance in temperature was ± 0.2 °C. The samples were kept in nearly saturated water vapor during the whole time of the measurements. Because the rheological properties of such systems are dependent on the shear–deformation history, the sample was gently inserted in the top of the plate of the sensor and then the plate was slowly elevated to its measuring position always with the same velocity. The sample squeezed out from the sensor system was removed gently. Measurements were carried out after a 10-min waiting (relaxation–thermostation–saturation) period.

The dynamic (oscillation) tests were done in the viscoelastic region on all the samples. First, the linear viscoelastic region was determined by measuring the complex modulus versus the stress at a given low frequency and then 2.5 Pa was chosen as a stress amplitude, which was found to be in the linear viscoelastic region in all cases.

The viscosity was measured as a function of the shear in a steady-state way. The viscosity was measured as a function of time at each stress value and the result was accepted if its change was lower than 0.05%/s.

For the small-angle X-ray scattering (SAXS) measurements the samples were transferred into thin-walled Mark quartz capillaries (Hilgenberg, Germany) of 1-mm diameter. The capillaries were closed with a two-component synthetic resin and were transferred in metal capillary holders into an aluminum block. This block was settled directly in the beamline and used as a thermal incubator for controlled annealing at different temperatures. The block was held at the desired temperatures by a thermostat. The actual temperatures were constant to within less than 0.05 °C as measured by a thermocouple. The windows of the block were covered with Mylar foil.

The SAXS camera was a Kratky camera (Anton Paar, Graz, Austria) attached to the line-focus window of a Cu tube at a generator (Seifert, Ahrensburg, Germany) operated at 40 kV and 20 mA. A Ni filter was used to eliminate $K\beta$ radiation. The scattering data were collected with a proportional counter. The scattering data curves were normalized and desmeared for the slit geometry.

Results

Structure of the lamellar phase

In the lamellar region the samples showed double refraction between crossed polarizers. In the texture many Maltese crosses, crosses which open if the sample position is rotating and stripes are visible. While Maltese crosses are characteristic of crystals with one optical axis, the opening crosses indicate the presence of crystals with two optical axes. From this fact it is clear that the sample is heterogeneous on a microscopic scale; it contains different types of lamellar domains: probably some of the carbon chains are in a frozen state and some are probably molten. Moreover, all the samples were pseudoisotropic; after some minutes of waiting large black regions were visible between crossed polarizers. When the slide was tilted, strong double reflection occurred again. This phenomenon is due to the alignment of lamellae parallel to the plane of the slide [26]. In the case of the pseudoisotropic samples, between the black regions some Maltese crosses were also visible. It is probable that the parallelly aligned lamellae with originally two optical axes cause these crosses, but the presence of some nonoriented domains is also possible.

In the case of A7, with increasing concentration a change in the structure occurs. While at lower concentration under a polarization microscope the sample shows big crosses and small more-or-less statistically aligned stripes, at 80% (w/w) A7 concentration the texture is strongly striped, the stripes are long and are aligned parallel to one another. This change in the texture may be due to a transition which causes a stronger ordering in the sample; probably the domains order more parallelly to one another. At this concentration the sample is also pseudoisotropic.

The change in the repeat distance of a nonionic lamellar surfactant–water system during the shearing was measured by SAXS [28] and it was found that the repeat distance decreased from 4.83 to 4.68 nm when the shear deformation was increased. After a 3-h shearing the ordering of the domain structure was increased but the average number of layers in a domain remained constant.

The change in the repeat distance of the lamellae was measured by SAXS for all the samples investigated. Typical results of the scattering experiments are shown in Fig. 1. It can be seen that in all cases a marked maximum occurs owing to the strong ordering of the lamellae. From the maximum of the peaks, repeat distances were calculated and these are summarized in Table 1. As is shown, the repeat distance decreases with increasing surfactant concentration, shows no dependence on temperature and is a function of the ethoxylation number of the surfactant. The interlamellar water layer thickness decreases with the increase in the surfactant concentration.

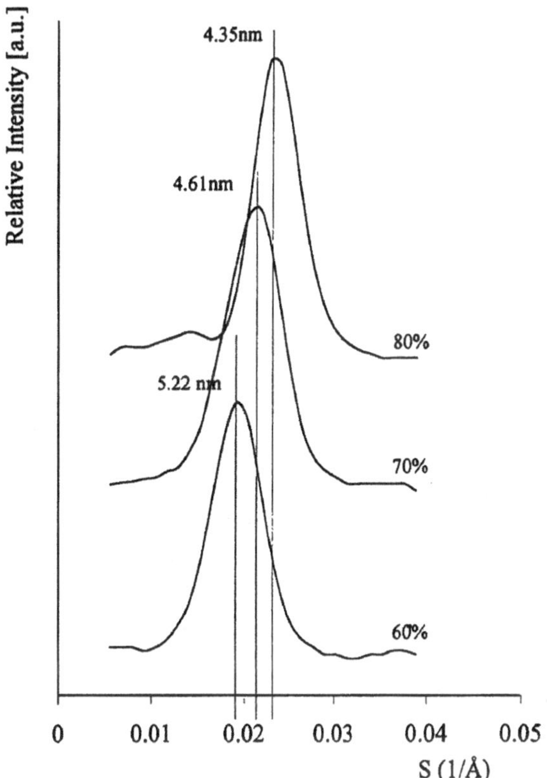

Fig. 1 Typical results of small-angle scattering measurements at 25 °C. The calculated repeat distance is also shown in the figure for samples containing A7 in different concentrations

Table 1 Repeat distance of lamellae of the samples investigated measured by small-angle X-ray scattering

Surfactant concentration (% w/w)	Temperature (°C)	Repeat distance (nm)
60	25	5.22
70	25	4.61
80	25	4.35
70	40	4.61
70	50	4.61

Rheological properties

Dynamic measurements

Typical complex modulus (G^*) versus stress curves at low frequency of the applied stress are shown in Fig. 2. G^* shows only a slight dependence on the applied stress up to about 10 Pa. This range is called the linear viscoelastic region. Here the applied stress does not cause a marked change in the sample or, in other words, we gain information on a slightly perturbed sample. Above 10 Pa, G^* decreases strongly with stress, which indicates that the structure of the sample is changed. The limit of the linear viscoelastic region is dependent on the composition and the temperature, but we found it to be higher than 2.5 Pa in all cases.

Typical results of the storage modulus, the loss modulus and the complex viscosity of the lamellar mesophase, formed from aqueous Synperonic A7, as a function of the frequency in the linear viscoelastic region can be seen in Fig. 3. The system is more elastic than

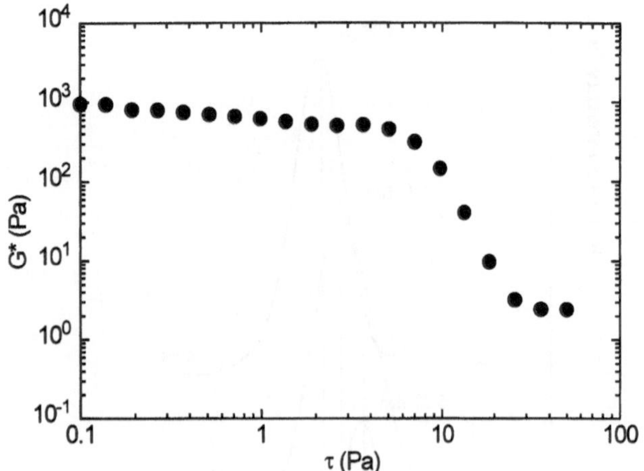

Fig. 2 Complex modulus (G^*) of a lamellar sample of 70% (w/w) A7 solution as a function of the applied stress (τ) at a constant (low) frequency (0.036 Hz) at 25 °C

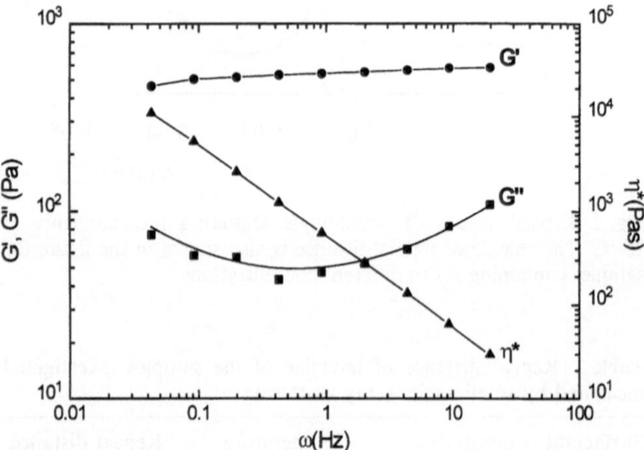

Fig. 3 Storage modulus (G'), loss modulus (G'') and complex viscosity (η^*) of the 70% (w/w) aqueous A7 lamellar system as a function of the frequency (ω) at a stress of 2.5 Pa at 25 °C

viscous in the range of frequency investigated, the storage modulus is higher by about 1 order of magnitude than the loss modulus throughout the whole frequency range. While the storage modulus has a weak dependence on the applied frequency, the loss modulus shows a minimum. The complex viscosity is strongly frequency dependent: the higher the frequency the lower the complex viscosity. The picture described here is characteristic only of the lamellar mesophase, both of the other phases (hexagonal phase and inverse micellar phase) which surround the lamellar phase in the phase diagram show completely different behavior as a function of frequency.

The effect of concentration on the storage and loss moduli was examined within the lamellar concentration region at 25 °C. The shape of the previously mentioned characteristics as a function of frequency does not change with the concentration of the surfactants. A summary of the results can be seen in Fig. 4, where we show the storage modulus, the loss modulus and their ratio (tan δ) as a function of the surfactant concentration at a given frequency. It can be seen that both the storage modulus and the loss modulus increase with increasing concentration, but the rate of their increase is different. As the system becomes more concentrated the sample behaves more and more elastically, as can be seen from the decrease in tan δ. At the highest concentration region both the storage modulus and the loss modulus begin to decrease. On investigating the structure of the sample with a polarization microscope a change in texture can be observed at 80% (w/w) concentration. While in the case of the concentration range 60–75% (w/w) Maltese crosses are seen, in the case of 80% (w/w) concentration, the texture is of the oily striped type. This change in texture indicates an orientation change of the domains, which is connected to the deviation of the trend of the storage modulus. These observations are in good agreement with the findings of Simitrova [27].

The effect of temperature was also examined for the 70 % (w/w) Synperonic A7 lamellar system. The shape of the storage modulus, the loss modulus and the complex viscosity as a function of frequency does not change with temperature until the phase boundary. Above this temperature the system shows rheological behavior characteristic of the system consisting of inverse micelles. This was found at the same temperature where the scattering experiments indicate the complete disappearance of the lamellar mesophase. In the lamellar region the quantitative results of dynamic tests are strongly temperature dependent (Fig. 5). Both the storage modulus and the loss modulus decrease with increasing temperature, but the rate of their change is different, the decrease of the elastic component being much more pronounced. This can be evaluated from tang δ, which

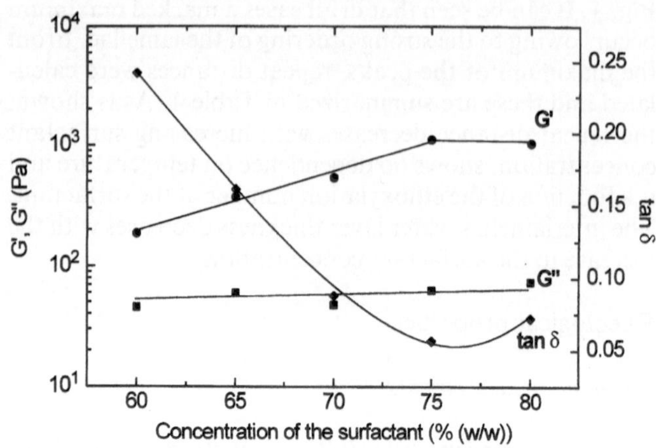

Fig. 4 The concentration dependence of G', G'' and their ratio (tan δ) at 1 Hz, measured with a stress amplitude of 2.5 Pa at 25 °C

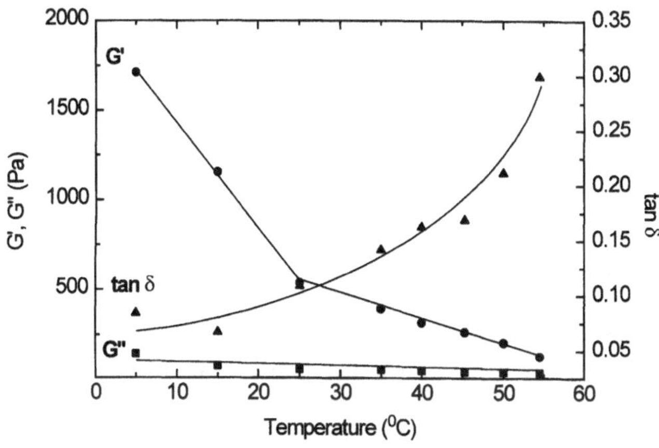

Fig. 5 The temperature dependence of G', G'' and their ratio (tan δ) at 1 Hz measured with a stress amplitude of 2.5 Pa

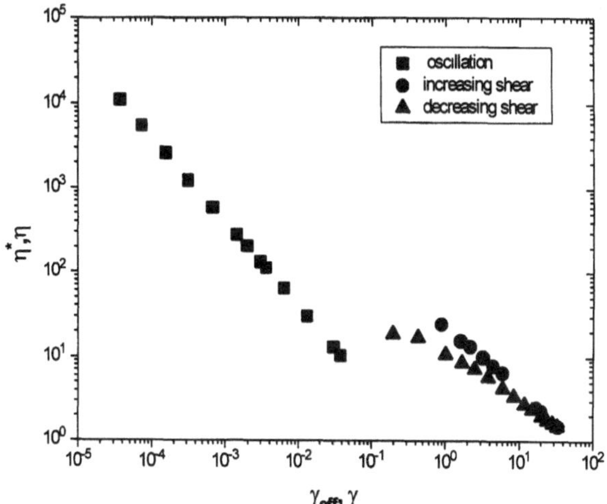

Fig. 6 Complex and shear viscosity versus frequency or effective strain of a 70% (w/w) aqueous A7 lamellar system at 25 °C

increases with temperature. The change in the storage modulus is more pronounced in the lower-temperature region (between about 5 and 25 °C). The elastic component of flow diminishes as the phase boundary is reached with increasing temperature. The change in the storage modulus can be characterized by two straight lines and the break point of the lines is at 25 °C. This value is very close to the melting point of the water-free surfactant. Assuming that the melting point of the carbon chains is not influenced strongly by the aqueous layers which are separated from these chains with the ethoxylate groups, this break point seems to be a transition temperature between the gel and the liquid-crystalline phases of the lamellar domains.

Steady-state viscosity measurements

A typical complex viscosity versus frequency curve for different surfactant concentrations is shown in Fig. 2. Both the shear viscosity and the complex viscosity are decreasing functions of the shear rate and the frequency, respectively; however, neither the Cox–Merz rule nor the modified Cox–Merz rule is followed (i.e. the complex viscosity is equal to the shear viscosity if the shear rate is the same as the frequency or the effective strain – which is equal to the strain amplitude multiplied by the frequency). Both the shear viscosity and the complex viscosity increase with an increase in the surfactant concentration in all regions of the shear rate and frequency, respectively. They decrease with an increase in temperature. as can be seen from Fig. 6 for the shear viscosity.

Discussion

The rheological properties of the lamellar mesophase formed from the nonionic Synperonic systems were investigated both dynamically (with oscillation tests) and statically (with creep tests and steady-state viscosity measurements). In the linear viscoelastic region, where the structure of the system does not change significantly during the rheological measurements, the lamellar mesophase behaves as an elastic gel; its storage modulus is higher by about 1 order of magnitude than its loss modulus.

Lamellar phases formed from other aqueous surfactant mixtures have a similar viscoelastic behavior to the system studied here, for example Robles–Vásquez and coworkers [13, 14] reported similar behavior for a lamellar mesophase of aqueous Aerosol OT; in that case the storage modulus was nearly independent of frequency, the loss modulus showed a minimum and the complex viscosity decreased with increasing frequency and did not obey the Cox–Merz rule. According to this we can conclude that this type of behavior is observed in the lamellar phases and is a characteristic of such phases.

The results of the rheological behavior in the linear viscoelastic range were evaluated by using the slip-plane theory developed by Jones and McLeish [18] for cubic phases. This theory assumes that parallelly aligned planes exist in the sample, and the shear stress acting tangentially results in a displacement between them (Fig. 7). When we apply a fixed strain to the sample, if no slipping occurs, the stress exerted at the top surface of the sample is

$$\tau = GP + \eta\dot{P} \ , \tag{1}$$

where G is the bulk modulus, η is the bulk viscosity of the liquid-crystalline phase, P is the gradient of displacement and $\dot{P}$ is its time derivative. Equation (1) is a simple Voigt–Kelvin equation. Once a slip has occurred, the strain in one layer is large but in all others is very small. The deformation in the slip plane is very large and is mainly viscous. The viscous interaction between domains

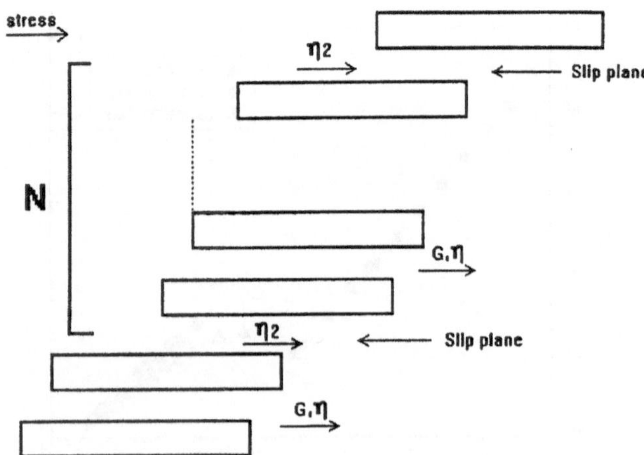

Fig. 7 Slip-plane model

(in the slip plane) is disturbed by an elastic lattice potential. The force balance in the slip plane is

$$GP + \eta \dot{P} - \frac{\eta_2}{a}(x_n - x_{n-1}) - G_2 \sin\frac{2\pi}{a}(x_n - x_{n-1}) = 0 \ , \tag{2}$$

where G_2 is an elastic modulus at the slip plane. If we neglect the effect of the lattice potential we can get for the components of the complex modulus

$$G'(\omega) = AG\frac{(\lambda\omega)^2 - (\lambda_1\omega)^2}{1 + (\lambda\omega)^2} \ , \tag{3}$$

$$G''(\omega) = AG\frac{\lambda\omega(1 + \lambda\lambda_1\omega^2)}{1 + (\lambda\omega)^2} \ , \tag{4}$$

where $\lambda_1 = \eta/G$ and $A = \frac{N\eta_2}{\eta + N\eta_2}$. $\lambda = \frac{\eta + N\eta_2}{G}$ is the relaxation time.

This linear model was extended by introducing a weak elastic force at the slip plane owing to the connectivity of the structure. The components of the complex modulus were considered as a series, and the higher terms were calculated by the perturbation method.

It was supposed that at low frequencies there is a slip–stick phenomenon, there is no slip below a critical stress value. In this region the two components of the complex modulus are

$$G' = \frac{G\varepsilon}{\gamma_0}\left[1 - \left(1 - \frac{\varepsilon}{\gamma_0}\right)^2\right]^{\frac{1}{2}} \approx \gamma_0^{-\frac{3}{2}} \ , \tag{5}$$

$$G'' = \frac{G\varepsilon}{\gamma_0}\left(1 - \frac{\varepsilon}{\gamma_0}\right) \approx \gamma_0^{-1} \ , \tag{6}$$

where ε is the critical value of the strain. Considering the shape of our modulus curves it seemed to us that the slip–stick model and the linear model may describe our curves. In Table 2 we summarize the calculated values for G' and G'' using the slip–stick model (assuming a ratio of 0.8 for ε/γ_0 and the G value was taken as the inverse of the instantaneous creep compliance.) The agreement between the calculated and measured values is rather good.

The fitted curves can describe the experimental curves rather well in the case of the storage modulus, but the agreement is not too good in the case of the loss modulus curves: the minimum values of the fitted curves are sharper than the experimentally measured ones. When we took into consideration the effect of the lattice potential by the perturbation method the results did not improve significantly and the agreement remained poor between the experimental and calculated results. As an example we show the fitted results as a function of frequency in the case of 70 wt% Synperonic A7 at 25 °C in Fig. 8.

The density of the slip planes calculated from λ with G (G is kept equal to AG, because $A \cong 1$ if λ and λ_1 are strongly different) strongly dependent on η_2, which is unknown. The two limits for η_2 are the viscosity of pure water and the calculated viscosity between two lamellae (η). With this we get a range of values for N between 10^3 and 10^7; this range seems to be too high.

Table 2 Measured and calculated moduli at 10^{-2} Hz for samples at 25 °C containing surfactant in different concentrations

Concentration (% w/w)	Storage measured (Pa)	Modulus calculated (Pa)	Loss measured (Pa)	Modulus calculated (Pa)
60	220	198	35	40
70	606	636	113	97
80	1,250	1,050	202	200

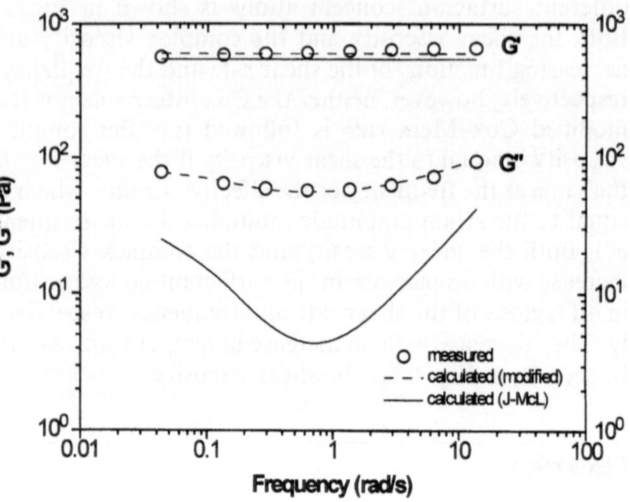

Fig. 8 G' and G'' of a 70% (w/w) aqueous A7 lamellar system as a function of the frequency at a stress of 2.5 Pa at 25 °C. Comparison of the measured and calculated values

G increases with increasing surfactant concentration and decreases with increasing temperature. It seems that G is a function of the lamellar–lamellar interactions. At a given temperature the repeat distance of the lamellae decreases and this means an increased interaction between lamellae, which causes an increase in G. At a given surfactant concentration the repeat distance is independent of the temperature, but naturally the interactions depend on temperature. In the case of lamellae formed from nonionic surfactants the interactions originate from the van der Waals forces and steric–hydration forces. The hydration repulsion is strongly temperature dependent. From these facts it seems reasonable to conclude that steric–hydration interactions play an important role in determining the interlamellae elastic modulus (G). The same conclusion (the importance of repulsion forces) was drawn by Robles-Vasquez and coworkers [13, 14].

λ shows a maximum as a function of temperature. It is dependent on the viscosity values, the elastic modulus and the rheological domain size (N). While the ratio of the viscosities against elastic modulus increases with increasing temperature, N may decrease (the enhanced temperature may increase the number of slip planes) and the superposition of these changes can cause the maximum in λ.

λ_1 (the ratio of the interlamellar viscosity and elasticity) is a monotonically decreasing function of the concentration. This is in good agreement with the the ratio of the macroscopical elasticity (G') and viscous component (G''), tang δ, which is a monotonic increasing function of the concentration.

If we use a modified form of Eqs. (3) and (4) in which a constant was added to the right side of the equations we obtain excellent agreement between the measured and the fitted values. The calculated parameters are summarized in Table 3. The data in the table show that the value of the constant is very close to the minimum value of the loss modulus.

The G^E term of the model might be explained by that part of the applied stress which is used to change the orientation of the domains (Frank stress) and which is not included in the theory of Doi, Ohta and McLeish. If we take into consideration the Frank stress, the left side of Eq. (1) consists of two terms: the Frank stress and the shear stress. Applying the common form of the Frank stress leads to

$$\tau_F = KE/a^2 \ , \qquad (7)$$

where K is a characteristic Frank elastic constant which has a typical value of 10^{-11}–10^{-10} N, E is an orientation-dependent dimensionless tensor and a is the domain size. Assuming that τ_F simply equals $\beta/G''_{min} = G''_E$ minus the Frank modulus and considering a β value of 10 as is typical for polymer systems we get a domain size of the order of 100 nm, which is a more acceptable estimation.

The steady-state flow curves show the yield value. The theory given by Doi et al. [20] predicted a yield value for block copolymer liquid crystals but their simple equation describing the steady-state viscosity versus the shear rate did not give a good approximation for our experimental results. In our system the shear rate causes some orientation and some change in the structure, which at higher shear rates should be different than in the linear viscoelastic region. The steady-state flow curves were described by a Carreau-type equation:

$$\tau - \tau_0 = \frac{\eta_a \dot{\gamma}}{1 + a(\lambda \dot{\gamma})^m} \ , \qquad (8)$$

where τ_0 is the yield value, η_a is the viscosity extrapolated to $\tau - \tau_0 = 0$, λ is a characteristic time and a and m are constants. Equation (8) can be simply derived if we assume breaking kinetics for the domain contact loss.

Equation (4) was used for the evaluation of the flow curves of the samples; the results can be seen in Table 4. η_a increases with increasing concentration and decreases with increasing temperature. The decrease in η_a with temperature is exponential and can be characterized by an Arrhenius-type equation. The activation energy was determined to be 2 J/mol for 70% concentration. The m value is independent of the A7 concentration; only the m value of the sample containing 60 wt% A7 is lower than the other values. m is a decreasing function of the

Table 3 Fitting results of the linear slip-plane theory for samples as a function of the surfactant concentration and sample temperature (G^E is the extra modulus needed for better fitting)

c (% w/w)	T (°C)	AG (Pa)	λ (s)	λ_1 (s)	G^E (Pa)
60	25	220	161	0.0317	48.2
65	25	360	85	0.0240	53.1
70	25	560	132	0.0184	46.7
75	25	1,010	122	0.0082	68.3
80	25	1,010	152	0.0103	75.0
70	5	2,040	24	0.0154	184.3
70	15	1,375	38	0.0086	89.1
70	40	320	65	0.0189	43.1
70	50	110	10	0.0346	41.0

Table 4 Fitted constants of Eq. (8) as a function of the surfactant concentration and the temperature

c (% w/w)	T (°C)	τ_o(Pa)	η_a (Pas)	b	λ (s)	m
60	25	4	19.03	8.988	0.111	0.446
65	25	14	80.38	7.277	0.810	0.633
70	25	28	144.70	3.926	1.544	0.782
75	25	33	220.4	2.109	0.227	0.751
80	25	39	250	16	0.388	0.758
80	30	35	78.6	0.392	48.967	0.667
80	40	25	6.5	2.246	321.6	0.308
80	50	8	0.6	8.905	985.9	0.022

temperature. From the experimental data it seems to us that in shear flow the lamellar system changes its orientation at low shear rate and at higher shear rates the structure of the system starts to change by changing the connectivity or the size of the domains. Equation (9) has a similar form as the temporary network equations [22].

Conclusions

The complex rheological behavior of a liquid-crystalline surfactant–water system was investigated. The storage modulus versus frequency was well described by the slip-plane theory but in the case of the loss modulus we obtained a significant deviation from the theoretical value. The modified version of the theory gave excellent agreement with the experiments. The necessity of the modification was estimated to be due to the existence of the Frank stress, which has the role to change the orientation of domains under shear. The steady-state flow curves were well described in the framework of slip-plane theory if we introduce the breaking kinetics of the domain structure and we can get a modified Carreau-type equation. The fact that the system does not obey the modified Cox–Merz rule indicates that breaking of the domains does not occur in the linear viscoelastic region.

References

1. (a) Brown GH (ed) (1975) Advances in liquid crystals, vol 1. Academic, New York; (b) Brown GH (ed) (1976) Advances in liquid crystals, vol 2. Academic, New York; (c) Brown GH (ed) (1978) Advances in liquid crystals, vol 3. Academic, New York; Brown GH (ed) (1979) Advances in liquid crystals, vol 4. Academic, New York; (e) Brown GH (ed) (1982) Advances in liquid crystals, vol 5. Academic, New York
2. Gray G, Winsor PE (1974) Liquid crystals and plastic crystals. Wiley, New York
3. Oswald A, Huang H, Huang J, Valint P (1984) US Patent 4,434,062
4. Brown GH, Walker JJ (1979) Liquid crystals and biological structures. Academic, New York
5. Krüssmann H, Bercovici R (1993) Tenside Surfactants Deterg 30:99
6. Suzuki T, Tsutsumi H, Ishida A (1982) 12th International Congress IFSCC Paris 1:117
7. Bohlin L, Fontell KJ (1978) Colloid Interface Sci 67:273
8. Oswald P, Allain MJ (1988) Colloid Interface Sci 126:45
9. Galleos C, Nieto M, Nieto C, Munoz J (1991) J Prog Colloid Polym Sci 84:236
10. Alcantara MR, Vanin JA (1995) Colloids Surf 97:151
11. Valiente M (1995) Colloids Surf 105:265
12. Penfold J, Staples E, Lodhi A, Tucker L, Tiddy GJT (1997) J Phys Chem B 101:66
13. Robles-Vasquez O, Corona-Galvan S, Sottero JFA, Puig JE (1993) J Colloid Interface Sci 160:65
14. Robles-Vasquez O, Sottero JFA, Puig JE, Monero O (1994) J Colloid Interface Sci 160:436
15. Hoffmann H, Thuing C, Schmeidel P, Munkert V (1994) Langmuir 10:3972
16. Hoffmann H, Thuing C, Schmeidel P, Munkert V, Ulbricht W (1994) Tenside Surfactants Deterg 31:389
17. Panizza P, Roux D, Vuillame V, Lu CYD, Cates ME (1996) Langmuir 12:248
18. Jones JL, McLeish TCB (1995) Langmuir 11:785
19. Radiman S, Toprakcioglu C, McLeish TCB (1994) Langmuir 10:61
20. Doi M, Harden JL, Ohta T (1993) Macromolecules 26:4935
21. Harden JL, Doi M (1996) J Rheol 40:187
22. Larson GR, Doi M (1991) J Rheol 35:539
23. Marucci M, Maffettone PL (1985) Pure Appl Chem 57:1545
24. Németh Z, Halász L, Pálinkás J, Bóta A, Horányi T (1998) Colloids Surf A 145:107
25. Németh Z, Pálinkás J, Halász L, Bóta A, Horányi T (1999) Tenside Surfactants Deterg 36:88
26. Tamamushi B (1976) Pure Appl Chem 48:441
25. Paasch S, Schambil F, Schwunger MJ (1989) Langmuir 5:1344
26. Halász L, Macskási L (1996) Plast Rubber 33:146
27. Dimitrova GT, Tadros TF, Luckham PF, Kipps MR (1996) 12:315
28 Bóta A (1999) J Hung Chem Soc 105:270

Progr Colloid Polym Sci (2001) 117: 167–171

É. Kiss
C. N. C. Lam
T. M. Duc
E. I. Vargha-Butler

Surface characterization of polylactide/polyglycolide copolymers

É. Kiss (✉)
Department of Colloid Chemistry,
L. Eötvös University, P.O. Box 32,
1518 Budapest 112, Hungary
e-mail: kissevak@ludens.elte.hu
Tel.: +36-1-2090555
Fax: +36-1-2090602

C. N. C. Lam · E. I. Vargha-Butler
Department of Mechanical and Industrial Engineering, University of Toronto,
Toronto, ON, M5S 3G8, Canada

T. M. Duc
Biophy Research S.A., Marseille, France

Abstract The wettability and the surface composition of biodegradable polymers used as drug delivery particles were determined and compared. Films of poly(lactic acid), poly(glycolic acid), and their copolymers with four different lactide/glycolide ratios of 85/15, 75/25, 65/35 and 50/50 were formed by solvent-casting on polar (silicon wafer, glass and mica) substrates. Advancing and receding water contact angles were measured goniometrically and also by using the automated axisymmetric drop shape analysis method. Both the advancing and the receding water contact angles of the copolymers were higher than was expected from the values obtained for the two homopolymers considering the increasing glycolide content of the copolymers. Surface analysis of the polyester films was performed by X-ray photoelectron spectroscopy at various takeoff angles to get information on the depth distribution of the components. Surface enrichment of the nonpolar (lactide) component in the outermost layer of the copolymer films was detected. The wettability behavior was in accordance with the surface composition results, suggesting a surface activity effect resulting in the orientation of the nonpolar segments of the copolymer chains towards the surface of the film and hence producing more hydrophobic character for the copolymer films than should be due to their bulk composition.

Key words Wettability · Biodegradable polymers · Drug carrier · X-ray photoelectron spectroscopy surface analysis · Lactide/glycolide copolymers

Introduction

Poly(lactic acid), DL-PLA, and its copolymers with poly(glycolic acid) (PGA), DL-PLGA, (with different copolymer ratios) are widely recommended [1–5] as polymeric drug delivery systems. The wettability of these biodegradable drug carriers influences pharmaceutically important processes such as drug encapsulation and release mechanisms. Furthermore, the hydrophobicity of the polymer determines the extent of the surface modification for the purpose of enhancing the biocompatibility of the microspheres and the nanospheres in pharmaceutical, and medical applications [6–8].

Considering the applications, it is important to establish whether the surface characteristics of the homopolymers and the copolymers are different from each other, because the interchange of these drug carrier polymers, owing to their similar wetting characteristics, could be beneficiary for different industries. In recent literature [9–11] only few wettability data on these copolymers can be found. In these studies, the advancing contact angles decreased with increasing PGA ratio, as expected.

The purpose of the present work was to study the wettability of DL-PLA and DL-PLGA copolymer surfaces. The other homopolymer of the series, PGA, was also studied for comparison only, since PGA is not a substance considered for the preparation of a microparticulate drug carrier owing to its nonsuitable physicochemical properties. The interpretation of the wettability results was carried out using the chemical composition data obtained by X-ray photoelectron spectroscopy (XPS) on the surface layer of the polymer films.

The outcome of this study might provide a guideline regarding the surface hydrophobicity of the lactide/glycolide-type polymers, and the data can be used to prepare drug carriers for successful and optimized drug delivery.

Experimental

Materials

DL-PLA (M_w: 106,000), PGA, and four of their random copolymers, DL-PLGA, with 85/15 and 75/25 (M_w: 90,000–126,000), as well as with 65/35 and 50/50 (M_w: 40,000–75,000) component ratios of lactide/glycolide obtained from Sigma-Aldrich Canada (Oakville, Ontario, Canada) were used for polymer film preparation. The structural compositions of the homopolymers, i.e., DL-PLA and PGA are illustrated in Fig. 1.

Dichloromethane (ACS reagent high-performance liquid chromatography grade) supplied by Sigma-Aldrich was used for sample film preparation. Double-distilled water was applied as the measuring liquid.

Silicon wafers ⟨100⟩ (Silicon Sense, Naschua, N.H., USA) with a thickness of 525 ± 50 μm, glass plates (microscopic cover glasses: 22 × 40 × 0.16 mm, Menzel-Glaser, Germany) and mica sheets (Mica Supply, UK) were used as substrates for the polymer films.

Sample preparation

The silicon wafers were cut into a rectangular shape of about 2.5 × 1.5 cm. For contact-angle measurements using the axisymmetric drop shape analysis profile (ADSA-P) a hole 1 mm in diameter was drilled in the centre of each wafer by using a diamond drill bit from Lunzer (New York, N.Y., SMS –0.027). The wafer surfaces and similarly the glass plates were then soaked in freshly prepared persulphuric acid (volume ratio of concentrated sulphuric acid: hydrogen peroxide, 30 w/v% solution of 2:1) at least for 2 h. The samples were thoroughly washed with double-distilled water and dried under an IR lamp for 15 min before the coating process. The mica plates were freshly cleaved before the polymer deposition in order to minimize carbonaceous contamination.

The polymer films (except PGA) were prepared by a solvent-casting technique on the hydrophilic substrates. The polymers were dissolved in dichloromethane (2 w/v%) and 200 μl of each polymer solution was deposited onto the relevant substrate with a pipette. To prevent fast evaporation of solvent (hence to enhance the film smoothness) the samples were covered for 20 min, allowing the solution to spread on the surface. Then, the polymer films were dried in a vacuum oven overnight. A PGA layer was formed from a polymer melt on a glass substrate at 230 °C, then the film was cooled slowly to room temperature and stored in a vacuum box until the measurements were performed.

poly(lactic acid), PLA

$-CH(CH_3)-C(=O)-O-\overset{2}{C}H(\overset{1}{C}H_3)-\overset{3}{C}(=O)-O-$

poly(glycolic acid), PGA

$-CH_2-\overset{3}{C}(=O)-O-\overset{2}{C}H_2-C(=O)-O-$

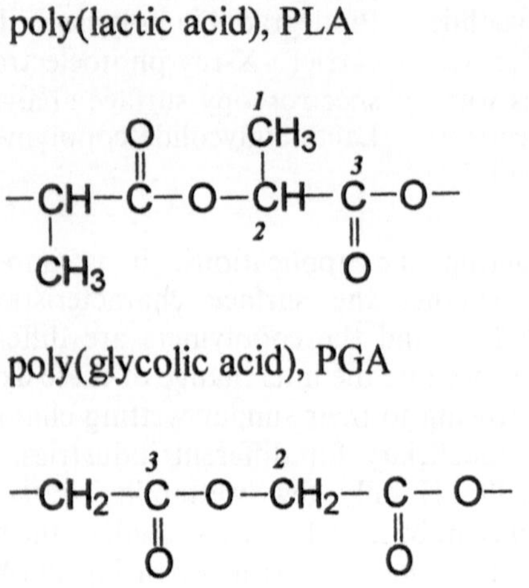

Fig. 1 Chemical composition of the homopolymers: poly(lactic acid) (*PLA*) and poly(glycolic acid) (*PGA*). The *numbers* mark the carbon atoms in different chemical environments resulting in C 1*s* electrons with various binding energies

Methods

Contact-angle measurement

Water contact angles were measured goniometrically using a Rame-Hart M2033 (Mt Lakes, N.J., USA) instrument. Water droplets on the polymer surface were formed in a closed chamber saturated with liquid vapour. The volume of the drop was increased step by step with a micropipette fixed to the measuring chamber and the static advancing angles were determined. Then, while withdrawing the liquid in a similar way, the receding angles were measured. Generally, 16–20 advancing and receding contact angles were determined at 25.0 ± 0.5 °C and were used to calculate the average values on each film.

The ADSA-P technique as an automated and more accurate dynamic method producing reliable contact-angle data was also applied to characterize the wetting properties of the polymer and copolymer surfaces. The principle of the method is the fitting of the experimental profile of a sessile drop by the Laplacian theoretical drop profile. Mathematical derivations and computational procedures can be found in previous work [12–15]. The volume of the droplet on the solid surface is increased and decreased continuously with a low rate, resulting in a contact line motion with a velocity of 0.1–1.0 mm/min for advancing and 0.3–2.0 mm/min for receding three-phase contact lines, respectively. The liquid drop is formed on the surface from below through a hole drilled into the solid sample before measurement. The shape of the drop is recorded by a microscope/charge-coupled-device camera and is processed by a computer operating at a rate of one picture in every 2 s. Individual sessile drops (4–10) were formed on each polymer and approximately 200 images were processed per drop. The measurements were performed at 23.0 ± 0.5 °C and at 45% relative humidity in a temperature-controlled room. Besides providing a large amount of accurate contact-angle data in an automated way, an additional benefit of the method is that the surface tension of the measuring liquid is monitored simultaneously during the contact-angle measurement.

The average advancing and receding contact angles were plotted with the corresponding 95% confidence limits which did not exceed 0.9° for the ADSA-P and 1.2° (with one exception: receding angle on PGA) for the goniometric method, respectively.

Surface analysis

Chemical analysis of the polymer (DL-PLA) and two selected copolymer (DL-PLGA 85/15 and DL-PLGA 50/50) films deposited on hydrophilic mica substrate was performed by XPS. Measurements were made using a SCIENTA 200 spectrometer using a monochromatized X-ray source (Al Kα at 1,486.6 eV) with a spot size of 4 mm × 1 mm. A flood gun with characteristics of 3 eV and 30% emission (0.3 mA) was necessary for charge compensation. The relative atomic composition of the surface layer of the polymer films was determined from the spectra, while the chemical structure of the surface layer was deduced using the chemical shift effect in the C 1*s* signal. In addition to the usually employed 90° takeoff angle relative to the sample surface, the measurement for each sample was repeated at a grazing 25° takeoff angle, in order to study the surface composition of the outermost (2.4 times thinner) layer of the polymer films as well, and hence to get information on the depth distribution of the components. All spectra were calibrated to the position of the aliphatic component of C 1*s* peak at 285.0 eV binding energy.

Results and discussion

Wettability of the polymer and copolymer surfaces

The mean advancing and receding contact angles of water measured on the polymer and copolymer surfaces are plotted in Fig. 2. The wettability is displayed as a function of the lactide content of the copolymers; thus 100% corresponds to the homopolymer PLA, while 0% represents the other homopolymer, PGA.

The wettability of the homopolymers indicates that PLA is rather hydrophobic ($\Theta_A \approx 80°$), but PGA seems to be considerably less hydrophobic ($\Theta_A \approx 57°$). The differences in the advancing and receding angles of water are about 20° and 30°, respectively; however, the advancing contact angles measured either by the ADSA-P technique or goniometrically on the copolymer surfaces do not change with the copolymer composition. The expected values are represented by dotted lines in Fig. 2 using the simple approximation that the hydrophobic character of the copolymers can be estimated from the properties of the homopolymers taking into account the component ratio. There is a significant deviation from this expectation; the advancing water contact angles of the copolymers do not differ from that determined for PLA. The results obtained by using organic liquids besides water also provided constant advancing angles on the copolymers [16].

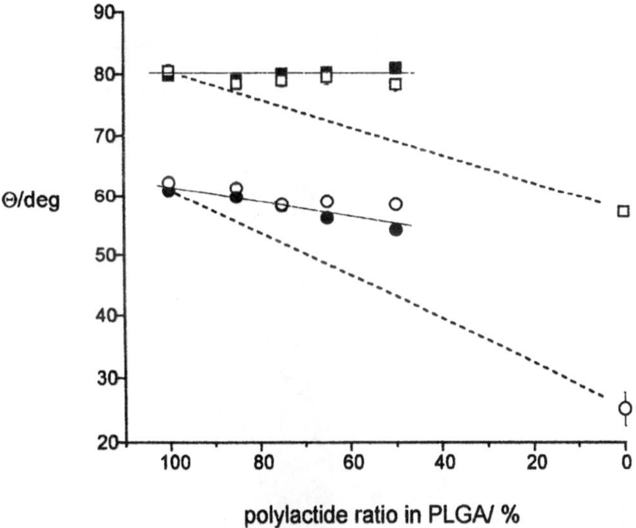

Fig. 2 Advancing (■) and receding (●) water contact angles measured with the axisymmetric drop shape analysis profile and goniometric (□, ○) methods on PLA/PGA films with various lactide contents. The *dotted lines* represent the expectable values of the contact angles considering the ratio of the components in the copolymers

Considering the receding contact angles there is also a notable deviation from the expected values. The measured angles are much higher than those predicted from the composition of the copolymers. In contrast to the advancing angles, the receding contact angles are not constant for the various copolymers but a slight but significant decrease was observed with decreasing polylactide ratio. The trend is clear in the receding angles measured by the ADSA-P technique.

From the wetting behaviour reported here we can conclude that the surfaces of PLGA copolymers do not differ or just slightly differ from that of a PLA film under the conditions of advancing contact with water.

Surface composition

The atomic compositions of the surface layer of one homopolymer (PLA) and two copolymer (PLGA with 85/15 and 50/50 component ratios) films were determined by the XPS technique. The overview spectra indicated the presence of only carbon and oxygen originating from the polymer films; no signal from the mica substrate was detected, which means that the polymers formed continuous films with a thickness greater than the analysis depth of the XPS experiment. The C/O atomic ratios characterizing the chemical composition measured at the two angles are summarized in Table 1. They differed slightly from the expected values of the stoichiometric ratio for PLA, while the C/O ratios obtained for PLGA85/15 and PLGA50/50 are in better agreement with the expected value. The XPS atomic ratios are to be considered only on a relative level and the most significant information is the surface carbon enrichment indicated by the higher C/O values at a 25° takeoff angle compared to those at 90°. This carbon enrichment increases relatively with the bulk PGA content in the polymer films.

Decomposition of the carbon signals into their different chemical components and their angular dependence provided further and more relevant information on the structure of the surface films [17–19]. Peak synthesis was performed on the C 1*s* spectra with the following chemical shifts: C–C and C–H at 285.0 eV,

Table 1 Chemical composition of the surface layer of poly(lactic acid) (*PLA*) and PLA/poly(glycolic acid) copolymer films (*PLGA*) determined by X-ray photoelectron spectroscopy (*XPS*)

Polymer	Stoichiometry C/O	Carbon components (%)		
		C–C (1)	C–O (2)	O–C=O (3)
PLA				
Theoretical	1.5	33.3	33.3	33.3
XPS at 90°	1.33	34.1	33.6	32.3
XPS at 25°	1.37	33.3	34.3	32.4
PLGA 85/15				
Theoretical	1.41	28.4	35.8	35.8
XPS at 90°	1.30	31.4	34.2	34.4
XPS at 25°	1.40	32.8	34.1	33.1
PLGA 50/50				
Theoretical	1.22	16.6	41.7	41.7
XPS at 90°	1.18	23.4	38.6	38.0
XPS at 25°	1.29	26.1	37.2	36.7

C–O at 287.0 eV, O–C=O at 289.0 eV (Fig. 3). The results of the peak synthesis as well with the numbers referring to the various carbons from Fig. 1 are collected in Table 1. From the analysis performed at a 90° takeoff angle the carbon component values found for PLA were in good agreement with the theoretical composition; however, the experimentally determined composition of the film reflected a higher C–C content at the expense of the more polar C–O and O–C=O components in the surface layer of the PLGA copolymers. This effect became more pronounced for higher PGA content (PLGA50/50). The component analysis was also performed at a 25° takeoff angle, which provided data corresponding to the chemical composition of the outermost layer owing to the sampling of a 2.4 times thinner layer than at 90°. The deviation of the XPS data obtained in this case from the theoretical values was above 15 and 50% for PLGA85/15 and PLGA50/50

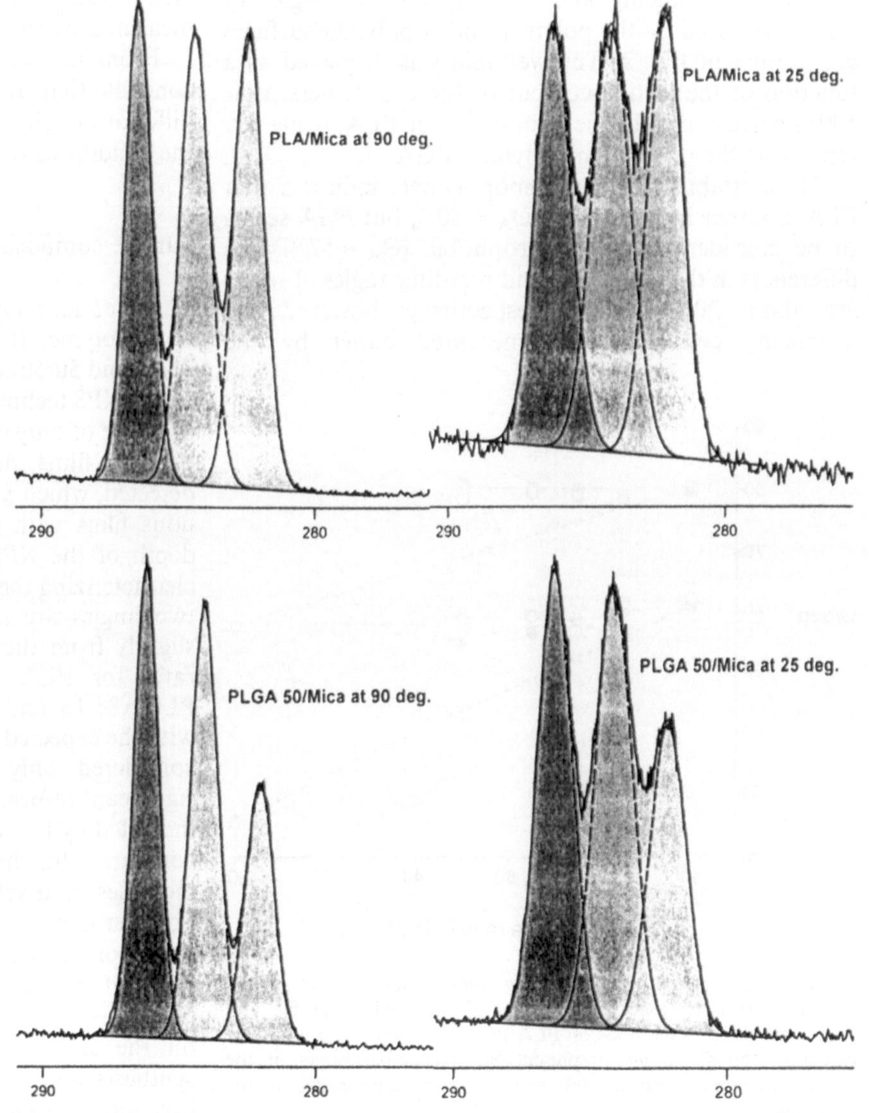

Fig. 3 Deconvoluted peaks of C 1*s* X-ray photoelectron spectroscopy signals obtained for PLA and PLGA50/50 copolymer films at 90°and 25° takeoff angles

copolymer films, respectively. This indicates a considerable alteration of the composition of the surface layer characterized by a surface enrichment of aliphatic carbon related to that of the "bulk" phase of the copolymer films.

Both the wettability and the surface composition studies revealed that DL-PLGA copolymer films represent a more hydrophobic surface than would be expected according to their bulk composition. A reasonable explanation for that might be the surface orientation of the nonpolar groups of the copolymers allowed by a certain segment mobility at room temperature. The effect of that kind of surface activity was proved to increase with higher glycolide content. The results show that DL-PLGA copolymer surfaces up to 50/50 component ratio are of similar hydrophobicity to the PLA homopolymer under dry conditions, when the polymer surface first contacts with water (advancing contact angles).

It is worth noting, however, that the receding water contact angles determined by the ADSA–P technique decreased with increasing glycolide (polar) component of the copolymers. Receding angles are formed in ADSA-P measurements on a solid surface in a controlled way following a well-defined contact with water, leading to hydrated state of the surface. During the contact time, the surface composition of the copolymers can change owing to the presence of a polar liquid at the interface if the segments have the freedom to orient to an energetically preferable position [20]. That change in the surface distribution or the orientation of the polymer segments was indicated by the lower receding water contact angles, but those values were still higher than the values expected from the bulk composition.

The behaviour of the copolymer surfaces, namely the ability to adapt to some extent the surface composition to the properties of the neighbouring phase further supports the hypothesis that the surface polarity of DL-PLGA) copolymer films is dominated by the nonpolar groups oriented at the solid/air interface.

Acknowledgements We give our special thanks to A.W. Neumann for using his facilities. This research was supported by the Ministry of Culture and Education (Hungary) through Research and Development Project FKFP 0156/1997, by the Natural Sciences and Engineering Research Council of Canada through grant no. OGPOO37393 and by a grant from Merck Frosst Centre for Therapeutic Research (Canada). The authors express their appreciation to Z. Policova, J.Y. Lu and L. Zhang (Department of Mechanical and Industrial Engineering, University of Toronto, Canada) and to Mrs. Hórvölgyi (Eötvös University, Budapest) for their valuable help in the measurements and technical evaluation.

References

1. Lewis DH (1990) In: Chasin M, Langer R (eds) Biodegradable Polymers as drug delivery systems. Dekker, New York, pp 1–42
2. DeLuca PP, Mehta RC, Hausberger AG, Thanoo BC (1993) In: El-Nokaly MA, Piatt DM, Charpentier BA (eds) Polymeric delivery systems. American Chemical Society Symposium Series 520. American Chemical Society, Washington, DC, pp 53–79
3. Park TG (1995) Biomaterials 16:1123
4. Brannon-Peppas L (1997) Med Plast Biomater Arch 1:11
5. Fu K, Pack DW, Klibanov AM, Langer R (2000) Pharm Res 17:100
6. Allemann E, Rousseau J, Brasseur N, Kudrevich SV, Lewis K, van Lier JE (1996) Int J Cancer 66:821
7. Coombes AGA, Scholes PD, Davies MC, Illum L, Davis SS (1994) Biomaterials 15:673
8. Stolnik S, Dunn SE, Garnett MC, Davies MC, Coombes AGA, Taylor DC, Irving MP, Purkiss SC, Tadros TF, Davis SS, Illum L (1994) Pharm Res 11:1800
9. Lu Z, Bei J, Wang S (1999) J Controlled Release 61:107
10. Norris DA, Puri N, Labib ME, Sinko PJ (1999) J Controlled Release 59:173
11. Lück M, Pistel K-F, Li Y-X, Blunk T, Müller RH, Kissel KJ (1998) J Controlled Release 55:107
12. Rotenberg Y, Boruvka L, Neumann, AW (1983) J Colloid Interface Sci 93:169
13. Cheng P (1990) PhD thesis. University of Toronto
14. Cheng P, Li D, Boruvka L, Rotenberg Y, Neumann AW (1990) Colloids Surf 43:151
15. Ortega JM, Rheinboldt WC (1970) Iterative solution of nonlinear equations in several variables. Academic press, New York
16. Vargha-Butler EI, Kiss É, Lam CNC, Keresztes Z, Kálmán E, Zhang L, Neumann AW (2001) Colloid Polym Sci (in press)
17. Beamson G, Briggs D (1992) High resolution XPS of organic polymers. The Scienta ESCA300 Database. Wiley, Chichester, UK
18. Duc TM (1995) Surf Rev Lett 2:833
19. Soletti JM, Botreau M, Sommer F, Brunat WL, Kasas S, Duc TM, Celio MR (1996) Langmuir 12:5379
20. Andrade JD (ed) (1988)Polymer surface dynamics. Plenum press; New York

Progr Colloid Polym Sci (2001) 117:172–181

H.-G. Kilian
M. Koepf
V. I. Vettegren

Model of reversible aggregation: universal features of fluctuating ensembles

H.-G. Kilian (✉) · M. Koepf
Experimentelle Physik, University of Ulm, Albert-Einstein-Allee 11, 89069 Ulm, Germany

V. I. Vettegren
AF Joffe Physico-Technical Institute, Russian Academy of Science, Polytechnichaja ul. 26, St. Petersburg, 194021 Russia

Abstract The model of reversible aggregation predicts universal reduced ensemble number or energy distributions described by Γ-functions of different classes, typified by the value of a single parameter, p. The universality of the model is evidenced by the description of ensembles of carbon black and liquids, of defects on the surface of loaded metals and of cells of bacteria and yeast.

Key words Aggregation · Ensemble structure · Fluctuation · Universality · Glasses

Introduction

Reversible aggregation of atoms, molecules or molecular units ("particles") [1–3] is typified by weak particle-to-particle contact energies in the range of the mean thermal energy. The configuration of aggregate ensembles and the intraaggregate states fluctuate. Such conditions are met in very different systems. The aggregate ensemble of carbon black [4–6] produced in a gas stream at very high temperatures (around 1800 K) within extremely short times (microseconds) is depicted in Fig. 1 [7]. In liquids, aggregates "collapse" forming densely packed patterns as it is evidenced by the surface structure of glassy poly(methyl methacrylate) [3] and of a monolayer of bottle-brush molecules [3, 8] depicted in Fig. 2. Since bottle-brush molecules have a diameter of about 6 nm the chains can be seen. The anisotropic intraaggregate patterns show that folded loops or chain ends are concentrated in the boundaries of the aggregates.

The model of reversible aggregation [3] allows us to describe the stationary ensemble properties and to discover universal features as demonstrated later for carbon black, liquids, defect ensembles at the surface of loaded metals and for bacteria and yeast.

The model of reversible aggregation

The key assumption [3] is that "units" form metastable aggregates with a liquidlike anisotropic internal structure. Equivalent ensemble configurations are generated consecutively. Described analogously to many simultaneously and reversibly running linear chemical reactions [3, 9] the thermodynamic interpretation of the reaction constant reveals that aggregation should mainly be controlled by the standard aggregation energy, Δu_p. For all the different systems studied until now Δu_p is constantly assigned to a positive value identifying aggregates as dynamic clusters. The energy of internal modes of motion overpowers the negative "contact energy" per unit. This is illustrated in Fig. 3.

This aggregation energy of an aggregate with y contacts may then be written as [3]

$$\Delta u_{py} = y\Delta u_p \quad . \tag{1}$$

Neglecting "extra contacts" as in branched configurations, the number of contacts, y, is $y = y_a - 1$ (y_a: number of units). The number of aggregates with y contacts, $n'_p(y)$, is then written as

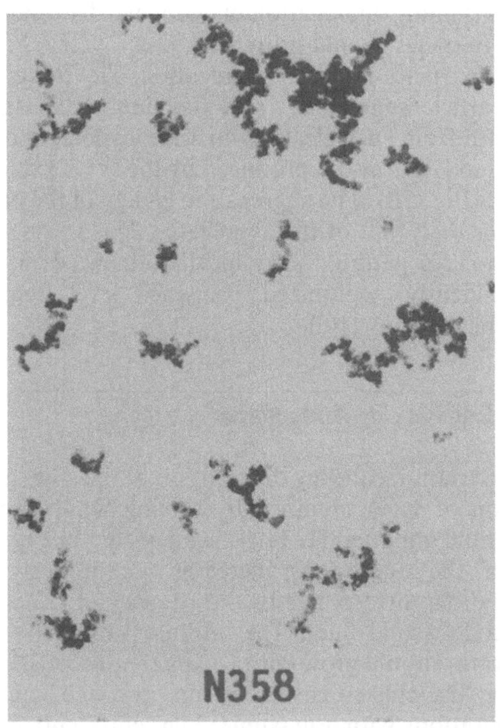

340 nm

Fig. 1 Transmission electron microscopic picture of carbon black N358 according to Ref. [7]

$$n_p(y) = n_0 y^p \exp(-\beta y \Delta u_p),$$
$$\beta = \frac{1}{k_B T}, \tag{2}$$

where n_0 is the normalization factor, k_B is the Boltzmann constant and T is the absolute temperature. The value of the exponent p of the front factor of Eq. (2) defines different "classes of aggregate ensembles". For isotropic aggregates p is zero. An ensemble of randomly oriented aggregates the anisotropy of which grows with y is characterized by $p=2$ owing to the additional orientation ensemble entropy. Δu_p may depend on the value of p.

The size distribution of aggregates ("mass distribution") is then given by

$$h_p(y) = y n_p(y) = n_0 y^{p+1} \exp(-\beta y \Delta u_p) \ , \tag{3}$$

so the energy distribution is equal to

$$h_{pu}(y) = n_0 y^{p+1} \Delta u_0 \exp(-\beta y \Delta u_p) \tag{4}$$

With $\Delta u_p > 0$ the mean size of the aggregates, $\langle y(T) \rangle$, should grow proportionally to the absolute temperature [3].

From Eqs. (2) and (4) the reduced number distribution, n_{pr}, and the conjugated energy distribution, h_{pur}, are deduced to be given by [3]

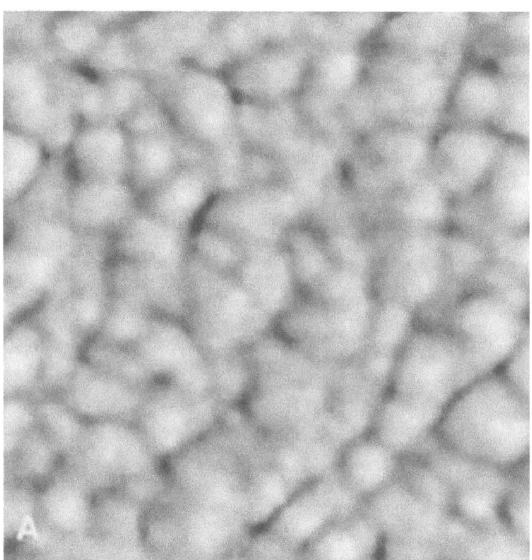

0 300 nm

0 312 nm

Fig. 2 **A** Pulsed force scanning microscopic picture of poly(methyl methacrylate) (*PMMA*) in the glassy state wetted with alcohol [3]. **B** Tapping-mode image of a monolayer of bottle-brush molecules depleted on mica [3]

$$n_{pr} = \frac{n_p(\eta_{py})}{n_{p\eta}} = \eta_{py}^p \exp(-\eta_{py}); \quad n_{p\eta} = C\frac{(k_B T)^p}{\Delta u_p^p},$$
$$h_{pur} = \frac{h_{pu}(\eta_{py})}{h_{p\eta}} = \eta_{py}^{p+1} \exp(-\eta_{py}); \quad h_{p\eta} = C\frac{(k_B T)^{p+1}}{\Delta u_p^p},$$
$$\eta_{py} = \beta y \Delta u_p \ . \tag{6}$$

Quasistationary aggregate ensembles of the same class (parameter p) are predicted to show the same reduced distribution.

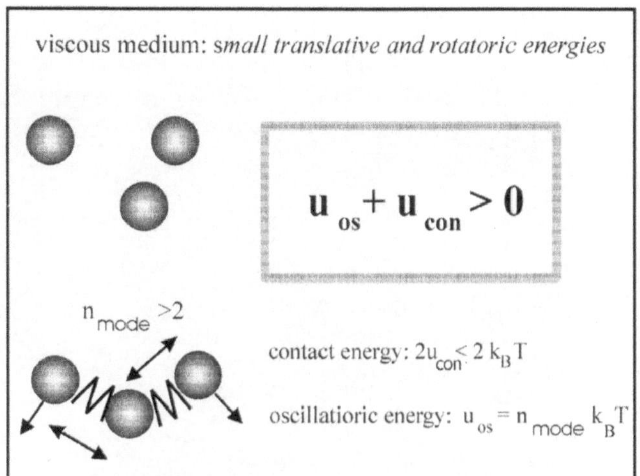

Fig. 3 The process of aggregation. The internal energy of all the modes of oscillation per unit, u_{os}, should exceed the weak contact energy, u_{con}

Experiments

The reduced frequency distribution of aggregates deduced from the images in Fig. 2 is shown in Fig. 4. The solid line is computed with the ($p = 1$) version of Eq. (3) ("two-dimensional!" layer). For each class this leads to the general conclusion that the mean aggregate size at the glass temperature should be identical [3]. The anisotropic intraaggregate structure growing with y should be fixed during the life time of an aggregate.

The frequency of particles of different types of carbon black [7] is plotted against the diameter in Fig. 5A. The ($p = 2$) version of Eq. (3) reproduces the data fairly well. The particle ensemble is optimized irrespective of the always running aggregation phenomena. The anisotropy of the particles should grow with y.

The particles undergo aggregation. The frequency of the "particle aggregates" as a function of the diameter (deduced from such electron microscopy images by Hess and Herd [7]) is also reproduced by the ($p = 2$) version of Eq. (3) (Fig. 5B). The aggregation energy of the particles is larger than that of the aggregates. This is a necessary condition for getting hierarchically organized ensembles independently "optimized" on each level (maximum ensemble entropy [10]).

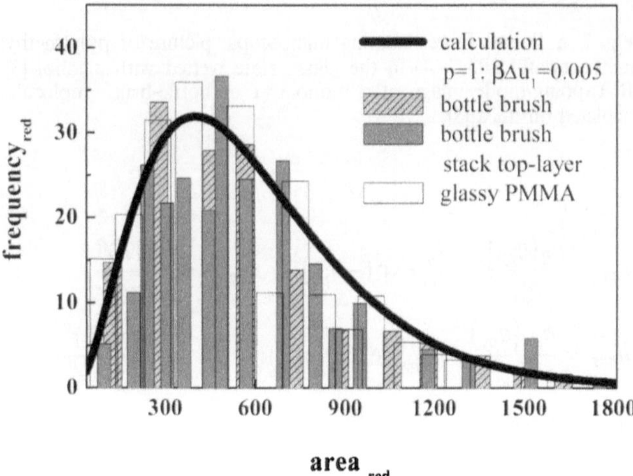

Fig. 4 Reduced frequency distribution of aggregates at the glassy surface of PMMA and for a frozen monolayer of bottle-brush molecules as a function of the reduced area

Quasistationary dynamic states

The invariant frequency distribution of the aggregates as function of $y/\langle y\rangle$ obtained by solving Smoluchovsky's differential equation [1, 11, 12] is depicted in Fig. 6 (the size of the aggregates increases continuously). The distribution corresponds to a stationary ($p = 2$) class of our model (solid line). The internal anisotropy of the aggregates should grow with y. Apparently, in the course of time the cluster ensemble runs through equivalent configurations with a maximized ensemble entropy. The entropy production is minimized [13]. This explains straightforwardly why, under the given conditions, carbon black shows the same reduced size distribution for particles and aggregates.

An important test

For $\Delta u_p > 0$ the size of the aggregates should grow proportionally to T (Eq. 5). This can be scrutinized by comparing the distribution of carbon black generated at about 1600 K with that after special treatment of cellulose acetate butyrate (CAB) chip dispersions at about 450 K [14]. "Process-induced activation" should be accounted for by introducing the "effective temperature" $T^* > T$.

Herd et al. [14] deduced the volume fraction, $V'(y_a)/V(y_a)$, of aggregates as a function of the "relative free volume" fraction, $v'(y_a)/v(y_a)$. $v'(y_a)/v(y_a)$ is described by [9]

$$\frac{v'(y_a)}{v} = \frac{v_{eq}(y_a)}{v} = \chi y_a^{2\kappa-1} \quad . \tag{7}$$

$V'(y_a)/V(y_a)$ is then written as

$$\frac{V'(y_a)}{V} = C_{V'} y_a^{3\kappa-1} y_a^2 \exp(-\beta y \Delta u_p) , \quad C_{V'} = \chi C_0 \quad . \tag{8}$$

The constants $C_{V'}$ and C_0 are linked by the parameter χ (Eq. 7). After the CAB treatment, the size of the aggregates is reduced as shown in Fig. 7, where the distribution $V(y_a)'/V(y_a)$ of a "dry ensemble" is com-

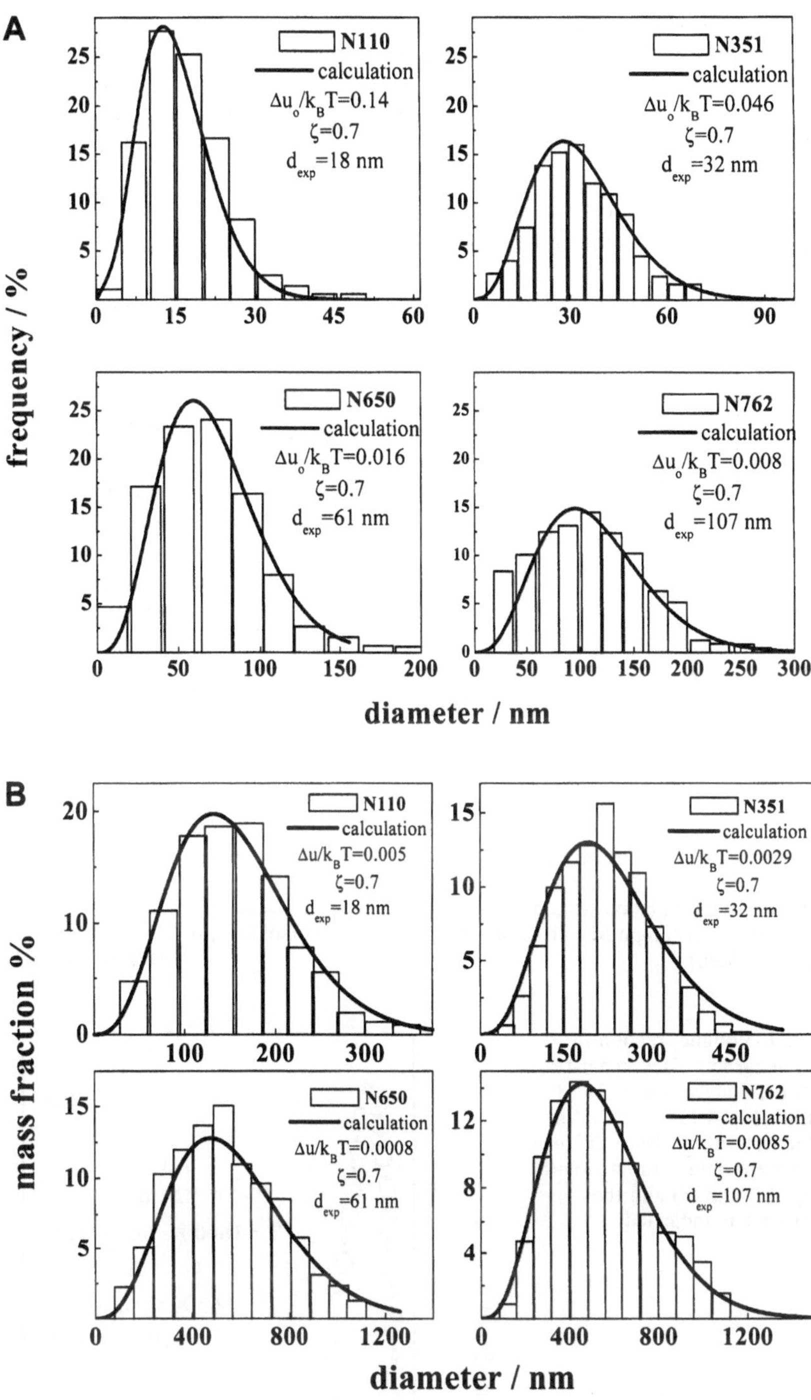

Fig. 5 **A** Frequency distribution of particles of different mean diameters (d_{exp}) of particles of carbon black against the diameter deduced from electron microscope pictures [7]. The *solid lines* represent data computed with the ($p=2$) version of Eq. (2). d_{ya} is defined by $d_{ya}=2y_a{}^{\zeta}l_o$ (l_o set equal to 1 nm). The parameter ζ is 1/3 for densely packed particles in a cubic lattice. For loosely arranged aggregates ζ increases. Parameters are indicated. **B** Mass fraction of aggregates plotted against the diameter deduced from electron microscope pictures [7]. The *solid lines* represent data computed with the ($p=2$) version of Eq. (3).

pared with the CAB pattern. The data can be fitted with Eqs. (7) and (8) by only replacing $T^*=1600$ K by $T^*=563$ K in full accordance with Eq. (5).

Defect ensemble of the surface of loaded metals

Scanning tunnelling microscopy reveals a fluctuating ensemble of submicroscopic defects on the surface of loaded metals (copper, gold, molybdenum and palladium) [15–18]. As a representative example, the pattern of copper under constant stress ($\sigma=350$ MPa) after 16 h is depicted in Fig. 8.

In loaded metals, dislocations can move at least within conservative sliding planes which are segregated at the surface. Fluctuating defect ensembles are generated (Fig. 8A), characterized by a mean life time of the order of 20 h. The data here are fitted with the ($p=1$) version of Eq. (2). The reduced size distributions of loaded metals (copper, gold, molybdenum and palladium) fall on a master curve (Fig. 9).

Since defects with the same shape are oriented (Fig. 8A, B) the $(p=1)$ class representation of Eq. (2) demands that defects should exhibit different randomly generated metastable internal structures. The ensemble entropy increases, and p becomes equal to 1.

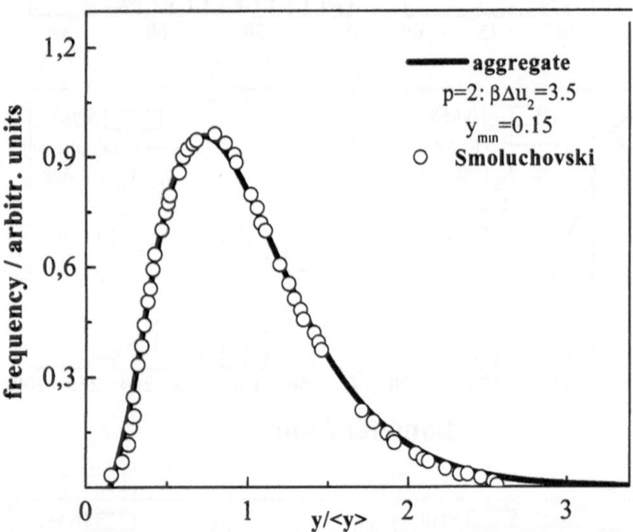

Fig. 6 Reduced number distribution of aggregates of different size as a function of $y/\langle y\rangle$ obtained by a numerical solution of the Smoluchovski equation according to Refs. [11, 12]. The *solid line* represents data computed with the use of the $(p=2)$ version of Eq. (6) $(\langle y\rangle \approx 1/\beta\Delta u_1)$

Bacteria and yeast ensembles

Images of *Escherichia coli* and of yeast ensembles [19] are shown in Fig. 10. Embedded in a culture medium the cells grow, whereby the internal properties and functions of each cell are at each moment entirely determined by the proteins made according to the genetic instructions stored in the DNA molecules [19]. The cytoplasm of the cells seems to be gelatinized since the constituents are linked to each other.

A key point is that the cells reproduce quickly by dividing in two [19]. Under optimum conditions, when food is plentiful, a bacterium can duplicate itself in as little as 20 min, depending, of course, on environmental conditions. If growth and decomposition rates are identical a stationary ensemble with a more-or-less broad cell size distribution should be constituted. The size should now be characterized by the number of "units". A unit is possibly represented by the chain-standing molecular group with a side-standing amino acid and the environment each unit is, on average, surrounded by. Yet, at the moment it is not necessary to identify these units. When a new unit is generated within the cell the energy should increase by the standard energy Δu_p, which is comparably large since it includes covalent peptide bonds. The state of reference is the "nonrealizable system" without contacts between the units. Δu_p should be positive, i.e. the energy of "internal modes of motion" is supposed to overpower the negative contact energy. Δu_p is likely to depend on the properties of the culture medium.

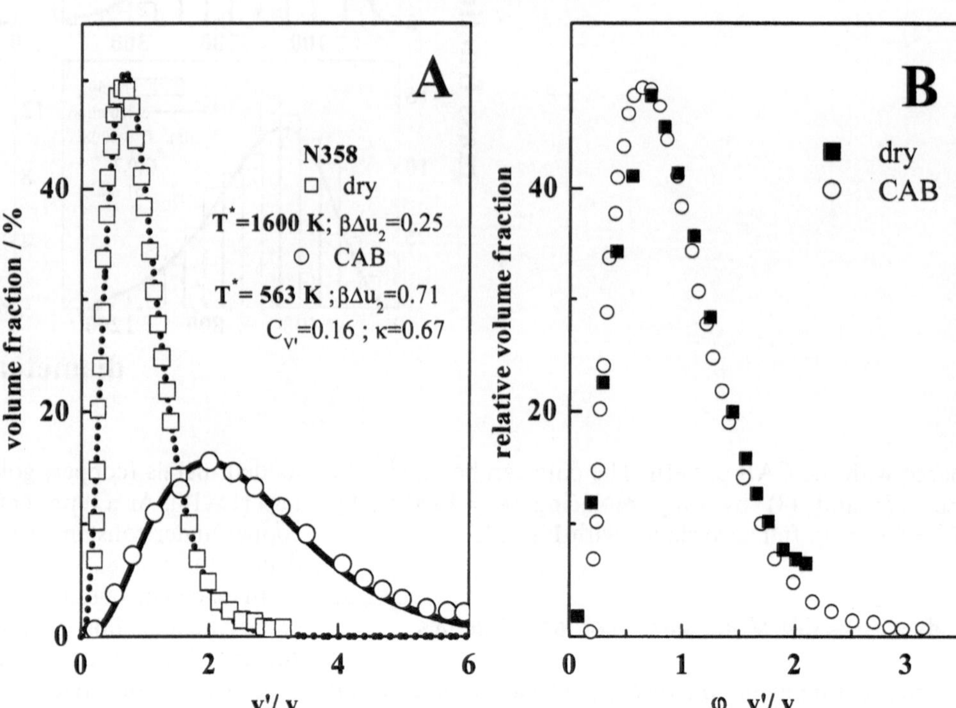

Fig. 7 Volume fraction of N358 carbon black against $v'(y)/v(y)$ according to Herd et al. [14]: **A** □ dry, ○ CAB, **B** master curve. The *solid lines* represent data computed with Eqs. (7) and (8) and the parameters indicated

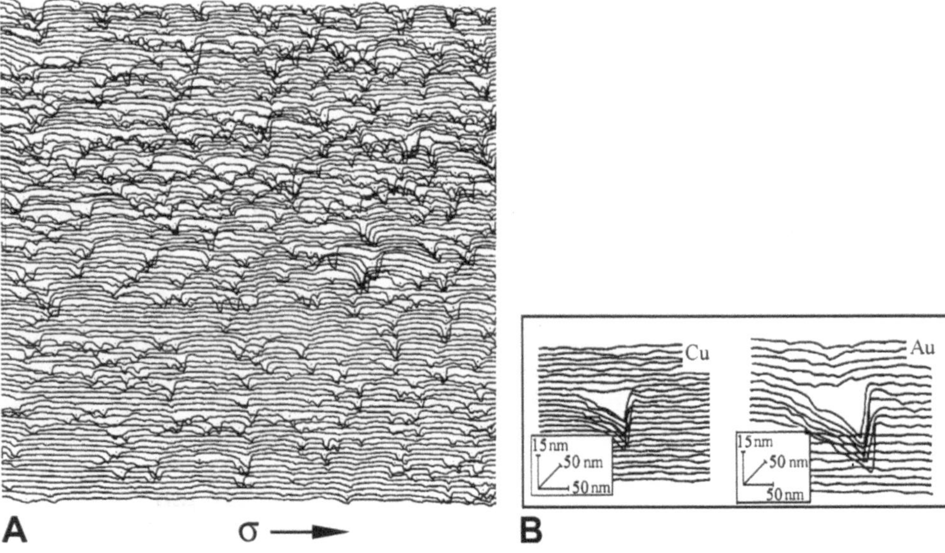

Fig. 8 A Tunnelling microscopic topogram of the defect ensemble of copper under a stress of $\sigma = 350$ MPa at $t = 16$ h according to Kilian et al. [18]. **B** Single defects at the surface of loaded copper and gold crystals [18]

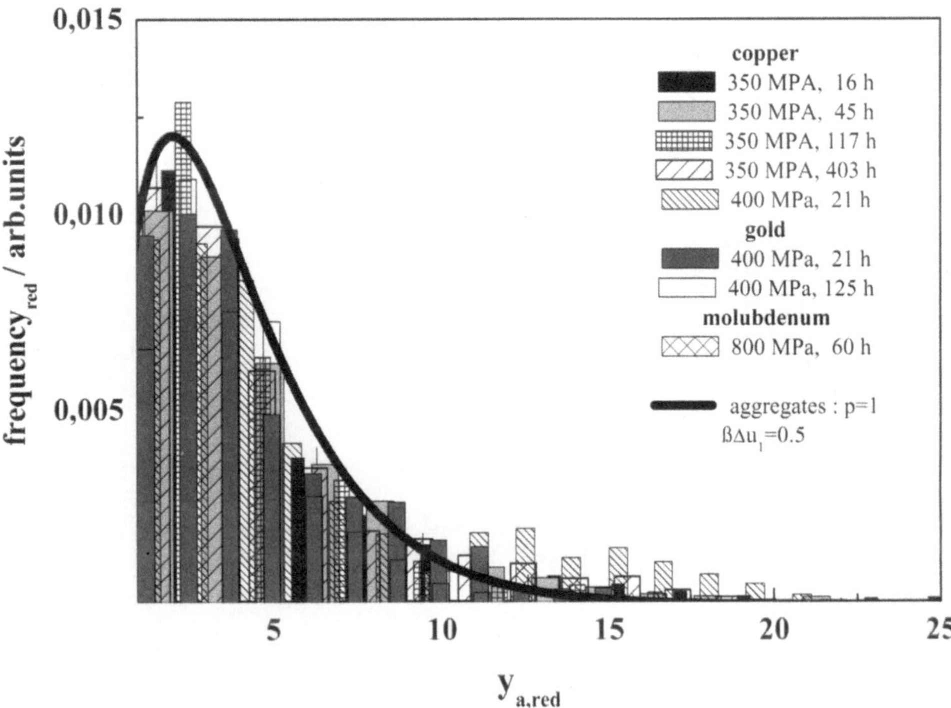

Fig. 9 Master curve of the number distributions of the defects at the surface of crystals of loaded copper, gold and molybdenum as function of the reduced depths, $y_{a,\text{red}}$ [15, 16, 17, 18]. The *solid line* represents data computed with the $(p = 1)$ version of Eq. (2)

Living cell ensembles show a "birth volume" (y_b), i.e. a smallest cell. The number of "contacts" between units in a cell set equal to $y = y_a - 1$ the quasistationary number distribution, $n_p(y)$, should be given by the modified version of Eq. (2)

$$n_p(y) = n_0(y - y_b)^p \exp[-\beta(y - y_b)\Delta u_p] \ , \tag{9}$$

where $1/\beta$ represents the mean energy stored in enzymatically excited intermediate states in the cell cytoplasm; it exceeds the standard energy Δu_p, so fluctuations can easily be activated. This distribution is a "maximum entropy pattern" [3]. The parameter p should not depend on environmental conditions. Rewriting Eq. (9) in terms of the reduced variable $\eta_{y\mathbf{b}}$ we are led to the universal relation

$$\begin{aligned} \eta_{\text{red}} &= \frac{\eta_{y\mathbf{b}}}{C} = \eta_{y\mathbf{s}}^p \exp(-\eta_{y\mathbf{b}}), \\ \eta_{y\mathbf{b}} &= \beta(y - y_b)\Delta u_p; \quad C = (\beta\Delta u_p)^{-p} \ . \end{aligned} \tag{10}$$

The proof of our conception is thus easily done by verifying that the reduced quasistatic number distribu-

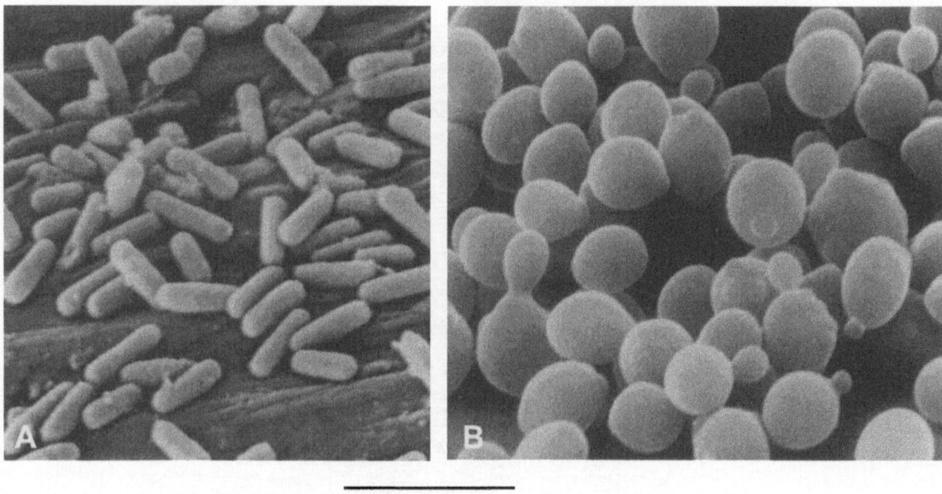

Fig. 10 A *Escherichia coli* according to Ref. [19]. **B** Yeast cells according to Ref. [19]

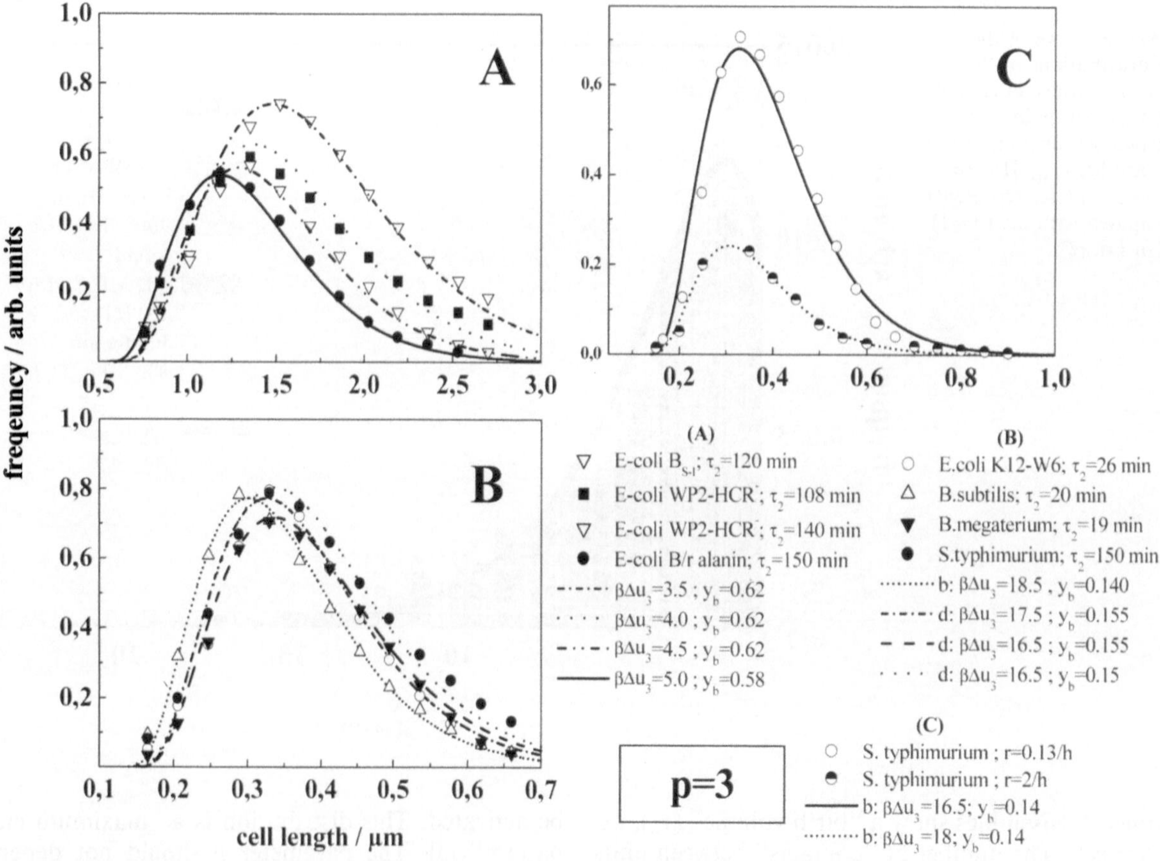

Fig. 11A–C Frequency distributions of the bacteria at different doubling times, τ_2 [20] and different growth rates, r [21]. The *lines* represent data computed with the $(p=3)$ version with the parameters in the micrometre representation of Eq. (2)

tions of bacteria or yeast cells of different sizes, typified by p, are universal.

The cell size distribution

We present in Fig. 11 the frequency of bacteria as a function of the length at different mean duplication times, τ_2 [20] i.e. under different growth rates, r [21] $(r=\ln 2/\tau_2)$. The data are fairly well reproduced with the $(p=3)$ version of Eq. (9). According to results not

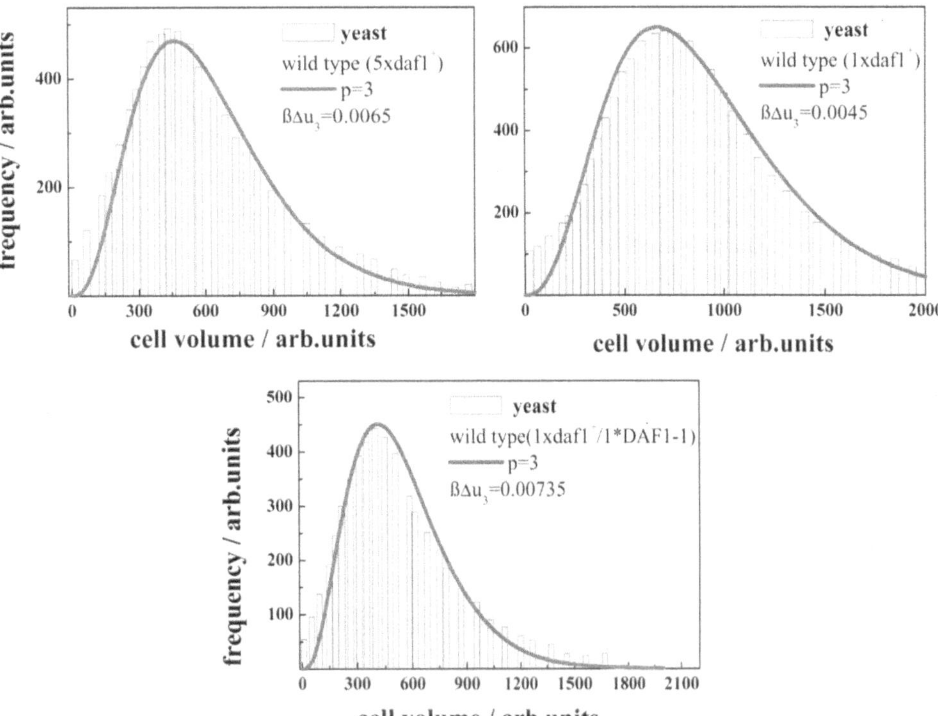

Fig. 12 Frequency distributions of different mutants of yeast cells as a function of the cell volume according to Ref. [22]. The *solid lines* represent data computed with the ($p=3$) version of Eq. (2) with the parameters in the nanometre representation as indicated

reported here bacteria establish quasistationary ensembles of this class even when irreversible processes run off.

Ensembles of mutants of yeast show analogous volume distributions of the ($p=3$) type [22] (Fig. 12). The data represented by the solid lines were computed with Eq. (9). Δu_3 changes slightly with a modification of the genetic code, i.e. the conformation within the cells of yeast depends on the individual structure of the various proteins. The $\beta\Delta u_3$ values cover the usual range.

The internal structure of bacteria is likely to be anisotropic showing an uniaxial fibre symmetry [19]. If the bacteria are randomly oriented and if the internal structure is fixed p should be equal to 2. Yet, according to the energy landscape of a protein molecule depicted in Fig. 13 [23] different "quasi-one-dimensional metastable configurations" of the intracellular protein complex can be generated thermally. The ensemble entropy increases and p goes to 3.

It was not expected that the frequency of the cell volume of yeast cells also belongs to the ($p=3$) class. The anisotropic properties of the cell should in both cases exhibit a fibre texture, the configuration of which should fluctuate topologically analogously to the behaviour of bacteria.

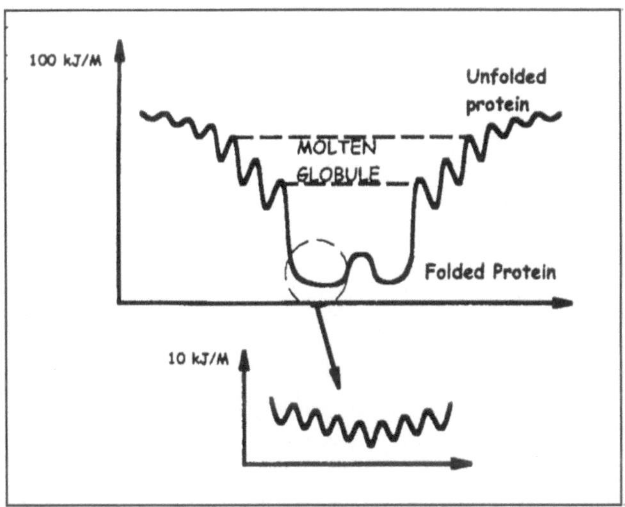

Fig. 13 Sketch of the energy landscape of a folded protein according to Fraunfelder et al. [23], which is typified by a pocket with a set of low-energy potential walls. They characterize the many metastable configurations (or conformations?) the folded protein with an invariant global structure is able to visit in the course of time (thermally activated)

The length distribution of the proteins

The proteins constitute most of the dry mass of a cell [19]. The prospective number of proteins of different lengths has been determined for bacteria and yeast cells by decoding the genome. The number of proteins of three different bacteria and for yeast (*Saccharomyces cerevisia*) is plotted against their length (i.e. the number of amino acids per strand) [19, 24] in Fig. 14. For both systems the data represented by the solid

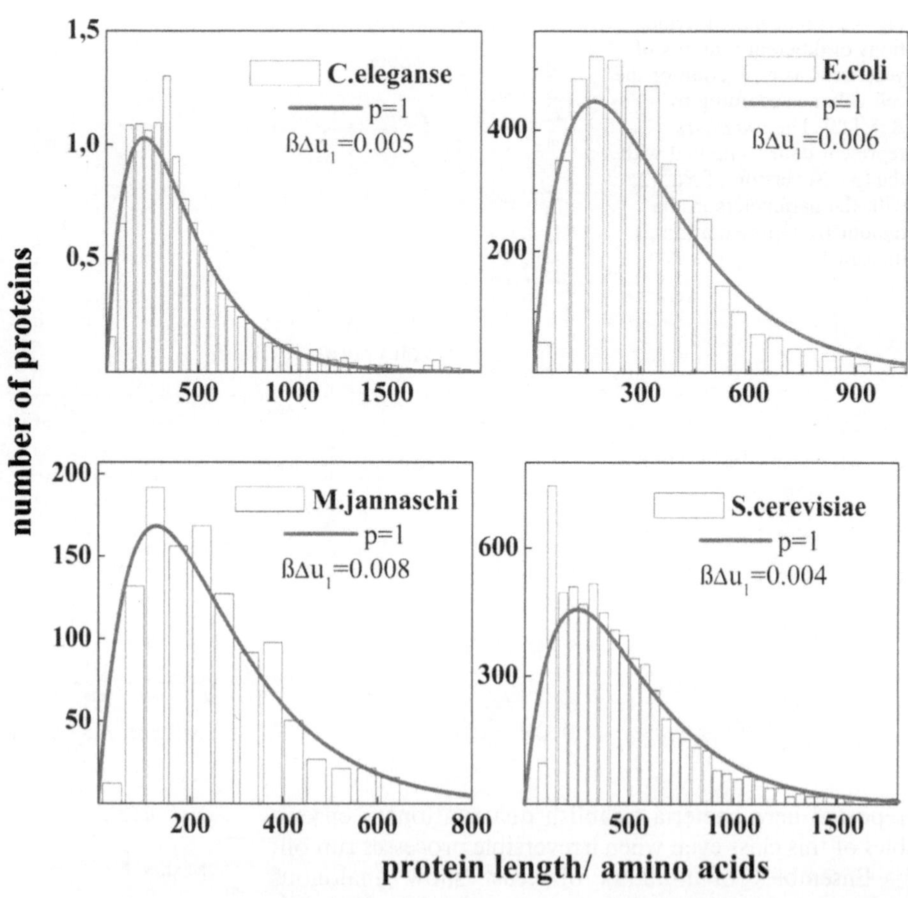

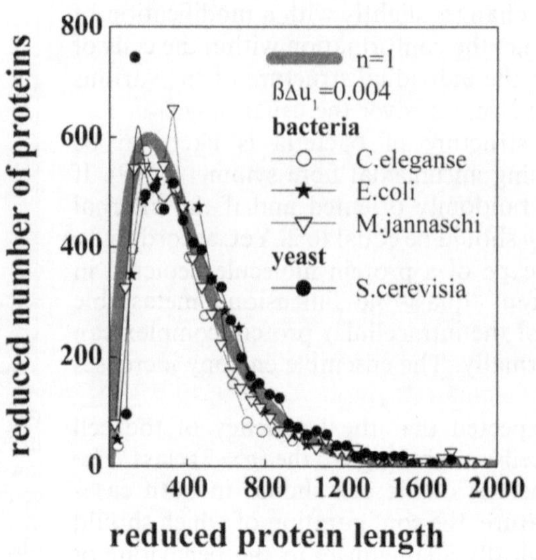

Fig. 14 Plot of proteins encoded by several chromosomes [19] of different bacteria and of the yeast *Saccharomyces cerevisia* showing the numbers of predicted proteins of different sizes according to Ref. [24]. The *solid lines* represent data computed with the ($p=1$) version of Eq. (2) with the parameters as indicated. The *lowest panel* shows the master curve of all the distributions shown above

lines were computed with the ($p=1$) version of Eq. (9). Rapid enzyme catalyzed generation or decomposition of proteins is necessary to permit optimization of the intracell protein sequences as observed here.

Despite the differences in the structure of bacteria or yeast cells the reduced size distribution of the intracell proteins [24] belongs to the same universal ($p=1$) class. Hence, in both cell types many statistically equivalent configurations of the proteins should continuously be generated or decomposed. Since the cytoplasm is gelatinized, any free orientation of the proteins is inhibited. The same argument holds for yeast cells. By having the proteins linked in an intracell scaffold, the thermally activated fluctuation of their "chainlike" configurations should be represented by quasi-one-dimensional random statistics ($p=1$). Δu_1 falls in the usual range of $0.004 \leq \beta\Delta u_1 \leq 0.008$.

Despite the remarkable differences of the individual properties (different molecular units, different intracellular structures, no nucleus in the cells of bacteria in contrast to yeast cells, different complex structures of the various folded proteins), bacteria and yeast cells seem to be able to optimize their ensemble structure adequately and independently on each level according to the extremum principles of thermodynamics: Both cell ensembles belong to the ($p=3$) class, while the "intracell ensemble" is of the ($p=1$) type. This is manifested by the master curve of the length distributions of both cell types (Fig. 14). The reduced cell size distributions of both systems are, of course, also identical.

Final comments

The results presented here identify universal features of fluctuating ensembles. Aggregates and their analogues should behave like metastable microphases, so their internal properties can be characterized by phenomenological parameters, the values of which do not fluctuate to measurable extents.

The mechanisms of reversible aggregation rely on having weak long-range interaction energies. The global properties are controlled by the finite size of the aggregates, while their individual characteristics are determined by the properties of the units. Aggregates are, in any case, metastable dynamic entities ($\Delta u_p > 0$!). The isoenergetic ensemble configurations fluctuate so as to maximize the ensemble entropy. Controlled by these boundary conditions, ensembles constituted by very different units (carbon black, liquids, loaded metals and simple living systems) generate their structure analogously. Each class (p) turns out to be characterized by generalized symmetries which are typical for the model of reversible aggregation. Moreover, the ensembles are, by definition, "perfect" and are able to "heal themselves" via structure-controlled relaxation processes. This guarantees unsurpassed perfection and reproduction, impressively proven with the description of ensemble properties of bacteria and yeast cells, which are "globally" controlled by the extremum principles of thermodynamics. This is an important matter since the statistical identity of cell ensembles is guaranteed. A genesis of complex individual ensemble structures, essentially as in ensembles of simple living systems, is not impeded at all. A unique correlation between ensemble structure and cell functions is likely to be individually installed in each case.

References

1. Smoluchovski M (1916) Phys Z 17:585
2. Botet R, Jullien R., Kolb M (1984) Phys Rev A 30:2150
3. Kilian HG, Zink B, Metzler R (1997) J Chem Phys 107:8697
4. Donnet JB, Bansal RC, Wang MJ (eds) (1993) Carbon black. Dekker, New York
5. Donnet JB, Voet A (1976) Carbon black. Dekker, New York
6. Wagner HG (1081) In: Siegela DC, Smitter GW (eds) Particle formation during combustion. Plenum, New York, pp 1
7. Hess WM, Herd CR (1993) In: Donnet JB, Bansal RC, Wang MJ (eds) Carbon black. Dekker, New York, p 89
8. Pieper B, Dulfer N, Kilian HG, Wolff S (1992) Colloid Polym Sci 270:29
9. Kilian HG (2001) Colloid Polym Sci (in press)
10. Montroll EW, Shlesinger MF (1983) J Stat Phys 32:209
11. Jullien R (1990) New J Chem 14:239
12. Botet R, Jullien R, Kolb M (1984) Phys Rev A 30:2150
13. Prigogine I (1961) Introduction to thermodynamics of irreversible processes. Wiley, New York
14. Herd CR, McDonald GC, Hess WM (1991) Rubber Chem Technol 65:1
15. Vettegren VI, Rakhimov SS, Svetlov VN (1996) Phys Solid State 38:323
16. Vettegren VI, Rakhimov SS, Svetlov VN (1997) Phys Solid State 39:1389
17. Vettegren VI, Rakhimov SS, Svetlov VN (1997) Proc SPIE 33:226
18. Kilian HG, Vettegren VI, Svetlov VN (2000) Physics Solid State 42:2083
19. Alberts B, Bray D, Johnson A, Lewis J, Raff M, Roberts K, Walter P (1998) Essential cell biology. Garlan, New York
20. Kubitchek HE (1969) Biophys J 9:792
21. Ecker RE, Schaechter M (1963) Ann NY Acad Sci 102:549
22. Gross FR (1989) J Cell Sci 12:117
23. Fraunfelder H, Wolynes PG, Austin RH (1999) Rev Mod Phys 71:419
24. Netzer WJ, Hartl WU (1997) Nature 388:343

Progr Colloid Polym Sci (2001) 117: 182–188

D. Pawlowski
B. Tieke

Change of structure and phase behaviour during homo- and copolymerisation of (2-methacryloyloxyethyl)dodecyldimethyl-ammonium bromide in a hexagonal lyotropic mesophase

D. Pawlowski · B. Tieke (✉)
Institut für Physikalische Chemie der Universität zu Köln,
Luxemburger Strasse 116,
50939 Cologne, Germany

Abstract The γ-ray initiated polymerisation of (2-methacryloyloxyethyl)dodecyldimethylammonium bromide in a hexagonal lyotropic mesophase was investigated. The binary mixture with water and the ternary mixture with water and acrylamide were studied. Especially the changes in the structure and the phase behaviour during polymerisation were investigated. Small-angle X-ray scattering indicates that the polymerisation of the binary mixture is accompanied by a discontinuous increase of the *d* spacing from 3.08 to 3.36 nm at about 5% conversion although the hexagonal order of the system is retained. The increase originates from a conformational transition in the headgroup region of (2-methacryloyloxyethyl)dodecyldimethylammonium bromide and leads to a strong acceleration of the reaction rate. Evidence for the thermodynamic stability of the system during the entire reaction process is presented. Admixture of acrylamide to the binary system results in a decrease in the transition temperature from the hexagonal phase to the isotropic solution until for a system with (2-methacryloyloxyethyl)dodecyldimethylammonium bromide and acrylamide in equimolar ratio the mesophase has completely disappeared at room temperature. However, upon γ-ray irradiation the transition temperature is raised again and for systems which are not liquid-crystalline in the monomeric state a mesophase is eventually induced. The polymerisation rate of the ternary system is lower because the acrylamide induces a different surfactant conformation in the mesophase. IR spectroscopic analysis of the copolymer formed in the ternary system indicates that acrylamide is enriched at low conversion.

Key words γ-ray irradiation · Polymerisation · Cationic surfactant monomer · Lyotropic phase

Introduction

Highly concentrated aqueous solutions of surfactants often exhibit lyotropic liquid-crystalline (LLC) phase behaviour [1–3]. In the LLC phase well-defined nanometer-scale architectures are present; these exhibit interesting material properties and are useful for a variety of applications [4]. Unfortunately, they are rather instable to perturbations such as a change of pressure, temperature and composition of the mixture. Therefore, several studies were carried out in order to prepare more stable LLC nanostructures, either by mixing amphiphilic polymers and water [5–7], by growth of polymers within the hydrophilic regions of LLC phases of commercial, nonpolymerisable surfactants [8–10] or by the in situ polymerisation of reactive surfactants in the LLC phase [11–20]. Especially the latter method appears to be very straightforward and elegant; however, recent studies indicate that the in situ polymerisation may proceed under a phase change [20], be incomplete [15–17], not occur at all [15] or the resultant polymeric nanostructures are the result of a phase separation and are not from the monomeric LLC arrays [9, 21, 22].

A big problem of the in situ polymerisation in the LLC phase is that a general understanding of the polymerisation process is still missing. Most of the studies only dealt with the characterisation of the system before and after polymerisation, but the systems were not examined during polymerisation. In order to elucidate the optimum reaction conditions and to prepare stable LLC materials suitable for applications, more detailed studies of the polymerisation process itself are necessary, for example, by investigating the phase behaviour and the structural changes during the polymerisation.

The purpose of our contribution is therefore to present a detailed study of the polymerisation process of a reactive surfactant in a LLC phase. (2-Methacryloyloxyethyl)dodecyldimethylammonium bromide (**1**) (Scheme 1) was chosen for this purpose as this compound

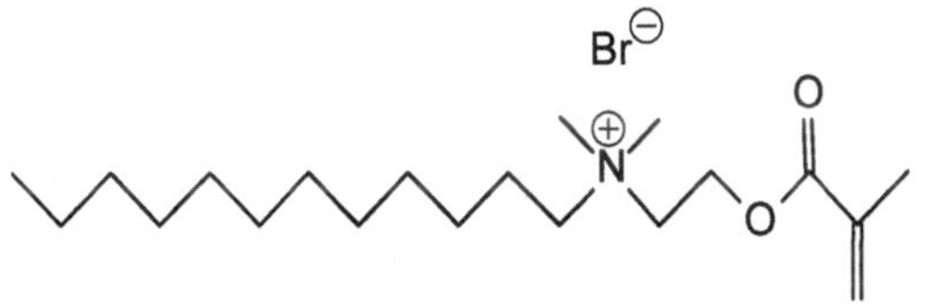

Scheme 1 Structure of (2-methacryloyloxyethyl)dodecyldimethylammonium bromide 1

is known to completely polymerise upon γ-ray irradiation in the hexagonal phase under retention of the macroscopic order [23]. Here the changes in the phase behaviour and the structure during the polymerisation as well as the reaction kinetics are described in detail. In addition, the copolymerisation of **1** and acrylamide in the ternary hexagonal phase is reported and is compared with the polymerisation behaviour of **1** in the binary mixture with water.

Experimental

Materials

1 was prepared by reaction of dimethylaminoethyl methacrylate with 1-bromododecane in acetone at 41 °C according to procedures described in the literature [23, 24]. A white solid was obtained which was recrystallized from ethyl acetate and dried in vacuum (m.p. 84 °C). Acrylamide was used as received from Fluka. Deionized water was used for all the experiments (MilliQ water).

Lyotropic solutions of the **1**/water binary system were prepared by mixing the surfactant with previously degassed water, followed by stirring for 10 min under purging with nitrogen to keep the oxygen content as low as possible. Subsequently, the samples were sealed and stored at 5 °C in the dark for 24–48 h. Lyotropic solutions of the **1**/acrylamide/water ternary system were prepared by mixing solid **1** and acrylamide in molar ratios of 1:1, 2:1 or 4:1 using a vibrating mill. Subsequently, the lyotropic solutions were prepared as described for the binary mixtures.

Polymerisation was carried out by subjecting the samples to ^{60}Co γ-ray irradiation (dose rate 0.388 kGyh^{-1}). The conversion to polymer was determined gravimetrically after leaching out the water-soluble monomer from the water-swellable (but not water-soluble) polymer and after subsequent drying of the polymer gel in a vacuum at room temperature until a constant weight was obtained.

Methods

The composition of the copolymer was determined using IR spectroscopy. The ratio of the band intensities of the C=O stretching modes at 1,650 cm^{-1} (ester group) and 1,720 cm^{-1} (amide I) was used to evaluate the composition.

The phase behaviour was studied using a standard Zeiss polarising microscope equipped with hot stage and a camera (Zeiss MC 80). To investigate the phase diagrams, a few drops of the aqueous solution were placed between two glass slides, heated to the isotropic state, sheared and recooled to room temperature. Subsequently, the phase-transition temperature was determined by viewing the samples in the polarising microscope during slow and constant heating at a rate of about 0.2 °Cmin^{-1}.

Small-angle X-ray scattering (SAXS) was carried out on sealed samples in fine-glass capillary tubes (1 mm diameter) using a Kratky compact camera (Anton Paar) with block collimation. Ni-filtered Cu K_α radiation with $\lambda = 0.154$ nm was used. The acceleration voltage was 40 kV at an anode current of 30 mA. Reflexes were monitored in continuous scan mode. The scattering curves were corrected for slit-smearing effects by a computational desmearing procedure.

Results and discussion

Polymerisation of the 1/water binary mixture

As previously described [23], **1** forms LLC phases in aqueous solution at a surfactant concentration above 55 wt%. The phase diagram in Fig. 1 (lower part) indicates the formation of hexagonal (H_α), cubic (Q_α) and lamellar (L_α) phases. In the following, we concentrate on a detailed study of the polymerisation in the hexagonal phase, the surfactant concentration being 70 wt%. As indicated in Fig. 1 (middle and upper part), the phase-transition temperature from the hexagonal phase to the micellar solution constantly increases upon exposure to γ-rays. At a dose of 3.9 kGy the maximum transition temperature has increased from 55 to 72 °C. This is a rather strong effect, especially if one takes into account that only 5 wt% of polymer has been formed in the sample, as can be concluded from the conversion versus dose curve shown in Fig. 2. The fact that the transition temperature goes up may be taken as the indication of the miscibility of monomer and polymer. Further indication of the miscibility results from the SAXS diagrams taken after different radiation doses of the sample. All the diagrams indicate a strong first-order peak of the hexagonal phase, which is retained during the entire reaction (Fig. 3). The half width of the peak also remains nearly constant. This indicates the presence of a

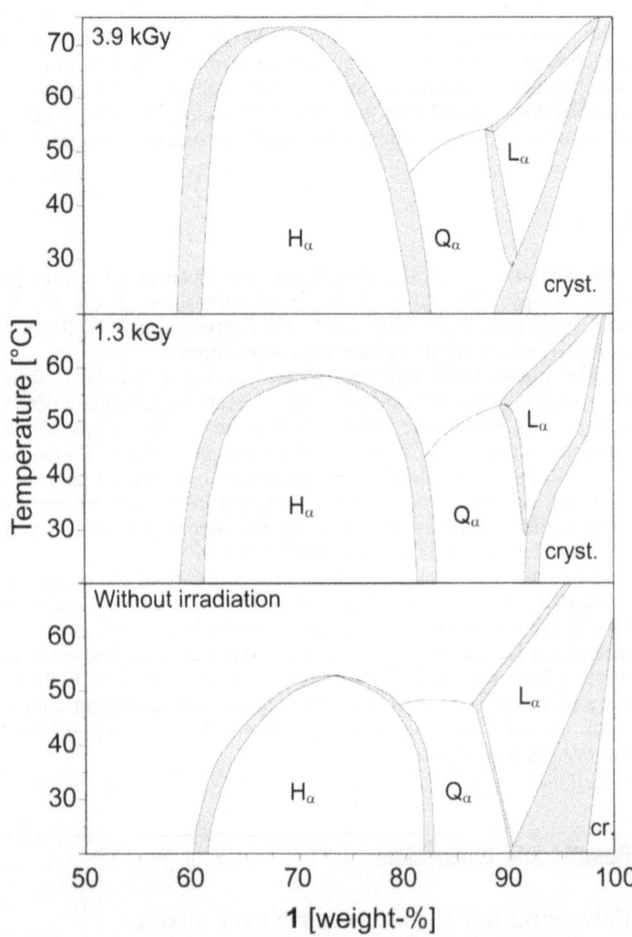

Fig. 1 Phase diagram of the (2-methacryloyloxyethyl)dodecyldimethylammonium bromide (**1**)/water binary system without γ-ray irradiation (*lower part*) and after exposure to 1.3 kGy (*middle*) and 3.9 kGy (*upper part*) (*hatched areas* indicate two-phase regions)

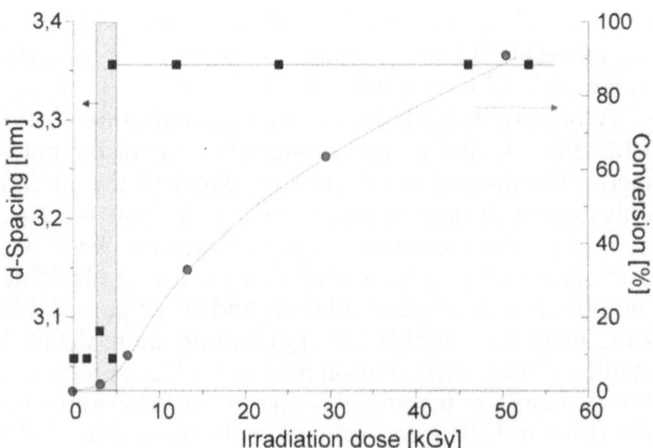

Fig. 2 Plot of the *d* spacing of the hexagonal phase and the conversion as a function of the γ-ray dose (surfactant concentration: 70 wt%)

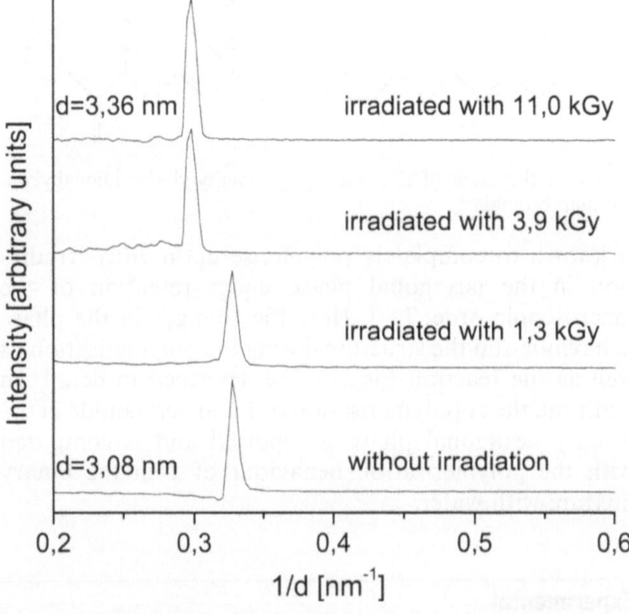

Fig. 3 Small-angle X-ray scattering patterns of aqueous mixtures of **1** in the hexagonal-phase region (concentration 70 wt%) before and after irradiation with various γ-ray doses

homogeneous hexagonal phase independent of the polymer content of the sample, in good agreement with the polarising micrographs reported earlier [23]. From the angular position of the first-order peak the *d* spacing of the hexagonal phase was evaluated as a function of the radiation dose. As shown in Fig. 2, the original *d* spacing of 3.08 nm of the monomer remains constant until a γ-ray dose of approximately 4 kGy is reached, which corresponds to a polymer content of about 5 wt%. Then, the *d* spacing abruptly changes to a value of 3.36 nm and again remains constant until the polymerisation has gone to completion. The abrupt change indicates a structural transition within the hexagonal phase, which we denote as a transition from hexagonal phase I to hexagonal phase II.

As the conversion versus dose curve in Fig. 2 indicates, the polymerisation proceeds very slowly in phase I, while in phase II an accelerated reaction which finally goes to completion is found. We believe that phase I becomes increasingly destabilized by the growing polymer (and/or oligomer) chains until finally the transition into the more stable phase II takes place. The sudden increase in the *d* value by almost 10% indicates a pronounced conformational transition of the surfactant molecules. In agreement with Hamid and Sherrington [25] we believe that the monomer molecules originally show intramolecular interaction of the methacryloyloxy group with the onium nitrogen centre, which could explain the small *d* spacing found in phase I. The exothermic polymerisation could easily provide the energy necessary to overcome this interaction so that phase II with a larger d spacing is formed, in which the molecules attain a more elongated shape. The rearrangement in the headgroup region leads to a different mutual orientation of the vinyl groups, which in phase II is sterically more favourable for polymerisation. This may explain the rapider reaction in phase II.

The monomer/polymer mixture formed upon irradiation might be either a thermodynamically stable or only a so-called frozen-in structure in which demixing is kinetically hindered by the high viscosity of the polymer. In the latter case, all attempts to prepare a homogeneous hexagonal phase by simply dissolving corresponding amounts of monomer and polymer in water would fail. Instead of a homogeneous mesophase, a two-phase system recognisable by SAXS peaks of either phase would result. In order to prove the thermodynamic stability, we first mixed solid monomer and polymer in different ratios and then prepared aqueous solutions with a solid content of 70 wt%. As shown in Fig. 4, the SAXS data of the monomer/polymer mixed systems exhibit nearly the same d spacing of 3.36 nm as the in situ polymerised system of identical composition. This indicates the spontaneous formation of hexagonal phase II and can be taken as proof of the thermodynamic stability of the in situ polymerised samples.

Copolymerisation of the **1**/acrylamide/water ternary system

In addition to the **1**/water binary system we also investigated the polymerisation of the **1**/acrylamide/water ternary system. This system was chosen because

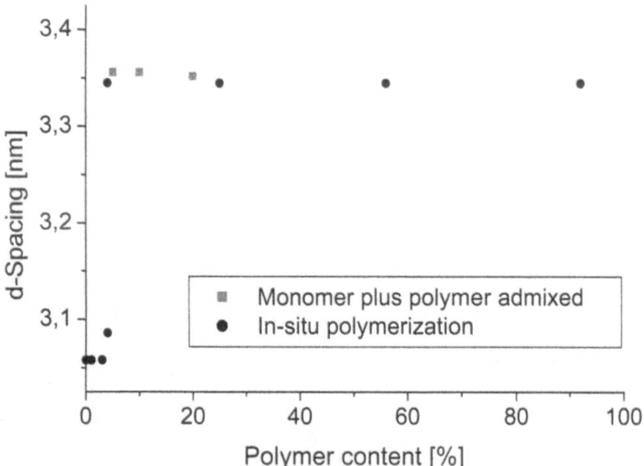

Fig. 4 Plot of the d spacing of the hexagonal phase as a function of the polymer content of the sample (the total concentration of monomeric plus polymeric **1** is always 70 wt%). Samples were prepared upon in situ polymerisation of the mesophase or upon admixture of polymer to the monomer phase

acrylamide and its polymer are highly water-soluble and thus in the case of copolymerisation with **1** an amphiphilic copolymer with potentially interesting self-organising properties could result and because the system is well comparable with the binary system described previously. Copolymerisation of water-soluble monomers and reactive surfactant monomers in LLC mesophases is still a new area and very little information is available in the literature.

As the partial phase diagram of Fig. 5a indicates, the addition of acrylamide to the **1**/water binary system with an initial surfactant concentration of 70 wt% leads to a continuous decrease in the stability of the hexagonal-phase region. As can be seen, the transition temperature from the hexagonal to the isotropic phase decreases from 55 °C for the acrylamide-free mixture to 35 °C for the comonomer mixture of **1** and acrylamide in a 2:1 molar ratio. However, as shown in Fig. 5b, the phase-transition temperature is raised upon γ-ray irradiation, quite similar to the binary mixture shown in Fig. 1. In the case of an equimolar aqueous mixture of **1** and acrylamide with a total monomer content of 75 wt% the hexagonal phase has completely disappeared at room temperature. Since we are presently not able to study samples below room temperature, it is not possible to decide whether the hexagonal phase has completely disappeared at all temperatures or whether the phase-transition temperature has dropped to a value below 20 °C. The interesting point is that a mesoscale structure reappears at 20 °C once the sample is exposed to γ-rays. Obviously the increasing polymer content stabilises the LLC phase and causes the system to form a hexagonal phase up to elevated temperature. Relevant polarising micrographs showing the ternary system before and after γ-ray rradiation are shown in Fig. 6.

The conversion to polymer of the ternary system was determined gravimetrically. The possible formation of the water-soluble acrylamide–homopolymer was checked, but no indications could be found. As shown in the conversion versus dose curves of Fig. 7, the polymerisation rate increases as the acrylamide content becomes higher, but the high rate of the binary system is not reached in any of the ternary mixtures (see also Fig. 2). The different reaction rates might be a consequence of interactions between the acrylamide and the surfactant molecules. Interactions are evident from the differences in the d spacing of the ternary and the binary system. While the d spacing of the ternary mixture containing **1** and acrylamide in a 4:1 molar ratio is 3.42 nm (Fig. 8), the binary system exhibits a value of 3.25 nm (Fig. 2). We therefore denote the new structure as hexagonal phase III. As also shown in Fig. 8, the d spacing first decreases to a value of 3.25 nm during polymerisation, and then increases to the final value of 3.31 nm. This value agrees with the value of the polymerised binary system denoted as hexagonal pha-

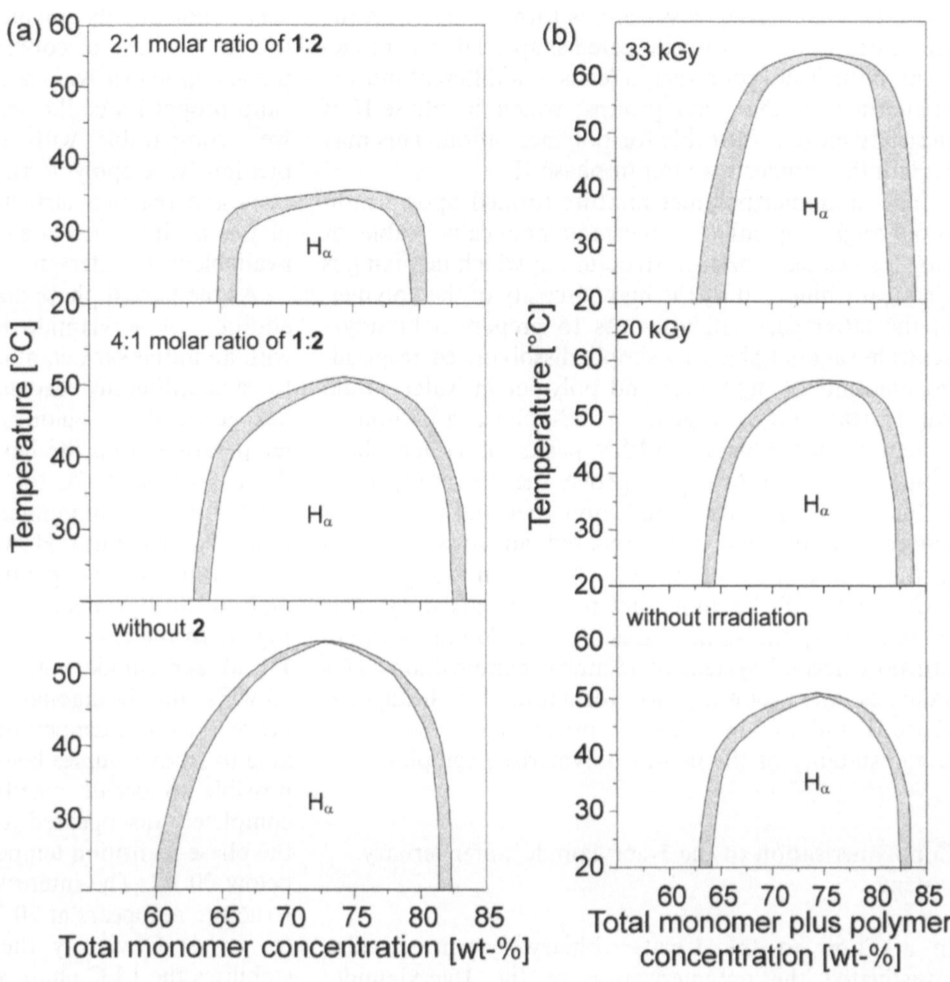

Fig. 5 Change of **a** the hexagonal-phase region of the **1**/water system upon addition of acrylamide (**2**) and of **b** the **1**/**2**/water system (with a 4:1 molar ratio of **1** and **2**) upon γ-ray irradiation (*hatched areas* indicate two-phase regions)

Fig. 6 Polarising micrographs of an aqueous mixture containing 75 wt% of an equimolar mixture of **1** and **2** before (*inset*) and after γ-ray irradiation with a dose of 110 kGy at room temperature

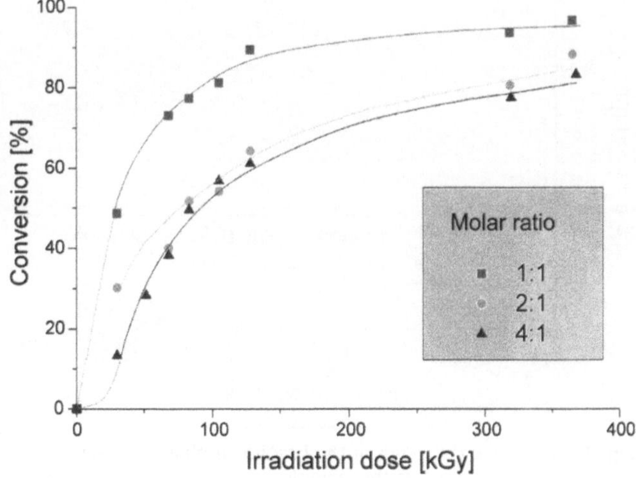

Fig. 7 Conversion versus dose curves of **1**/**2**/water ternary mixtures containing 75 wt% **1** and **2** in different molar ratio

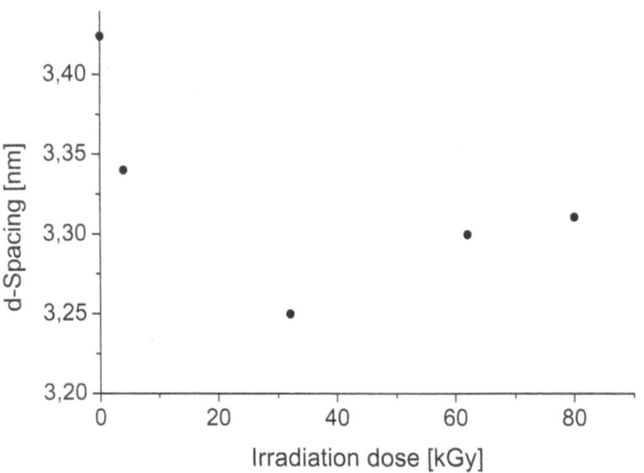

Fig. 8 Plot of the *d* spacing of a **1**/**2**/water ternary system containing 75 wt% **1** and **2** in a 4:1 molar ratio versus the γ-ray irradiation dose

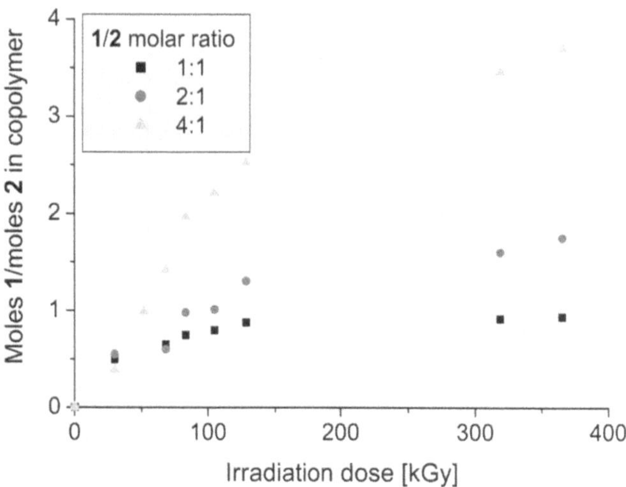

Fig. 9 Composition of the copolymer formed in the **1**/**2**/water ternary system containing 75 wt% **1** and **2** in different molar ratios as a function of the irradiation dose

se II. It is likely that a transition from phase III to II occurs at a dose of 35 kGy, when the lowest *d* value of 3.25 nm is reached. The half width of the first-order X-ray peak does not significantly change upon the transition, which is an indication that the order of the system is retained.

The different polymerisation rates of the binary and the ternary systems can be ascribed to a different molecular conformation of the surfactant resulting in a lower reactivity in case of hexagonal phase III. The fact that in the ternary mixture the reaction rate increases with the acrylamide content might be due to the preferential polymerisation of acrylamide in the aqueous phase so that copolymers with a higher polyacrylamide content are formed. A detailed analysis of the copolymer composition by IR spectroscopy indeed indicates a preferential incorporation of acrylamide monomer in the initially formed copolymer (Fig. 9), while the copolymer obtained at high conversion has the same composition as the original monomer mixture.

Conclusions

Our study demonstrates that the polymerisation of surfactant monomer **1** in a hexagonal mesophase proceeds without loss of the mesoscale order. Structural studies during the polymerisation indicate that the polymerisation is accompanied by a discontinuous change in the *d* spacing at about 5% conversion, while the hexagonal order is completely retained. The change in the *d* spacing can be ascribed to a conformational transition in the headgroup region of the surfactant molecules. The new conformation is sterically more favourable for polymerisation, so the reaction rate strongly increases after the transition has taken place. The low initial reaction rate combined with the sudden acceleration once a certain polymer content is formed generally renders the control of the polymerisation of **1** difficult. This may explain the inconsistent results previously reported on the polymerisation of this compound in mesoscale systems [23, 26]. The fact that a simple mixing of monomer, polymer and water leads to the same hexagonal order as obtained in the in situ polymerisation of the system can be taken as proof of the thermodynamic stability of the mesophase during the entire polymerisation process.

Our study also indicates that the addition of a water-soluble comonomer significantly alters the phase and polymerisation behaviour and is even able to suppress LLC phase behaviour completely. In contrast, γ-ray polymerisation increases the stability of the lyotropic phase and even induces liquid-crystalline phase behaviour in systems which in the monomeric state are not liquid-crystalline.

The study also shows that the change in structure and phase behaviour of **1** during polymerisation is quite different from other surfactant monomers and makes it difficult to draw more general conclusions from the present results. Besides the pressure, the temperature and the composition of the system, the molecular conformation, the mutual arrangement of the reactive units and the conversion to polymer are important parameters controlling the phase behaviour, the structure and the reactivity of the system in a complex manner. Clearly, the study of other systems of reactive monomers in a LLC phase is necessary to fully understand the in situ polymerisation in a mesoscale solution.

Acknowledgement The authors thank Astrid Haibel, II. Physikalisches Institut der Universität zu Köln, for the SAXS measurements.

References

1. Winsor PA (1968) Chem Rev 68:1
2. Gray GW, Winsor PA (eds) (1974) Liquid crystals and plastic crystals, vol 1, 1st edn. Wiley, New York
3. Kelker H, Hatz R (1980) Handbook of liquid crystals, 1st edn. Verlag Chemie, Weinheim
4. Engels T, v Rybinski W (1998) J Mater Chem 8:1313
5. Finkelmann H, Lühmann B, Rehage G (1982) Colloid Polym Sci 260:56
6. Hill RM, He M, Liu Z, Davis HT, Scriven LE (1993) Langmuir 9:2789
7. Wang W, Lieser G, Wegner G (1993) Liq Cryst 15:1
8. Laversanne R (1992) Macromolecules 25:489
9. Antonietti M, Caruso RA, Göltner CG, Weissenberger MC (1999) Marcomolecules 32:1383
10. Candau F, Zekhini Z, Durand JP (1986) J Colloid Interface Sci 114:398
11. Herz J, Reiss-Husson F, Rempp P, Luzatti V (1963) J Polym Sci Part C 4:1275
12. O'Brien DF, Armilage B, Benedicto A, Bennett DE, Lamparski HG, Lee YS, Srisiri W, Sisson TM (1998) Acc Chem Res 31:861
13. Meier W (1998) Macromolecules 31:2212
14. Miller SA, Ding JH, Gin DL (1999) Curr Opin Colloid Interface Sci 4:338
15. McGrath KM, Drummond CJ (1996) Colloid Polym Sci 274:316
16. McGrath KM (1996) Colloid Polym Sci 274:399
17. McGrath KM (1996) Colloid Polym Sci 274:499
18. Lester CL, Guymon CA (2000) Macromolecules 33:5448
19. Srisiri W, Benedicto A, O'Brien DF, Trouard TP (1998) Langmuir 14:1921
20. Thundatil R, Stoffer JO, Friberg SE (1980) J Polym Sci Part A Polym Chem 18:2629
21. Antonietti M, Hentze HP (1996) Colloid Polym Sci 274:696
22. Göltner CG, Antonietti M (1997) Adv Mater 9:431
23. Pawlowski D, Haibel A, Tieke B (1998) Ber Bunsenges Phys Chem 102:1865
24. Nagai K, Ohishi Y (1994) J Polym Sci Part A Polym Chem 32:445
25. Hamid SM, Sherrington DC (1987) Polymer 28:325
26. McGrath KM, Drummond CJ (1996) Colloid Polym Sci 274:612

Progr Colloid Polym Sci (2001) 117: 189–194

R. Borbás
É. Kiss
M. Nagy

Elastic properties of protein gels obtained by three-phase partitioning

R. Borbás · É. Kiss (✉) · M. Nagy
Loránd Eötvös University,
Department of Colloid Chemistry,
Budapest 112,
P.O. Box 32, Hungary-1518
e-mail: kissevak@ludens.elte.hu
Tel.: +36-1-2090555
Fax: +36-1-2090602

Abstract A special combination of alcoholic precipitation and salting out of proteins was described as three-phase partitioning [1]. Proteins are accumulated in a coherent middle phase in the partitioning system composed of *tert*-butanol, ammonium sulphate and water. The mechanical properties of the middle phase were determined in model systems using bovine serum albumin (BSA) or ovalbumin. The stress–deformation relationship obtained by uniaxial compression showed the elastic behaviour of the middle phase. It was found that the elastic modulus of the protein gel changed with stress. The middle phase became harder upon compression, possibly owing to some change in structure. The viscoelastic properties were determined by measuring the deformation as a function of time. The BSA gels showed higher elastic and viscous deformation and lower viscosity than the ovalbumin gels. The rheological behaviour of protein gels is consistent with the previous results obtained from the analysis of middle-phase components, which indicate that the middle phase formed in three-phase partitioning is an emulsion gel.

Key words Three-phase partitioning · Rheology of bovine serum albumin and ovalbumin gels · Emulsion gel · Protein separation

Introduction

Three-phase partitioning (TPP) is a protein separation method based on heterogeneous systems consisting of *tert*-butanol, ammonium sulphate and water. *tert*-Butanol has limited miscibility with water in the presence of ammonium sulphate; the mixture separates into two liquid phases [1, 2]. Applying this type of heterogeneous system to a protein solution leads to the appearance of a coherent protein-rich disc (a third phase) between the two liquid phases [1–4]. It was proved to be an effective method in the prepurification of proteins [3, 4] or the preparation of high activity horseradish peroxidase [5].

Several parameters affecting the behaviour of proteins during TPP have been studied so far in order to exploit the full potential of the method, but there are still many aspects uncovered in this area. This article focuses on the mechanical properties of the gel-like middle phases of TPP.

Gels can be classified either by the forces acting in the junction points or by the basic elements forming the network. Physical and chemical gels can be distinguished on the basis of the forces. The basic forming elements of the network can be either macromolecules or other kinds of colloidal particles. A special type of particle gel is a protein-stabilised emulsion gel, made of protein-coated droplets aggregated in a liquid which is immiscible with the interior of the droplets [6, 7].

Mechanical measurement is a typical tool for the characterisation of protein gels. The knowledge of the mechanical properties gives indirect information on the strength of the forces acting in the junction zones of the gel network, the flexibility of network elements and the concentration of cross-links. A gel is properly characterised rheologically by the elastic modulus and viscous components of deformation [7]. The mechanical measurements monitor network continuity over macroscopic dimensions on a supramolecular level, surveying the arrangement of molecules [8].

There are many fundamental works on the rheological characterisation of gels. Both the theoretical [8–13] and the experimental [11, 12, 14, 15] aspects of this subject have been widely studied.

The theoretical approach is based on a model with ideally elastic Gaussian chains. By considering the elasticity of one chain, the force acting between the two ends of the network chain, the Brownian motion of the segments, and using physical and statistical methods the following basic equation can be deduced:

$$\sigma = K_1 kT \frac{\overline{r_{cr}^2}}{r_{mp}^2} \frac{v}{V} \left(\lambda - \lambda^{-2}\right) , \tag{1}$$

which provides a connection between stress and deformation. σ stands for the applied force per unit undeformed area, K_1 is a constant, k is the Boltzmann factor, T is the absolute temperature, $\overline{r_{cr}^2}$ is the mean-square end-to-end distance, r_{mp}^2 is the most probable distance of the ends of network chains, v is the number of elastically active network chains in the volume, V, of the gel and λ is defined as the extension ratio of the gel: the actual length of the gel under σ stress per the initial length (l_0) of the gel.

The rearrangement of Eq. (1) results in the following expression:

$$[\sigma] = \frac{\sigma}{\lambda - \lambda^{-2}} \propto \frac{K_1 kT}{l_0} v , \tag{2}$$

where l_0 replaces $V^{1/3}$ in the case of uniaxial compression. This expression defines the reduced stress, $[\sigma]$, which is independent of the extension ratio and the stress.

The connection between stress and the deformation parameter can be written in another form [8]:

$$\sigma = \frac{E}{3} \left(\lambda^2 - \frac{1}{\lambda} \right) , \tag{3}$$

where E stands for Young's modulus.

Equations (1) and (3) work well for many synthetic polymer gels or heat-set protein gels, but in several cases nonideality has also been observed where this equation is no longer valid. For example, the Mooney–Rivlin formula or the equation of Blatz, Sharda and Tschoegl (which can be found in Refs [8, 11]) can be used in this cases.

The previously discussed model of rubber elasticity and the equations derived from it have been applied to biopolymer gels [8]; however, this model can be considered as a rough approximation of the behaviour of biopolymer gels. The network-forming protein is not a random coil and with the exception of a few cases, its end-to-end distance, conformations are regulated by the secondary, tertiary and quaternary structures of the protein, which are determined by the amino acid sequence of the molecule. In addition, the conformation of protein molecules can change in the presence of an interface (owing to partial unfolding), and this effect may contribute to the mechanical properties of the interfacial film. An adsorbed protein layer at the interface of droplets dispersed in an immiscible liquid controls the stability of the emulsion and, hence, the structure of an emulsion gel [6, 7].

Experimental

Bovine serum albumin (BSA) fraction V (from Reanal) and chicken egg ovalbumin (OVA) (from Sigma) lyophilised and with purity above 98% were used. *tert*-Butanol, high-performance liquid chromatography grade, was obtained from Sigma. All other reagents were of analytical grade and each reagent was used without further purification.

Five partitioning systems were used during the measurements, which differed in the ratio of *tert*-butanol, ammonium sulphate and water. The compositions in mass ratio are (listed in the manner of tert-butanol:ammonium sulphate:water) system 1 0.16:0.18:0.66; system 2 0.58:0.09:0.33; system 3 0.56:0.06:0.38; system 4 0.71:0.04: 0.25; system 5 0.18:0.11:0.71. The interfacial tension of the protein-free heterogeneous systems were 1 (system 5), 2 (systems 1 and 3) and 4.5 mNm^{-1} (systems 2 and 4).

Aqueous protein solution was mixed with aqueous ammonium sulphate solution to obtain a protein concentration of 3.5 gdm^{-3}, and finally *tert*-butanol was added to the system after 1 h. The system was thoroughly mixed by turning the tubes upside down 20 times and was centrifuged after standing for 1 h. The centrifugation was performed at 2,000 rpm (672*g*) for 10 min. The system was kept at 25 °C during the whole procedure.

The gel-like middle phase was separated, and then two deformation tests were performed using an apparatus developed in our department [16]. This equipment allowed the vertical compression of the sample with measurement of deformation and force within a wide range, from 10^{-5} to 2×10^{-3} m and from 10^{-4} to 2 N, respectively. The gel sample was placed and compressed between parallel glass plates. The samples were loaded and the deformation was measured on a scale proportional to the change in sample height with an accuracy of 10^{-5} m. The samples were either loaded step by step in the range 0–5 g and the deformation was measured as a function of stress (method 1) or they were loaded with a given weight (0.5 and 1 g) and the deformation was measured as a function of time (method 2).

Information could be obtained on the reversibility of deformation from both types of methods. The gel size was measured when the sample was loaded with the smallest possible weight, 0.01 g, (which corresponds to a stress of ~ 0.15 Nm^{-2}). This condition is mentioned as unloaded in the following. The reversibility of deformation was checked twice on the sample during the whole procedure of method 1, once after reaching a stress of ~ 2.2 Nm^{-2} and once after ~ 75 Nm^{-2}. In method 2 the reversibility was checked by unloading the sample at 600 s and measuring the gel size for 300 s.

Results and discussion

Proteins accumulate into a coherent third phase when treated with a heterogeneous two-liquid-system of TPP.

A gel-like phase is formed, which contains all four components of the two-liquid-phase system. The protein content of the gel is low (3–9 w/w% in the case of BSA, 0.4–2 w/w% in the case of OVA), and the *tert*-butanol/ammonium sulphate/water ratio in the gel is a function of the initial composition of the system [17].

The coherent phase produced by TPP is a physical gel because the mild conditions during which the gel was formed do not induce chemical reaction, i.e. no oxidising or reducing agents were present in the system, ammonium sulphate and *tert*-butanol are protein friendly and the system was held at standard temperature. These conditions permit the formation of physical connections only between the network elements.

The mechanical properties of protein gels were studied in two ways. The deformation of the sample was measured as a function of stress (method 1) and time (method 2). The stress–deformation curves of BSA and OVA gels developed in five different systems are displayed in Fig. 1a and b, respectively. The abscissa shows the extension ratio (λ) while the ordinate shows the stress (σ), which is negative because the applied force was compressive.

Figure 1 shows that Hooke's law is valid for the gels in the initial (see insert of Fig. 1b) and the final sections of the stress–deformation curves, i.e. stress is linearly proportional to deformation. The initial region was within 0.96 or $0.94 \leq \lambda \leq 1$ (depending on the protein studied) with stress values above $-2\ \mathrm{Nm^{-2}}$; the final region was below a stress of $-15\ \mathrm{Nm^{-2}}$. There was a turn between the two linear regions.

The hardness of the gel is indicated by the slope of the linear region of the curves. The slopes of the second linear section are displayed beside the curves in Fig. 1 in units of newtons per square metre. The higher the slope the harder the gel because a larger step in stress caused a smaller step in deformation. The slope of the second linear region is 1 magnitude higher than that of the first region, which means that the gels became considerably harder during compression. It can be seen that the OVA gels were not so hard as the BSA gels: the slopes of OVA curves were slightly smaller than those of BSA. The differences between the hardness of the two types of protein gels were not so high as was expected from the differences in their protein content.

The comparison of the gels developed from a single protein in five different systems led to the conclusion that BSA gels were harder if they were formed in a system with higher interfacial tension. The system with an interfacial tension of 1 $\mathrm{mNm^{-1}}$ produced a curve with the lowest value of the slope ($431 \pm 17\ \mathrm{Nm^{-2}}$), while the highest slopes (952 ± 72 and $796 \pm 49\ \mathrm{Nm^{-2}}$) were measured in the system with the highest interfacial tension. No such tendency could be observed in the case of OVA.

There should be a linear relationship between stress and $\lambda - \lambda^{-2}$ for macromolecular gels according to Eq. (1). The gel-like middle phase of TPP had an elastic behaviour similar to the gel in the model of rubber elasticity, even though it possessed a more complex structure. A linear relationship could be detected between stress and $\lambda - \lambda^{-2}$ around stress values above $-2\ \mathrm{Nm^{-2}}$ and below $-15\ \mathrm{Nm^{-2}}$ (data not displayed). There was a break between these two regions similarly to the stress–deformation curves.

The rearrangement of Eq. (1) results in the reduced stress, $[\sigma]$, defined by Eq. (2). The reduced stress should be independent of σ and λ, but in this case the relation

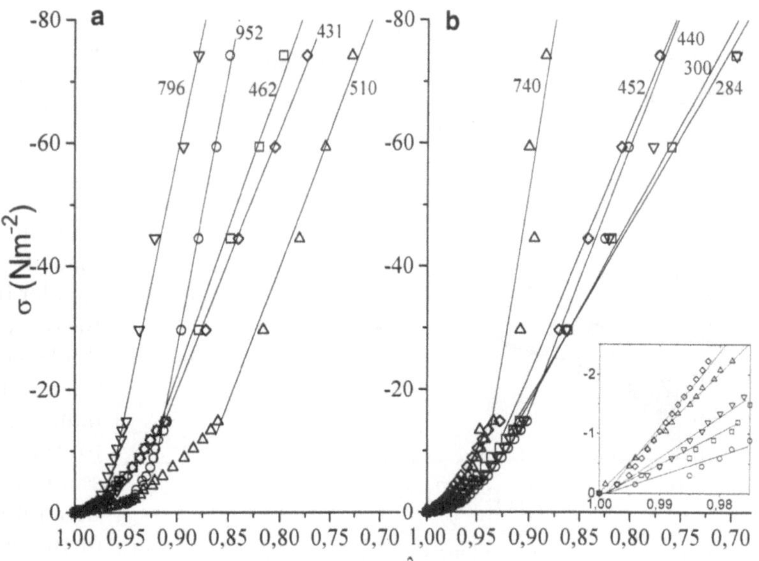

Fig. 1 The applied stress (force per unit undeformed area) as a function of extension ratio in the case of **a** bovine serum albumin (*BSA*) and **b** ovalbumin (*OVA*) gels, obtained by applying five different partitioning systems: systems 1 (□), 2 (○), 3 (△), 4 (▽), 5 (◇). The initial section of the curves of the OVA gels is shown as an *insert*. The slopes of the regressed lines are displayed next to the lines

seemed to be not generally valid. The actual relationship between the reduced stress and the stress for protein gels developed by TPP is displayed in Fig. 2: $[\sigma]$ grew with the load (as σ became lower). In the case of OVA this value became almost constant in the region $-30\ \mathrm{Nm}^{-2} \geq \sigma \geq -75\ \mathrm{Nm}^{-2}$ (except in system 5). The right-hand side of Eq. (2) contains a few constants (K_1, k, T, l_0) beside the factor v. There was no detectable change in the volume of the gels during the measurements, so the growing value of the reduced stress can be due to the change in the concentration of physical bonds relevant to the elastic property. A possible explanation could be that new interparticle interactions are formed because of the compression, and the new cross-links harden the gel. The syneresis (which could be observed to a small extent) can also induce some change in the structure of the gels.

The reduced stress is a modulus-type quantity which can approximate the value of the shear modulus of the sample [10] (if λ does not differ much from 1). Young's modulus of the protein gels calculated from Eq. (3) is almost equal to the 3 times the value of the reduced stress, which is expected from the literature [18]. (Greater deviations from a ratio of 3 were calculated for $\lambda < 0.9$.) Hence, it could be concluded that the elastic modulus of the protein gels became higher as the sample was loaded with a higher weight.

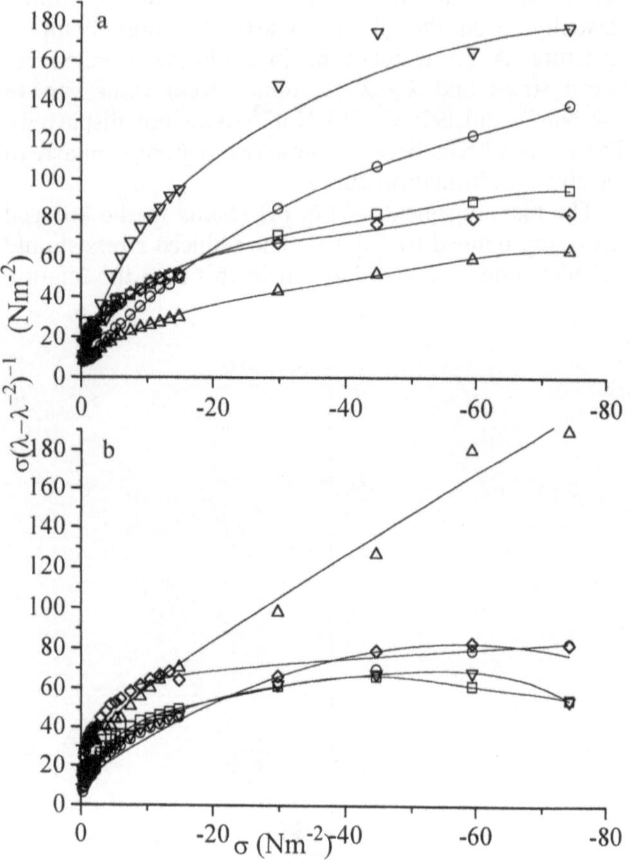

Fig. 2 The reduced stress as a function of stress in case of **a** BSA and **b** OVA gels, obtained by applying five different partitioning systems: systems 1 (□), 2 (○), 3 (△), 4 (▽), 5 (◇)

The viscoelastic properties of the middle phases were also studied by measuring the deformation of the sample loaded with a constant weight as a function of time. The results for BSA and OVA gels developed in TPP system 1 are shown in Fig. 3. The instant deformation value $(1-\lambda_1)$ was measured directly after loading the protein (BSA or OVA) gel. The elastic region of the gel extended to the point where the deformation versus time curve became linear; the value characterising this region is the elastic deformation, $1-\lambda_2$ (calculated by subtracting $1-\lambda_1$ from the intercept of the regressed linear line). The viscous component of deformation was active from 0 s, but when the elastic deformation vanished, this was the only component present. The extent of viscous deformation at the end of the period of loading $(1-\lambda_3)$ can be calculated by subtracting $1-\lambda_1$ and $1-\lambda_2$ from the maximum value of $1-\lambda$. The load was removed from the sample at 600 s, and after that the instant, elastic and viscous deformations could be measured in a similar way to the loaded section (signed by primes), giving information on reversibility [19]. These values of the deformations are summarised in Table 1.

The comparison of the different deformation values leads to interesting observations. It can be generally concluded that the higher the load the higher the deformation in the case of a single protein. No great difference was measured between the $1-\lambda_1$ values of OVA and BSA if they were loaded with the same weight; however, differences between the deformation of the two proteins appeared both in the elastic and in the viscous region: the elastic and viscous deformations of BSA gels significantly exceeded those of OVA gels. The elastic component of deformation of BSA gels was about 3 times higher than the viscous component (considering the measuring period), while the $1-\lambda_2$ and $1-\lambda_3$ values of the OVA gels were quite similar.

In the unloaded section $1-\lambda_1'$ did not differ much from $1-\lambda_1$ in the case of both protein gels loaded previously with 0.5 g. The $1-\lambda_1'$ value of the sample loaded previously with the heavier weight was much smaller than the instant deformation at loading. Possibly the original structure of the sample could not be regained if a larger force compressed the sample. In the unloaded section the elastic deformation decreased and the remaining deformation increased in every case compared to the corresponding values observed in the loaded section. The reversibility of the elasticity of protein gels was not high (92–79%), owing to either the compression or the flow of the gel.

The reversibility of the protein gels was also measured during method 1. It was found that the

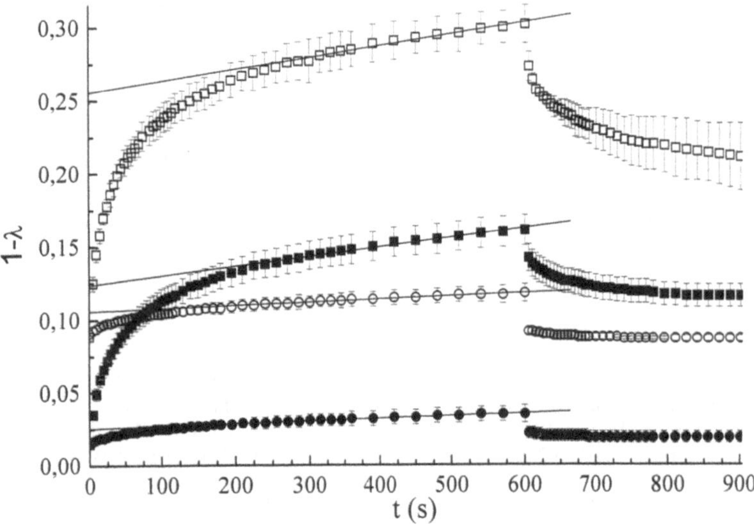

Fig. 3 The deformation $(1-\lambda)$ as a function of time in case of BSA (□) and OVA (○) gels. *Solid* and *open* symbols correspond to the samples loaded with 0.5 and 1 g, respectively

Table 1 The deformation parameters of bovine serum albumin (*BSA*) and ovalbumin (*OVA*) gels

Protein	Load (g)	Loaded			$\frac{d(1-\lambda)}{dt}\times 10^5$(s)	Unloaded		
		$1-\lambda_1$	$1-\lambda_2$	$1-\lambda_3$		$1-\lambda_1'$	$1-\lambda_2'$	$1-\lambda_3'$
BSA	0.5	0.014	0.107	0.039	7	0.019	0.026	0.116
BSA	1	0.089	0.165	0.048	8	0.029	0.062	0.211
OVA	0.5	0.015	0.011	0.009	2	0.013	0.003	0.019
OVA	1	0.091	0.015	0.012	2	0.025	0.006	0.087

gels were reversibly deformed up to $\sigma = -2.2\ \mathrm{Nm}^{-2}$ ($\lambda \geq 0.95$), and the degree of reversibility decreased to $\lambda = 0.88$ or 0.95 for BSA and OVA after the turn in the stress–deformation curves. The comparison of these values and the $1-\lambda_3'$ values in Table 1 suggest that both compression and the viscous deformation played an important role in the alteration of gel structure.

The rate of flow of the gels was also acquired from Fig. 3. The slope of the linear region of the deformation–time curves, $\left(\frac{d(1-\lambda)}{dt}\right)_{\sigma,T}$, was calculated form the measurements and is listed in Table 1. The rate of flow is inversely proportional to viscosity. The comparison of BSA and OVA showed that OVA was more viscous than BSA. There was no great difference between the rate values of BSA loaded with 0.5 and 1 g, and OVA behaved similarly to BSA. This suggested that the gel showed higher viscosity at higher compressing force. The increased viscosity upon compression also suggested that the protein gel underwent a certain change in structure as a result of deformation.

The way in which the middle phases were formed by TPP advocated that the middle phases are emulsion gels. Enzyme-set emulsion gels were reported to have an increasing shear modulus with increasing deformation [7]. Hence, the mechanical behaviour suggests that the gels are really a kind of particle gel.

Conclusions

An aqueous protein solution treated with *tert*-butanol, ammonium sulphate and water (in a suitable ratio) results in the enrichment of the protein as a coherent disc. The gel appears between two immiscible liquid phases. The protein disc was supposed to be an emulsion gel because it is formed after the emulsification of the two immiscible liquid phases of the heterogeneous TPP system. Mechanical characterisation of the middle phases was carried out to study the nature of the protein gel.

Stress–deformation curves were obtained for two different proteins, BSA and OVA, in five different partitioning systems. It was found that the gels became harder when they were compressed, which was shown by the increasing slope of the curves and also from the fact that the reduced stress increased with stress. A possible reason for hardening is that the structure of the gel changes owing to deformation.

The deformation–time measurements provided information on the elastic and viscous deformations, on the flow of the gels and on the reversibility of deformation. BSA gels showed higher elastic and viscous deformations compared to OVA gels, but their reversibility was lower. The viscous component of the deformation of BSA gels was higher, but the viscosity was calculated to be lower than that of OVA gels, which reflects that BSA lost more of its network structure as a consequence of compression.

It can be concluded from the mechanical characterisation of the TPP middle phases that there might be a certain alteration in the structure of the gels as a result of compression and also that the gel is supposedly a concentrated emulsion. Investigations of the structure are planned in order to gain further information on the formation of the middle phase and its connection with the interfacial properties of proteins, probably by microscopic and other techniques.

Acknowledgements The work was supported by the Hungarian Research Foundation (OTKA) project no. T 033065. The authors are indebted to Mrs. Hórvölgyi for helpful technical assistance.

References

1. Kiss É, Szamos J, Tamás B, Borbás R (1998) Colloids Surf A 142:295
2. Szamos J, Kiss É (1995) J Colloid Interface Sci 170:290
3. Tan KH, Lovrien R (1972) J Biol Chem 247:3278
4. Lovrien R, Goldensoph C, Anderson PC, Odegaard B (1987) In: Burgess R (ed) Protein purification: micro to macro. Liss, New York, pp 131–148
5. Szamos J, Horschke Á (1992) Acta Alimen 21:253
6. Dickinson E (1994) J Chem Soc Faraday Trans 90:173
7. Dickinson E (1998) J Chem Soc Faraday Trans 94:1657
8. Clark AH, Ross-Murphy SB (1987) Adv Polym Sci 83:57
9. Ferry JD (1948) Adv Protein Chem 4:1
10. Treolar LRG (1975) The physics of rubber elasticity. Clarendon, Oxford
11. Nagy M (1985) Colloid Polym Sci 263:245
12. Nagy M (1993) Magy Kem Foly 1:8
13. Nagy M, Nagy A (1996) Magy Kem Foly 10:427
14. Van Kleef FSM, Boskamp JV, van der Tempel M (1978) Biopolymers 17:225
15. Palusson M (1990) PhD thesis. University of Lund
16. Horkay F, Nagy M, Zrinyi M (1980) Acta Chim Acad Sci 103:387
17. Borbás R, Turza S, Szamos J, Kiss É (2001) Colloid Polym Sci (in press)
18. Grosberg AY, Khokhlov AR (1997) Giant molecules. Academic, San Diego, p 103
19. Rohrsetzer S (1996) In: Rohrsetzer S (ed) Kolloidika. Tankönyvkiadó, Budapest, pp 306–308

Progr Colloid Polym Sci (2001) 117: 195–199

B. Kolarić
S. Förster
Regine v. Klitzing

Interactions between polyelectrolyte brushes in free-standing liquid films: influence of ionic strength

B. Kolarić · Regine v. Klitzing (✉)
Iwan-N.-Stranski-Institut für
Physikalische und Theoretische Chemie,
Technische Universität Berlin,
Strasse des 17. Juni 112,
10623 Berlin, Germany
e-mail: klitzing@chem.tu-berlin.de
Tel.: +49-30-31426774
Fax: +49-30-31426602

S. Förster
Institut für Physikalische Chemie,
Universität Hamburg, Bundesstrasse 45,
20145 Hamburg, Germany

Abstract Foam films of charged amphiphilic diblock copolymers are investigated using a thin-film balance. At low polymer concentrations the disjoining pressure shows a continuous exponential decay as a function of the thickness This indicates film stabilization by electrostatic repulsion. At higher polymer concentration stratification occurs with increasing pressure. This is due to the formation of a layer of block copolymer micelles which is pressed out of the film in analogy to usual foam films containing micelles of small surfactant molecules. With increasing salt concentration the film becomes thinner as the charged part becomes more and more coiled and the surface charges more screened.

Key words Polymer brushes · Foam film · Amphiphilic diblock copolymer · Stratification · Disjoining pressure

Introduction

In aqueous solutions diblock copolymers consisting of a hydrophobic and a hydrophilic part can form many different kinds of aggregates, such as micelles, vesicles, wormlike or networklike structures [1–3]. The spherical micelles have a core of an insoluble block and a shell of a soluble block. The shape of the aggregates is affected by the polymer concentration, the ionic strength or the pH of the solution or in the case of adsorbed aggregates by the hydrophobicity of the substrate [4]. The variety of geometrical structures is similar to that of surfactants of low molecular weight but on a nanometer scale. If the hydrophilic block is a polyelectrolyte the quality of the solvent provides additional facilities to tune the shape and the interactions between the aggregates. This leads to new approaches to biological systems and new technical applications (e.g. stabilization of microemulsions, nanocasting); therefore, the aggregation has to be controlled in solution. In this context, more information about the interactions between the aggregates is required. The diblock copolymers are assumed to form a brushlike structure at the interface of the aggregate. The interaction between polyelectrolyte brushes [5, 6] or polyelectrolytes [7–9] adsorbed at solid interfaces has already been investigated using a surface force apparatus. The difference to liquid interfaces (liquid/liquid or air/liquid interface) is the constant packing density after the adsorption process is finished [10]. At liquid interfaces the density of surface-active molecules is variable depending on the different parameters, such as concentration, ionic strength or pH. Especially, in foam films (two opposing liquid interfaces) the surface density increases with decreasing film thickness [11].

Investigations of the structure at one liquid interface (air/water) show that amphiphilic diblock copolymers present polymer brushes which are anchored by the hydrophobic block on the interface [12, 13]. In the present study first disjoining pressure measurements between two opposing charged polymer brushes between liquid interfaces are described. An amphiphilic copolymer consisting of a hydrophobic and a polyelectrolyte block is investigated in the confined geometry of free-standing foam films. These films give the opportunity to control the thickness and, therefore, the degree of confinement and the interaction between the brushes. The term "geometrical confinement" means that the film thickness is of the same order of magnitude as the diameter of a micelle (several tens of nanometers). The film consists of two air/water interfaces and the question arises if the second interfaces influence the formation of polymer brushes which is assumed at a single air/water interface. Furthermore, it should be clarified if the amphiphilic block copolymer behaves in a foam film like a "big surfactant" and what influence the polyelectrolyte part has. For this reason the solvent quality was impaired for the polyelectrolyte part by the addition of salt.

Experimental

The amphiphilic diblock copolymer used for the present studies consists of a poly(styrene sulfonic acid) (PSSH) block of 144 monomers and a hydrophobic poly(ethylethylene) (PEE) block of 136 monomers. Synthetic details are described elsewhere [3]. All solutions were prepared with MilliQ water and the salt (NaCl) was purchased from Merck. The measurements were carried out with a thin-film balance using the porous-plate technique. This method was developed by Mysels and Jones [14] and subsequently enhanced by Exerova and coworkers [15, 16] for measuring the disjoining pressure, $\Pi(h)$, in the film as a function of the film thickness, h (disjoining pressure isotherm). The film is formed from an aqueous polymer solution over the (1 mm) hole drilled in a porous plate. The film holder is enclosed in a cell and the pressure inside the cell is changed using a syringe pump. The film thickness is determined by an interferometric method [17]. The equilibrium thickness is reached when the intensity is constant over a period of 20 min. Further details are described elsewhere [18].

Results

Influence of polymer concentration

Foam films of pure PSSH–PEE are stable without any additional surfactant. The disjoining pressure isotherms at PSSH–PEE concentrations of 1.8 and 3 g/l are shown in Fig. 1. At a concentration of 1.8 g/l the film drains continuously and the data points can be fitted with an exponential function:

$$\Pi_{el} = \Pi_0 \exp[-\kappa(h - 2h_0)] \ , \qquad (1)$$

with the Debye–Hückel length $1/\kappa$ and the interface layer thickness h_0. Π is connected to the surface potential, Ψ_0, by the relation [19]

$$\Pi(h) = 64kT\rho_\infty\gamma^2 \exp(-\kappa h) = (1.59 \times 10^8)[c_{el}]\gamma^2 \exp(-\kappa h), \qquad (2)$$

$$\gamma = \tanh(ze\Psi_0/4kT) \ . \qquad (3)$$

The second part of Eq. (2) is valid at 298 K for monovalent ions and the unit of the pressure is pascals. The thickness of the imterface h_0 was estimated with 22.5 nm as found by Helm et al. for the thickness of one ($PSSH_{144}$-PEE_{136}) brush at the air/water interface (private communication). The fit results in a Debye length of 15 nm and a Π_0 of around 2,200 Pa. The Debye

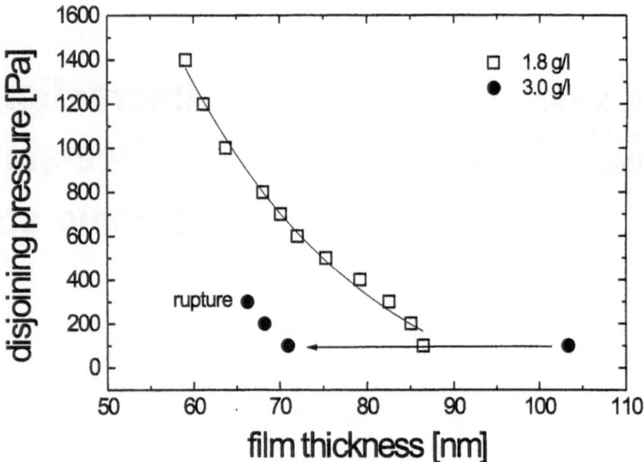

Fig. 1 Disjoining pressure isotherms of a poly(styrene sulfonic acid)–poly(ethylethylene) (*PSSH–PEE*) at two different concentrations: 1.8 and 3 g/l. The *solid line* corresponds to an exponential fit of the experimental data. For details see text

length corresponds to a counterion concentration of 4.4×10^{-4} mol/l. Using this concentration the fitted value of Π_0 results in a surface potential of around 20 mV (using Eqs. 2, 3). The sign of the potential cannot be determined by this method. The exponential decay indicates an electrostatic repulsion between the two film interfaces.

At a higher concentration of 3 g/l the film shows a discontinuous stratification. The isotherm shows a step in film thickness of approximately 32 nm (from 103 to 71 nm). This step size is in good correlation with the size of PSSH–PEE micelles in solution, which have a diameter of approximately 30 nm [4]. After the step the film is thinner than in the case of the lower concentration. The film is less stable and breaks at a pressure of about 300 Pa. The transition from a thicker film (h~100 nm) to a thinner film (h~70 nm) starts in the form of darker spots (Fig. 2) which extend to the "new" (smaller) thickness.

Influence of ionic strength

The disjoining pressure isotherms at a polymer concentration of 1.8 g/l without additional salt and at a NaCl concentration of 50 mmol/l are shown in Fig. 3. After addition of salt the film becomes thinner and the isotherm is steeper. This is an additional proof that the film is stabilized by electrostatic repulsion at low ionic strength. The repulsive interactions are screened partially at higher salt concentrations. The influence of salt at the higher polymer concentration of 3 g/l is shown in Fig. 4. With increasing ionic strength the film becomes thinner after the step and the film becomes less stable. After the addition of salt an experimental problem occurred: Since

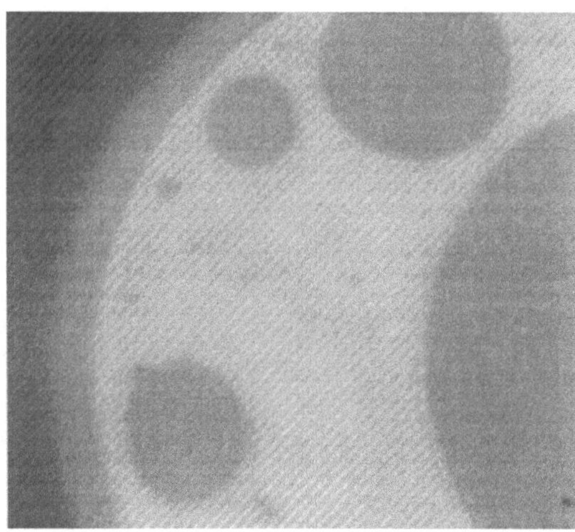

Fig. 2 Photograph of a PSSH–PEE film at a concentration of 3 g/l during the discontinuous transition from a thicker to a thinner film. In the area of the *darker spots* the film has a thickness of around 70 nm, in the *bright part* the film is around 100-nm thick. The *dark grey area* on the *left side* is the meniscus of the film but is not considered in this work

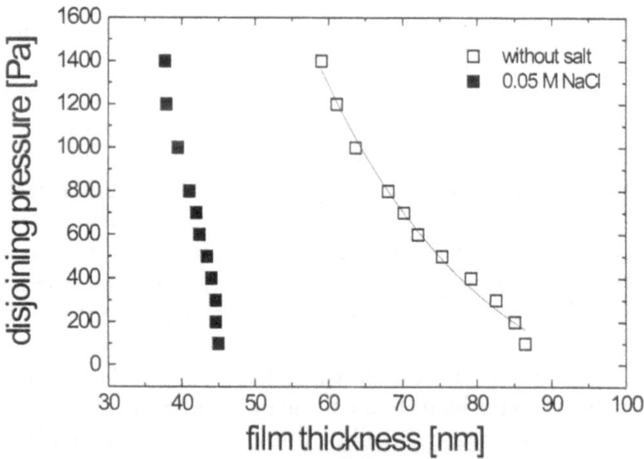

Fig. 3 Disjoining pressure isotherms of PSSH–PEE at a fixed concentration of 1.8 g/l without additional salt and at a NaCl concentration of 0.05 mol/l. The curve without salt is the same as in Fig. 1

the film was not in equilibrium before the step we were not able to estimate precisely the size of the jump. So, only the part of the disjoining pressure isotherm after the step is shown in Fig. 4. Without any additional salt this problem was less pronounced and we were able to determine the complete isotherm as shown in Fig. 1.

Discussion

At lower polymer concentrations the foam film drains continuously and it seems to be stabilized by electrostatic

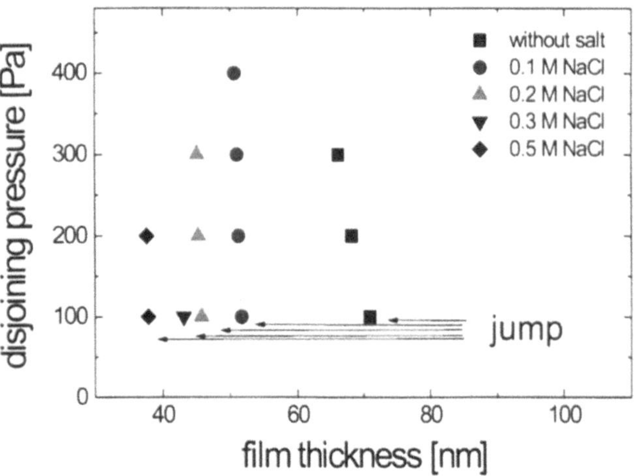

Fig. 4 Part of the disjoining pressure isotherm after the step in film thickness at a fixed PSSH–PEE concentration of 3 g/l and at different NaCl concentrations: 0, 0.1, 0.2, 0.3, 0.5 mol/l

interaction between the interfaces. The 22.5-nm-thick interface layer indicates the formation of polymer brushes at the opposing interfaces which are anchored at the film surface with the hydrophobic PEE block (Fig. 5a).

At the air/water interface X-ray measurement results of a similar system ($PSSH_{83}$–PEE_{114}) are interpreted in the way that the hydrophobic PEE block forms a collapsed 1-nm-thick layer at the interface and the hydrophilic PSS chains are extended into the solution. So, the copolymer molecules form a brush at the interface [12, 13]. X-ray measurements at vertical free-standing films of amphiphilic blockcopolmers (Pt-PSS) also led to the conclusion that polymer brushes are formed at the film interfaces [20]. In these studies the film is drained by gravitation forces at atmospheric pressure. With increasing pressure the opposing brushes are pressed together. In the present study they do not seem to interdigitate. If they were in contact, the electrostatic repulsive force would increase with decreasing h as $1/h$ instead of the exponential increase found [21].

The data points of the measurement at a lower concentration (1.8 g/l) can be fitted by an exponential function, which indicates an electrostatic repulsion between the brushes grafted at the interfaces. The "surface potential" of around 20 mV corresponds to the potential at the brush/solution interface. After Odijk [22] the Debye length consists of a part which is due to the counterions of the polylectrolyte and to the ions of additional salt. In the present case there is no salt; therefore, we can only argue with the counterion concentration of the PSS part to discuss the screening length of the electrostatic repulsion between the brushes. The experimentally determined Debye length is rather large with respect to the counterion concentration.

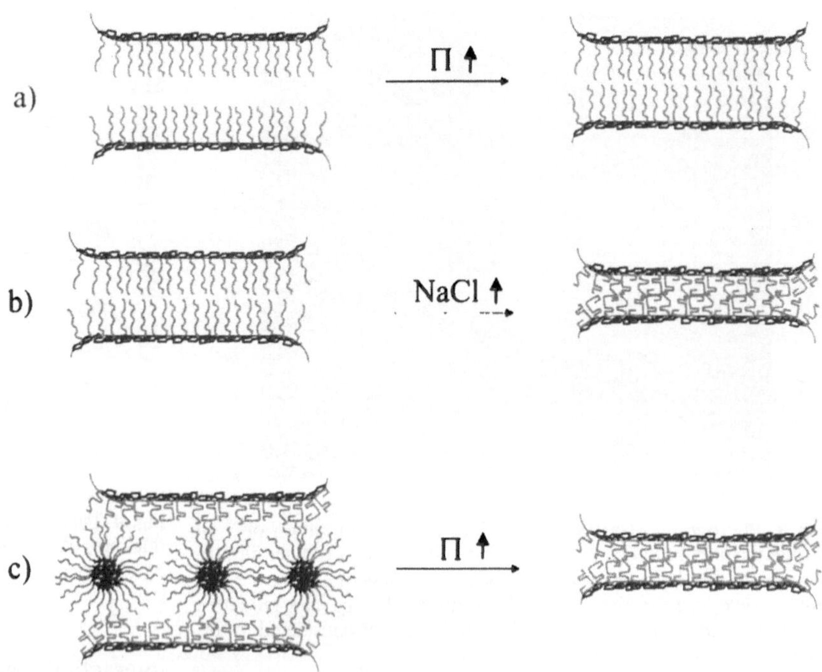

Fig. 5 Assumed structures of a PSSH–PEE film: **a** approach of two brushes at low polymer concentration, **b** influence of ionic strength and **c** expulsion of one layer of micelles at high polymer concentration

A PSSH–PEE concentration of 1.8 g/l corresponds to 7.4×10^{-3} mol/l monomer units of PSS. The distance between two charges along the PSS chain is about 2.5 Å and is smaller than the so-called Bjerrum length, l_B (7.1 Å in water). In this regime counterion condensation must be taken into account [23, 24]. This gives a concentration of free counterions of about one-third of the total number of counterions, i.e. 2.5×10^{-3} mol/l, which results in a Debye length of about 6 nm. The deviation from the experimentally determined value of 15 nm could mean that almost all counterions are in the brush (osmotically swollen brush) and that the counterion concentration in the film core between the brushes is much lower: $1/\kappa = 15$ nm is equivalent to a counterion concentration of about 4×10^{-4} mol/l.[1]

At a salt concentration of 50 mmol/l the Debye length is about 1.4 nm. (This value is smaller than the precision of the fit.) So, the isotherm becomes steeper (Fig. 3). Additionally, the conformation of the polyelectrolytes changes with the ionic strength and becomes more coiled (Fig. 5b). Both effects result in a reduction in film thickness. This salt influence is observed at both polymer concentrations. A decrease in film thickness, after addition of salt, was also observed in vertical films of charged diblock copolymers where the drainage is driven by gravitation [20].

Investigations of horizontal foam films of nonionic poly(ethylene oxide)–poly(propylene oxide)–poly(ethylene oxide) triblock copolymer (Synperonic) results in the transition from Derjaguin–Landau–Verwey–Overbeek (DLVO) forces to steric interactions as forces that stabilize the film. This indicates a brush-to-brush contact at higher capillary pressure [25]. In the present study, the films were not stable at high pressure. Consequently, it is not possible to state whether a comparable transition would also occur in films of charged block copolymers.

A film thinning is also measured after the increase in PSSH–PEE concentration (Fig. 1). The free counterions introduced by the PSS part lead to a screening of electrostatic repulsion between the brushes and along one brush. This leads to a coiling of the PSS part and also causes a decrease in film thickness with increasing polymer concentration. With respect to small surfactant molecules, this thinning of the interface layer after increasing the polymer concentration is counterintuitive since usually the packing of surface-active molecules increases with increasing concentration.

At a higher concentration (3 g/l) the isotherm is not continuous anymore: a step in film thickness occurs. The step size is similar to the diameter of the micelle in an aqueous solution [4]; therefore, a layer of micelles is assumed to be embedded in the film (Fig. 5c, left). In analogy to the solvation forces between spherical particles [19] the interfaces induce a lateral ordering of the micelles. The micelles are squeezed out of the film (Fig. 5c, right) into the surrounding bulk phase, which leads to a lower concentration in the film. This results in attractive depletion forces and in oscillation in the disjoining pressure. In the thin-film balance only the repulsive parts of a pressure oscillation can be measured,

[1] The pH of the water was 5.5. So, the concentration of hydronium and hydroxide ions was much lower than 4×10^{-4} mol/l

resulting in steps in film thickness. A similar picture exists for foam films of small surfactant molecules [26]. Foam films made from surfactant solutions above the critical micelle concentration shows several steps in film thickness which cannot be explained by DLVO forces. This stratification is explained by layer-by-layer expulsion of the micelles. In the present study a maximum of one step can be observed. A possible explanation could be that the pressure barrier which has to be overcome to squeeze out one layer is too small to observe a multilayer ordering between the interfaces. This could be due to a "softer" ordering of micelle layers than in the case of small surfactant micelles. So, just one oscillation could be measured resulting from a single layer of micelles in the film core. Since the step in film thickness and the diameter of the micelle (in solution) are of similar size it is assumed that the micelle is not compressed in the film. In this picture the amphiphilic PSSH–PEE molecule can be considered as a "big surfactant", which leads to a similar film stratification behavior as in the case of small surfactant but on a larger length scale.

Foam films consisting of hydrophilic polyelectrolytes also show stepwise drainage on a mesoscopic length scale [18, 27, 28] but the reason is different. In contrast to the foam films of amphiphilic diblock copolymer, films of hydrophilic polyelectrolytes are not stable. Surfactant has to be added and the interfacial layer is formed by a layer of surfactant or surfactant/polyelectrolytes complexes. The step sizes are independent of the surfactant used and it is assumed that they are caused by a mesoscopic ordering of nonanchored polyelectrolyte chains in the film core.

Conclusions

Foam films of charged amphiphilic diblock copolymers have a kind of "sandwich" structure of two opposing polymer brushes and are stabilized by electrostatic repulsion between the brushes. Further, the exponential decay of the $\Pi(h)$ isotherm indicates that the brushes do not interdigitate. The film thickness decreases with increasing salt concentration for two reasons:

1. As in films of small surfactant molecules the electrostatic repulsion between the two film interfaces is screened.
2. In contrast to foam films of small surfactant molecules the thickness of the interfacial (stagnant) layer shrinks in the case of polyelectrolyte brushes with increasing ionic strength.

In analogy to films of small surfactant molecules, micelles are embedded which can be squeezed out of the film. This is noticeable as steps in the film thickness.

Acknowledgements We thank the DFG (Schwerpunkt Polyelektrolyte mit definierter Molekülarchitektur) for financial support and Deven D. Parghi for reading the manuscript.

References

1. Gast A (1997) Curr Opin Colloid Interface Sci 2:258
2. Alexandridis P (1997) Curr Opin Colloid Interface Sci 2:478
3. Förster S, Hermsdorf S, Leube W, Schnablegger H, Regenbrecht M, Akari S (1999) J Phys Chem B 103: 6657
4. Regenbrecht M, Akari S, Förster S, Möhwald H (1999) J Phys Chem B 103:6669
5. Kurihara K, Kunitake T, Higashi N, Niwa M (1992) Langmuir 8:2087
6. Abe T, Higashi N, Niwa M, Kurihara K (1999) Langmuir 15:7725
7. Klein J, Luckham PF (1984) Colloids Surf 10:65
8. Komiyama Y, Israelachvili J (1992) Macromolecules 25:5081
9. Dahlgreen MAG, Hollenberg HCM, Claesson PM (1995) Langmuir 11:4480
10. Steitz R, Leiner V, Siebrecht R, v. Klitzing R (2000) Colloids Surf A 163:63
11. Exerowa D, Nikolov A, Zacharieva M (1981) J Colloid Interface Sci 81:419
12. Ahrens H, Förster S, Helm CA (1997) Macromolecules 30:8447
13. Ahrens H, Förster S, Helm CA (1998) Phys Rev Lett 81:4172
14. Mysels KJ, Jones MN (1966) Discuss Faraday Soc 42:42
15. Exerova D, Schedulko A (1971) Chim Phys 24:47
16. Exerova D, Kolarov T, Khristov KHR (1987) Colloids Surf A 22:171
17. Schedulko A (1967) Adv Colloid Interface Sci 1:391
18. v Klitzing R, Espert A, Asnacios A, Hellweg T, Colin A, Langevin D (1999) Colloids Surf A 149:131
19. Israelachvili J (1997) Intermolecular and surface forces. Academic, London
20. Guenoun P, Scalchli A, Sentenac D, Mays JW, Benattar JJ (1995) Phys Rev Lett 74:3628
21. Ohshima H (1999) Colloid Polym Sci 277:535
21. Odijk T (1979) Macromolecules 12:688
22. Manning GS (1969) J Chem Phys 51:924
23. Ray J, Manning GS (1997) Macromolecules 30:5739
24. Sedev R, Exerowa D (1999) Adv Colloid Interface Sci 83:111
25. Bergeron V, Radke CJ (1992) Langmuir 8:3020
26. Asnacios A, Espert A, Colin A, Langevin D (1997) Phys Rev Lett 78:4974
27. Kolarić B, Jaeger W, v Klitzing R (2000) J Phys Chem B 104:5096

Progr Colloid Polym Sci (2001) 117:200–203

H. Cölfen
L. Qi

The mechanism of the morphogenesis of $CaCO_3$ in the presence of poly(ethylene glycol)-*b*-poly(methacrylic acid)

H. Cölfen (✉)
Max Planck Institute of Colloids and Interfaces, Colloid Chemistry, Research Campus Golm, Am Mühlenberg 2, 14424 Potsdam, Germany
e-mail: coelfen@mpikg-golm.mpg.de
Tel.: +49-331-5679513
Fax: +49-331-5679502

L. Qi
College of Chemistry
Peking University
Beijing 100871, China

Abstract The morphogenesis of crystalline $CaCO_3$ in calcite modification was systematically investigated in the presence of poly(ethylene glycol)-*b*-poly(methacrylic acid) with respect to pH, polymer and mineral concentration. A variety of different particle morphologies was observed and at least two different mechanisms for the formation of the final $CaCO_3$ structures are proposed. At low polymer concentration or pH, the basis of the mechanism is that the polymer could selectively adsorb onto specific crystal surfaces leading to morphologies elongated along the crystal *c*-axis. At low pH, macroscopic elongated crystals were found resulting from a slow growth process, whereas at high pH or high polymer concentration, spherical structures dominated originating from a more complicated growth scenario: polymer adsorption onto the nanocrystals with subsequent defined aggregation into a superstructure. Here, the elongated crystal is stabilized on the nanometer scale with a growth self-termination and an acting electric multipole field arranges later nucleated particles into the observed superstructures.

Key words Morphogenesis · $CaCO_3$ · Block copolymers · Crystal design · Aggregation

Introduction

Living organisms are well able to synthesize highly optimized organic–inorganic hybrid materials with often complex forms by biomineralization (e.g., seashells, bone, teeth, diatomea and many others). For this, biological systems use organic molecules (mostly proteins and polysaccharides) that function variously as nucleators, cooperative modifiers and matrices or molds for minerals so that they exert exquisite control over the crystallization processes [1–5]. The strategy that organic additives and/or templates with complex functionalization patterns are used to control the nucleation, growth, and alignment of inorganic crystals has been universally applied for the biomimetic synthesis of inorganic materials with complex form [6]. A recently developed class of functional block copolymers as crystal growth modifiers, the so-called double-hydrophilic block copolymers [7], turned out to be very effective for the morphogenesis of inorganic crystals, such as calcium carbonate [7–9], calcium phosphate [10], barium sulfate [11, 12] and zinc oxide [13], owing to the separation of binding and solvating moiety. For $CaCO_3$, different morphologies were reported for the same polymer additive but for various experimental conditions [8, 9], suggesting the importance of experimental parameters such as pH, ionic strength and polymer as well as mineral concentration in a complex overall morphogenesis scenario. These parameters were varied in a recent study on the $CaCO_3$ crystallization in presence of poly(ethylene glycol)-*b*-poly(methacrylic acid) (PEG-*b*-PMAA) and their influence on the $CaCO_3$ morphology was elaborated [14]. In this work, we propose two mechanisms for the observed morphogenesis.

Experimental

The experimental conditions for $CaCO_3$ crystallization with PEG-*b*-PMAA as a growth modifier as well as the characterization methods have been described in previous work [14].

Results and conclusion

At pH 9 $CaCO_3$ crystallization occurs more slowly compared to at higher pH owing to the lower supersaturation which is determined by the pH-dependent carbonate–hydrogen carbonate equilibrium. Here, in the presence of PEG-*b*-PMAA (0.2 g/l), elongated $CaCO_3$ single microcrystals are formed and are shown in Fig. 1a and c.

These crystals show the calcite −1, 1, −4, −1, 0, 4 and 0,1,4 faces (or the three symmetry-related ones at the other end) at their tips and are elongated along the *c*-axis, which is the growth axis (see also Fig. 1c). This morphology is remarkably different from the normal calcite rhombohedrons and is not the obvious result of a preferential polymer adsorption on one or more of the 1, 0, 1; 1, −1, −1 or 0, 1, −1 (or the three symmetry-related) faces of rhombohedral calcite which show a quite even distribution of Ca^{2+} and CO_3^{2-} at the surface. However, a matching of the functional polymer block to the 0, 16, −1, 16, 0, 1 and 16, −16, −1 faces would result in the preferential adsorption to these faces, thus making those unusual faces the most stable, hindering further crystal growth and leading to the observed elongated structures. However, it must be stated that the observed crystal morphologies show some diversity, so the morphology control by polymer adsorption does not seem to be perfect. It is remarkable that these three faces are identical and expose dense, regular arrays of Ca^{2+} ions at their surface, which explains the preferred interaction of the polyanionic polymer block with these faces. At the crystal tips, however, the CO_3^{2-} triangles are exposed to the surface, leading to a negative charge. The Ca^{2+} ions are distributed on a regular lattice on the elongated faces; their distances in perpendicular directions being 499 and 405 pm. The distance between the carbon atoms on the edge of a tetrahedron in the PMAA backbone is 252 pm, so the COOH groups attached to every second carbon on the PMMA backbone favourably match the crystal lattice; we call this soft epitaxy. The elongated structures grow slowly to their final microcrystal size (Fig. 2, mechanism c).

At higher polymer concentrations, however, a different scenario is observed as the polymer is now well able to stabilize the higher surface area of nanoparticles (Fig. 2, mechanisms a, b).

The adsorption does not become selective anymore and the polymer adsorbs on all crystal faces, which leads to a stabilization of the nanoparticles, preventing them

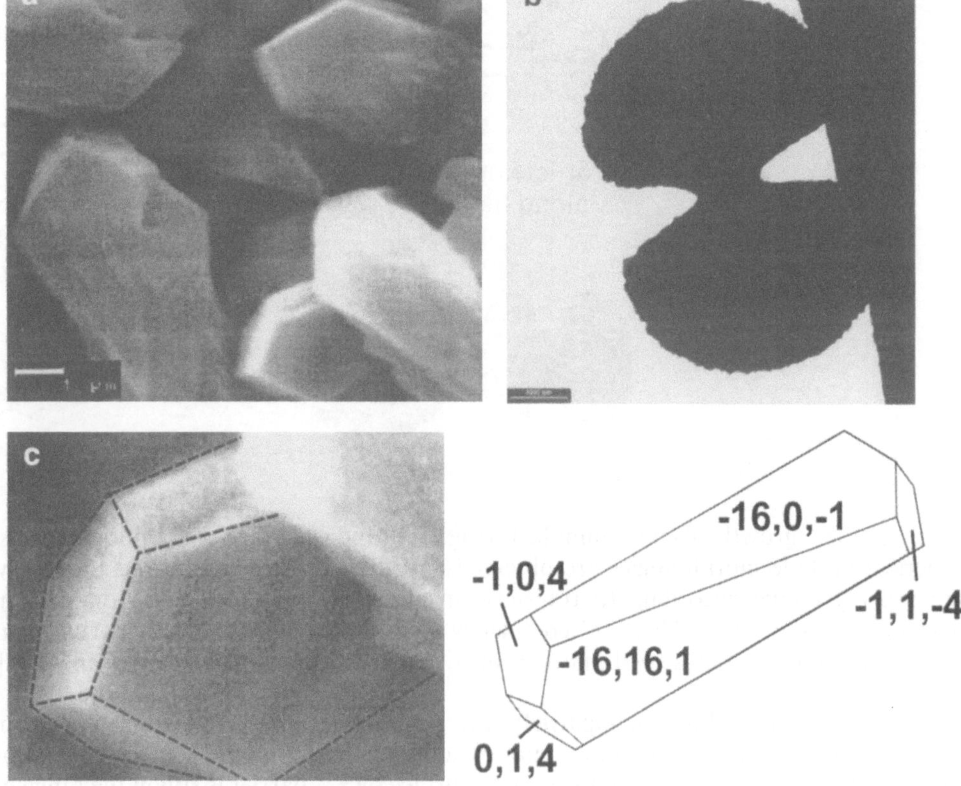

Fig. 1 $CaCO_3$ particles obtained in the presence of poly(ethylene glycol)-*b*-poly(methacrylic acid). $[CaCO_3] = 8$ mM (*a*) [polymer] = 0.2 g/l, pH 9; *b* [polymer] = 0.5 g/l, pH 10; *c* zoom in of the crystal in the lower-right-hand corner of *a* with the corresponding calcite Groth/Krantz crystal model 267 [18]

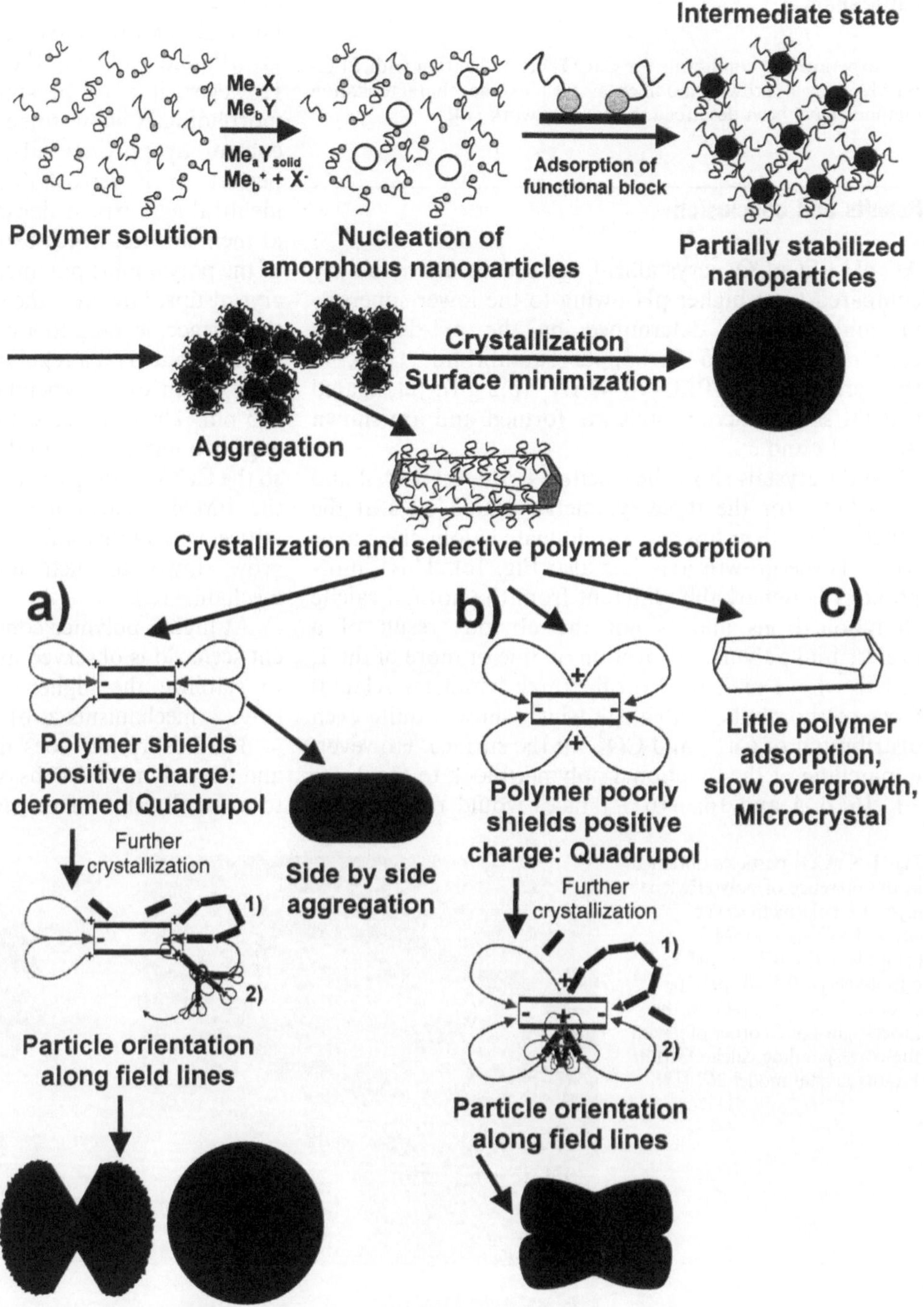

Fig. 2 Proposed mechanisms for the formation of various $CaCO_3$ morphologies resulting from the defined superstructuration of the nanoparticles

from further growth. Here, dumbbell (Fig. 1b) and spherical particle morphologies are observed and these are polycrystalline according to transmission electron microscopy thin cuts. Here, a completely different growth mechanism is observed and is shown schematically in Fig. 2.

As soon as amorphous nanoparticles are nucleated [14], the block copolymer becomes surface-active with the PMAA block adsorbing onto the particle surface by a surface-recognition process and the PEG extending into the solution and thus providing partial steric stabilization depending on the amount of polymer adsorbed, which is mostly not sufficient to prevent aggregation [14]. From the results in Ref. [14], these nanoparticles are amorphous. As the PEG block is rather short (3000 g/mol) extending approximately 3–4 nm into the solution, the steric stabilization is insufficient as the steric barrier is still in the range of the attractive van der Waals forces, which have a range of 5–10 nm. The particles are not completely stabilized and dipolar interactions can range through the stabilization layer and lead to directed particle aggregation. As a consequence of the solution ionic strength of about 0.01 mol/l, there is only little electrostatic shielding, so overall, the attractive van der Waals forces prevail, leading to particle aggregation. Owing to surface-energy minimization as well as to diffusion effects, a spherical superstructure shape is favoured. The particle size, on the other hand, is determined by the minimum of the free enthalpy where the enthalpy decrease owing to surface minimization is balanced by the entropy loss owing to the superstructure formation, so remarkably monodisperse spherical particles are often obtained [8].

The dumbbell particle morphology is regarded as an intermediate case of the sphere formation; however, it cannot be explained by such mechanism. Here, a mechanism suggested by Kniep and coworkers [15, 16] for the fractal growth of fluorapatite in a gelatin matrix is closely related to the observed morphology. If elongated particles are nucleated as the first crystalline species from the pool of the amorphous $CaCO_3$ nanoparticles and the polymer adsorbs specifically to the cationic elongated surfaces, charge and field shielding is obtained in a tensorial manner as outlined in Fig. 2, mechanisms a and b. In any case, the selective adsorption generates an electrical quadrupole field which increases with increasing crystal size so that a self limitation of crystal growth becomes operational and new nucleation events finally become favoured. The quadrupole field lines lead to a positioning of the later nucleated and thus smaller aggregating nanoparticles with respect to the primary crystals along these field lines. It is now the interplay between the van der Waals and dielectric forces and the quadrupole field which determines the superstructure of the compound particle. If attractive forces prevail, a side-by-side aggegation is favoured (for instance by lowering the pH and thus decreasing the negative charge at the crystal tips). If the quadrupole field is strong, it depends on the shielding of the field by the polymer on the elongated faces (Fig. 2, mechanisms a, b), which morphology is favoured. Poor shielding results in the reproduction of the polar field [17], whereas sufficient shielding results in peanutlike particles (Fig. 2, mechanism a, structure 1) or dumbbells (Fig. 2, mechanism a, structure 2) depending on whether further nanoparticles are structured along the field lines of the primary polar field (Fig. 2, mechanism a, structure 1) or if the field of the previously deposited particle is at least partly responsible for the orientation of the next deposited particle (Fig. 2, mechanism a, structure 2).

This results in the observed dumbbell shape where both ends of the dumbbells can grow together at later stages resulting in a spherical particle [15]. Whether or not this mechanism is valid for the $CaCO_3$ case could not be unambiguously elaborated as a result of the difficulty to experimentally observe the primary rodlike particle.

Acknowledgements The authors would like to thank Th. Goldschmidt AG, Essen, Germany, for the supply of the PEG-*b*-PMAA block copolymer and the Max Planck Society for financial support. H.C. also acknowledges the Dr. Hermann Schnell Foundation for financial support.

References

1. Mann S, Webb J, Williams RJP (eds) (1989) Biomineralization, chemical and biochemical perspectives. VCH, Weinheim
2. Mann S (1993) Nature 365:499
3. Mann S (1995) Biomimetic materials chemistry. VCH, Weinheim
4. Weiner S, Addadi L (1997) J Mater Chem 7:689
5. Fritz M, Morse DE (1998) Curr Opin Colloid Interface Sci 3:55
6. Mann S, Ozin GA (1996) Nature 382:313
7. Sedlak M, Antonietti M, Cölfen H (1998) Macromol Chem Phys 199:247
8. Cölfen H, Antonietti M (1998) Langmuir 14:582
9. Marentette JM, Norwig J, Stockelmann E, Meyer WH, Wegner G (1997) Adv Mater 9:647
10. Antonietti M, Breulmann M, Göltner CG, Cölfen H, Wong KK, Walsh D, Mann S (1998) Chem Eur J 4:2493
11. Qi L, Cölfen H, Antonietti M (2000) Angew Chem Int Ed Engl 39:604
12. Qi L, Cölfen H, Antonietti M (2000) Chem Mater 12:2392
13. Öner M, Norwig J, Meyer WH, Wegner G (1998) Chem Mater 10:460
14. Cölfen H, Qi L (2001) Chem Eur J 7:106
15. Kniep R, Busch S (1996) Angew Chem Int Ed Engl 35:22
16. Busch S, Dolhaine H, DuChesne A, Heinz S, Hochrein O, Laeri F, Podebrad O, Vietze U, Weiland T, Kniep R (1999) Eur J Inorg Chem 1643
17. Sedlak M, Cölfen H (2001) Makromol Chem Phys 202:587
18. Burchard U (1994) The P. Groth and F. Krantz crystal model collection. Freising

Progr Colloid Polym Sci (2001) 117:204–210

M. Benna
N. Kbir-Ariguib
C. Clinard
F. Bergaya

Card-house microstructure of purified sodium montmorillonite gels evidenced by filtration properties at different pH

M. Benna · C. Clinard · F. Bergaya (✉)
CRMD CNRS-Université d'Orléans 1b,
Rue de la Férollerie 45071.
Orléans Cédex 2, France
e-mail: f.bergaya@cnrs-orleans.fr

M. Benna · N. Kbir-Ariguib
Institut National de la Recherche
Scientifique et Technique. BP 95,
2050 Hammam-Lif, Tunisia

Abstract The effect of pH on the static filtration properties of purified sodium montmorillonite dispersions depends on the applied pressure. At 1.5×10^5 Pa, the double-layer repulsion resists the applied pressure at natural pH, which is close to neutral pH, and at basic pH. At acidic pH, where the cake is the thinnest and the least permeable, the edge-to-face attractions act as hinges. At natural pH, the thickest and the most permeable cake is obtained. It retains less water than the acidic one because of the absence of edge-to-face contact between particles. These interpretations are confirmed by transmission electron microscopy photographs. At basic pH, the cake retains more water than at lower pH, but it has an intermediate permeability, possibly because of adsorption of water on the negatively charged edges of the clay mineral layers. At 5.7×10^5 Pa, the permeability of the cakes obtained decreases as the pH increases, which probably means the breakdown of the double-layer repulsion at natural pH and the breakdown of the "repulsive network" at basic pH. The results also show that the water retention is the highest in the acidic cake and the lowest in the neutral cake, which probably means that the edge-to-face attractions in the acidic cake resist the pressure more strongly than the face-to-face repulsions in the cake at the natural pH. The X-ray diffraction patterns of the acidic cake do not show a basal reflection, whereas the other cakes show interparticular distances ranging from about 13 to about 39 Å (natural cake) and from about19 to about 33 Å (basic cake).

Key words Clays · pH and filtration properties · Microtexture

Introduction

The use of bentonite clays in drilling fluids is based on their ability to present specific rheological and filtration properties (under dynamic and static conditions). During the drilling operation (even when it is stopped) a thin and impermeable cake is formed on the wall of the hole. Low permeability and water loss are required macroscopic parameters for the suitability of a clay to be used in drilling fluids. Some clays become suitable for use in drilling fluids after acidic or basic activation. The filtration properties are thus dependent on the microtexture of the clay, which depends on two main factors: the water content and pH. The effect of clay content was studied by Benna et al. [1] and the effect of pH is the subject of the present article.

The objective of this work is to determine the influence of the clay microtexture on the static filtration properties of a purified sodium montmorillonite. The filtration properties are followed, as a function of pH variation, under two different constant pressures.

Literature review

Recent literature data on the static filtration properties of bentonite aqueous suspensions are very scarce [1], particularly those on the effect of pH. In studies related to the petroleum industry, Loeber [2] and Li [3] reported results for filtration properties of water/crude montmorillonite/electrolyte/polymer systems at the natural pH of the clay. The static filtrations were done over 0.5 h in an "API standard filter press". Scanning electron microscopy micrographs showed that the cake obtained from a 3% (w/w) crude bentonite aqueous suspension was homogeneous and its permeability was of 4.6×10^{-18} m^2 at 6×10^5 Pa [2]. For the water/clay system, this homogeneity was confirmed by the results of Li [3] which showed that the size, shape and mode of association of clay particles have a great effect on the texture and the permeability of the cake. For the same clay content (3% w/w) the permeability was 14.3×10^{-18} m^2 at 2×10^5 Pa and 4.5×10^{-18} m^2 at 7×10^5 Pa [3].

The effect of clay content in aqueous clay suspensions on the filtration properties was recently studied at natural pH [1]; however, the time of filtration was not fixed and the filtration experiment was stopped when the filtrate did not flow at all. The variation of permeability and/or thickness of the cakes obtained at two different applied pressures allows the sol–gel transition to be detected and the importance of the clay texture to be pointed out.

The pH variation was correlated to the microtexture of the clay in many rheological studies (see Ref. [4] for references). In the rheological investigations of aqueous clay suspensions, the association of particles was the subject of different interpretations [5]. The major question is the existence or the nonexistence of the edge-to-face interaction. It is very important to note that all authors who demonstrated the face-to-face association between clay particles by rheological investigation or textural studies [6–9] used montmorillonite suspensions at the natural pH of the clay, without any variation of pH. However, the existence of what is called a card-house [10] microtexture between clay particles in aqueous suspensions owing to attractive edge-to-face associations in acidic media, or owing to repulsive interparticle interactions in basic media, has been discussed in our previous work [4].

The water loss and the permeability of crude bentonite/cement slurries have been investigated in function of high pH values, from 8.1 to 12.66, in the presence of added calcium ions [11]. For the crude bentonite suspension (3% w/w) without added cement and at a Ca^{2+} concentration of 10^{-2} mol/l, the permeability is 14.5×10^{-18} m^2 at 6×10^5 Pa. The water loss remains constant until pH 12, but at pH 12.66, the permeability increases and the water loss decreases.

Experimental

Materials

The Wyoming bentonite used in this study was provided by Comptoir des Minéraux et Matières Premières, France (ref. SPV) and was purified by the classical NaCl exchange [4, 12].

The granulometry and the method of preparation of the suspensions have a great influence on the degree of dispersity of the clay and on the filtration properties. So, it is important to state that all samples were prepared in the same way. The granulometry of the dry purified sodium bentonite was controlled by sieving the dry clay powder to 80 μm. During the sample preparations, it was particular care was taken to prevent attack of the clay mineral structure by avoiding direct contact of acidic or basic compounds with the dry bentonite powder. The clay content of 5.5% (w/w) used corresponds to the lowest amount of this bentonite which exhibits gel behaviour.

Preparation of the suspensions

The acidic (HCl) or basic (NaOH) solution (50% by weight of the final suspension) was added to the aqueous clay suspension, which was shaken by hand for 5 min (5.5% clay and 44.5% bidistilled water by weight of the final suspension). The suspension was shaken on a roller shaker for 24 h, then left to rest for 24 h before the experiments were carried out.

pH measurements

Three values of pH were chosen according to previous results [4]: the natural pH of the aqueous bentonite suspension (pH 7.50) and pH 4.23 and 11.75 corresponding to the highest values of the yield stress in acidic and basic media. All pH values were measured with a high-precision Ingold combined glass electrode using a Bioblock Scientific 93313 pH meter.

Filtration

The suspension was filtered in a stainless steel Millipore filtration cell as already described [1]. The filtrate volume was measured as a function of time. The water retention, which is the complement of the water loss (water loss = 100–water retention), was determined by measuring the final filtrate volume ($V_{f.f.}$). As the volume of water in the initial suspension ($V_{i.w.}$) is known, the water retention is defined by

$$\text{Water retention} = [(V_{i.w.} - V_{f.f.})/V_{i.w.})] \times 100 \ . \quad (1)$$

The coefficient of permeability (k), in square metres, is calculated assuming that the microstructure of the cake is uniform, which means that k is constant in all the bulk volume of the cake during the filtration. This allows the use of the Darcy law:

$$\mathrm{d}V_{t.f.}/\mathrm{d}t = (kSP)/(\mu e) \ , \quad (2)$$

where $V_{t.f.}$ is the filtrate volume at time t, S is the filtration surface ($S = 13.2 \times 10^{-4}$ m^2), P is the experimental applied pressure, μ is the filtrate viscosity (in our experiments $\mu = 0.001$ Pas = viscosity of water) and e is the thickness of the cake.

Integration of this equation assuming volume conservation gives the final form of the Darcy law [1]:

$$V_{t.f.}^2 = (2kSP/\mu e)V_{f.f.}t \quad (3)$$

Thus, the permeability coefficient, k, can be determined experimentally from $V_{t.f.}^2 = f(t)$ curves, where the linear slope is $2kSPV_{f.f.}/\mu e$.

For each pH value studied, the static filtration experiments were carried out at 1.5×10^5 and 5.7×10^5. The flow rate of filtration was relatively high at the beginning and decreased near the end when the filtrate did not flow at all.

The cake removed from the filtration cell was immediately divided into two parts: one part was immediately used to measure the thickness of the cake; the second part of the cake, which was obtained at 1.5×10^5 Pa, was prepared for transmission electron microscopy (TEM) observation. When the applied pressure was 5.7×10^5 Pa, the second part of the cake was immediately used, without grinding, in a Siemens D500 diffractometer to obtain the X-ray patterns.

Results

Filtering properties

Darcy law

The linear variation of the square of the filtrate volume is shown in Fig. 1a and b as a function of time for all the sodium montmorillonite suspensions, indicating the validity of the Darcy law. At high applied pressure the time required for filtration increases as the pH increases. This behaviour was not observed at lower pressure.

Water retention

The water retention is probably the simplest macroscopic parameter which gives an idea of the microtexture of the cake; its variation with pH at 1.5×10^5 and 5.7×10^5 Pa is shown in Fig. 2. The cakes retain more water at the lower pressure. Whatever the applied pressure, the water retention of the cake is the lowest at natural pH. At basic pH, the water retention is higher than at acidic pH at the lower pressure. At a higher pressure of 5.7×10^5 Pa, the opposite behaviour is observed: the acidic cake retains more water than the basic one.

Thickness and permeability

The variations of the thickness and the permeability with pH depend on the applied pressure (Figs. 3, 4). At the lower pressure, the thickness and the permeability are highest at the natural pH and are higher at basic pH than at acidic pH; however, at 5.7×10^5 Pa the thickness and the permeability decrease as the pH increases and have their highest values at acidic pH.

Microstructure of the cakes

X ray diffraction of cakes obtained at 5.7×10^5 Pa

The X-ray diffraction patterns (Fig. 5) show a broad distribution of peaks between 13 and 39 Å for the cake at

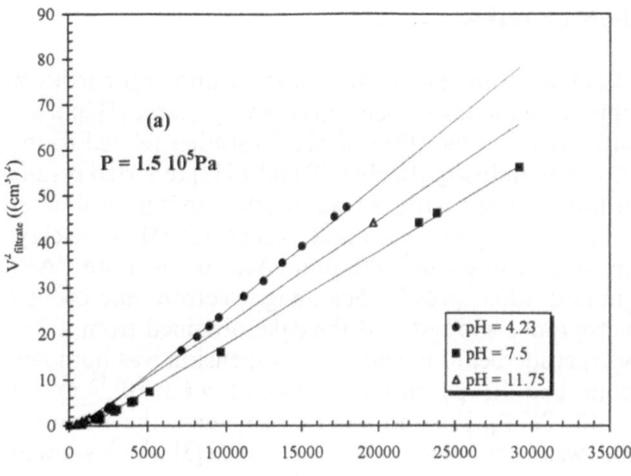

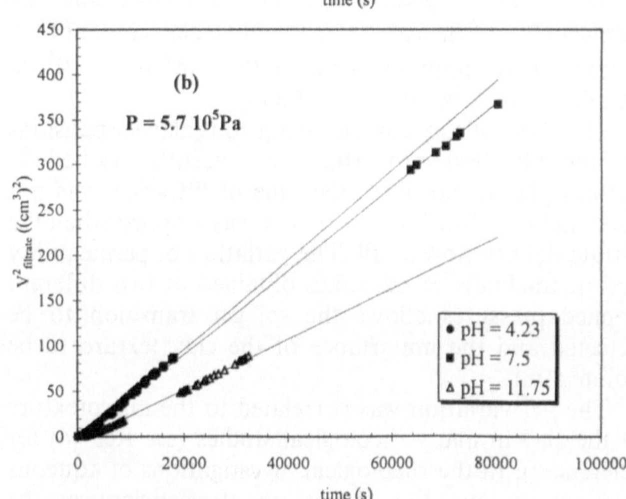

Fig. 1a, b Darcy law. Linear variation of the square of the filtrate volume with time

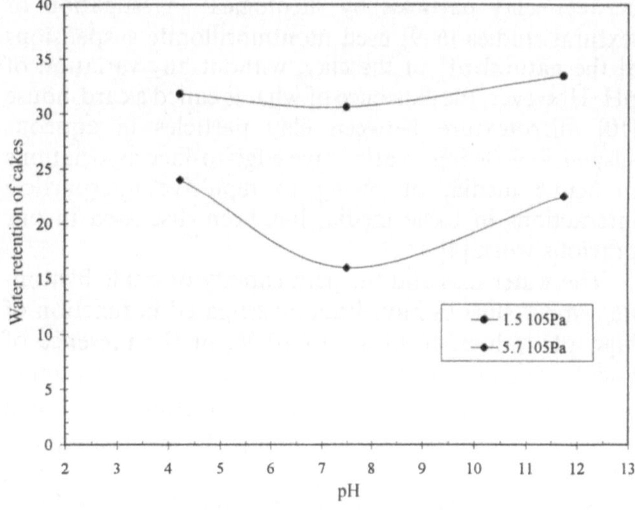

Fig. 2 Water retention of the cakes as a function of pH

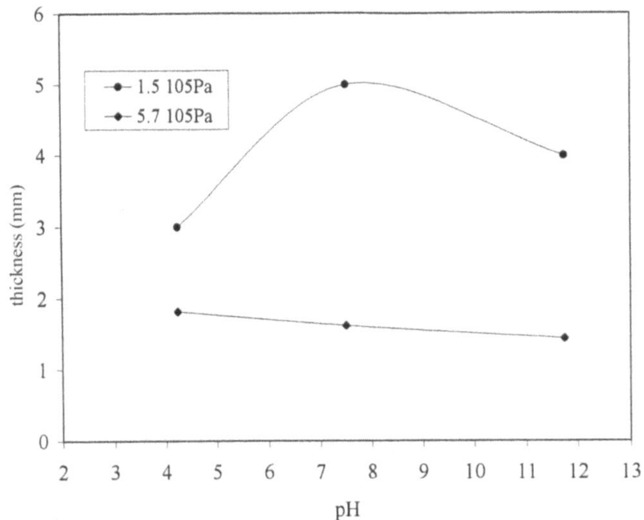

Fig. 3 Thickness of the cakes as a function of pH

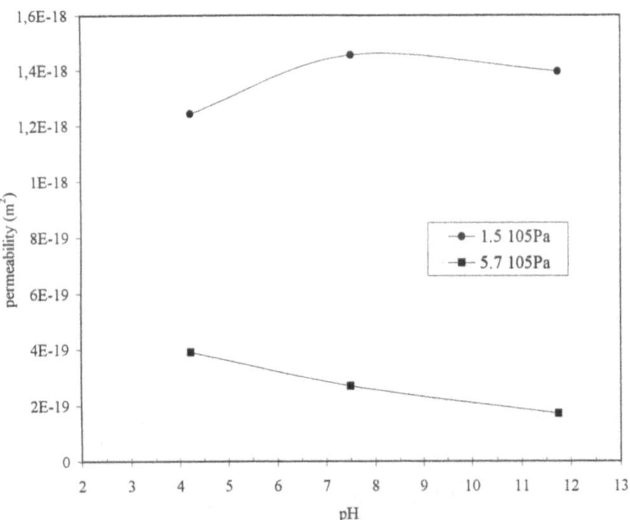

Fig. 4 Permeability of the cakes as a function of pH

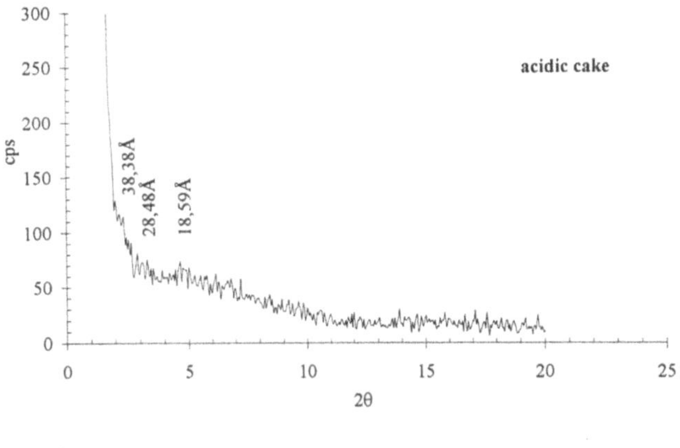

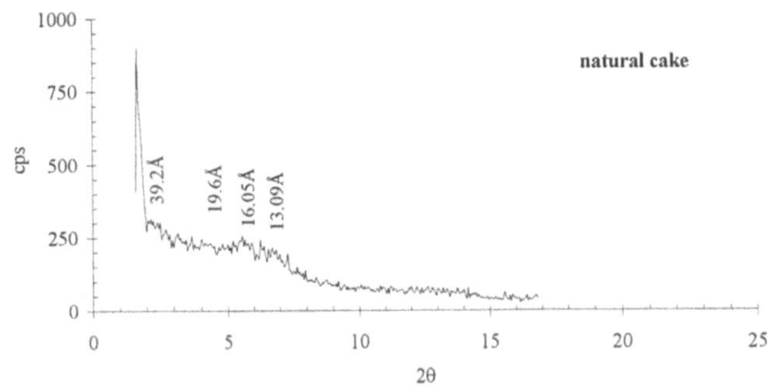

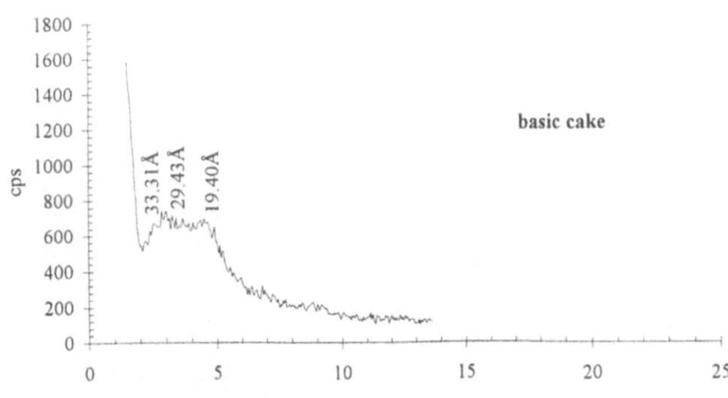

Fig. 5a–c X-ray diffractograms of the cakes obtained at 5.7×10^5 Pa

natural pH and between 19 and 33 Å for the cake at basic pH. The X-ray diffraction pattern of the acidic cake does not show significant peaks in the region of small distances and shows only a relatively weak peak at about 38 Å.

TEM of cakes obtained at 1.5×10^5 Pa

Observations at low magnification give an idea of the texture (homogeneity, orientation and porosity) of the cakes obtained.

- In the acidic cake (Fig. 6a) a pronounced disorientation of the particles is observed, and there are many zones that are very dense. The porosity essentially consists of closed pores with broad size distribution.
- The basic cake (Fig. 6b) is less disoriented than the acidic one. There is a preferential orientation which gives a majority of open pores. There are some zones that are denser than the others.
- The natural cake (Fig. 6c) is the most-oriented sample. A few zones are denser than the others. We have already observed this phenomenon in the initial natural gel. The porosity essentially is related to open pores.

Observations at high magnification are shown in Fig. 7.

- The acidic cake (Fig. 7a) is the most disoriented and the densest cake; however, a preferential orientation of the long particles is detectable. The smaller

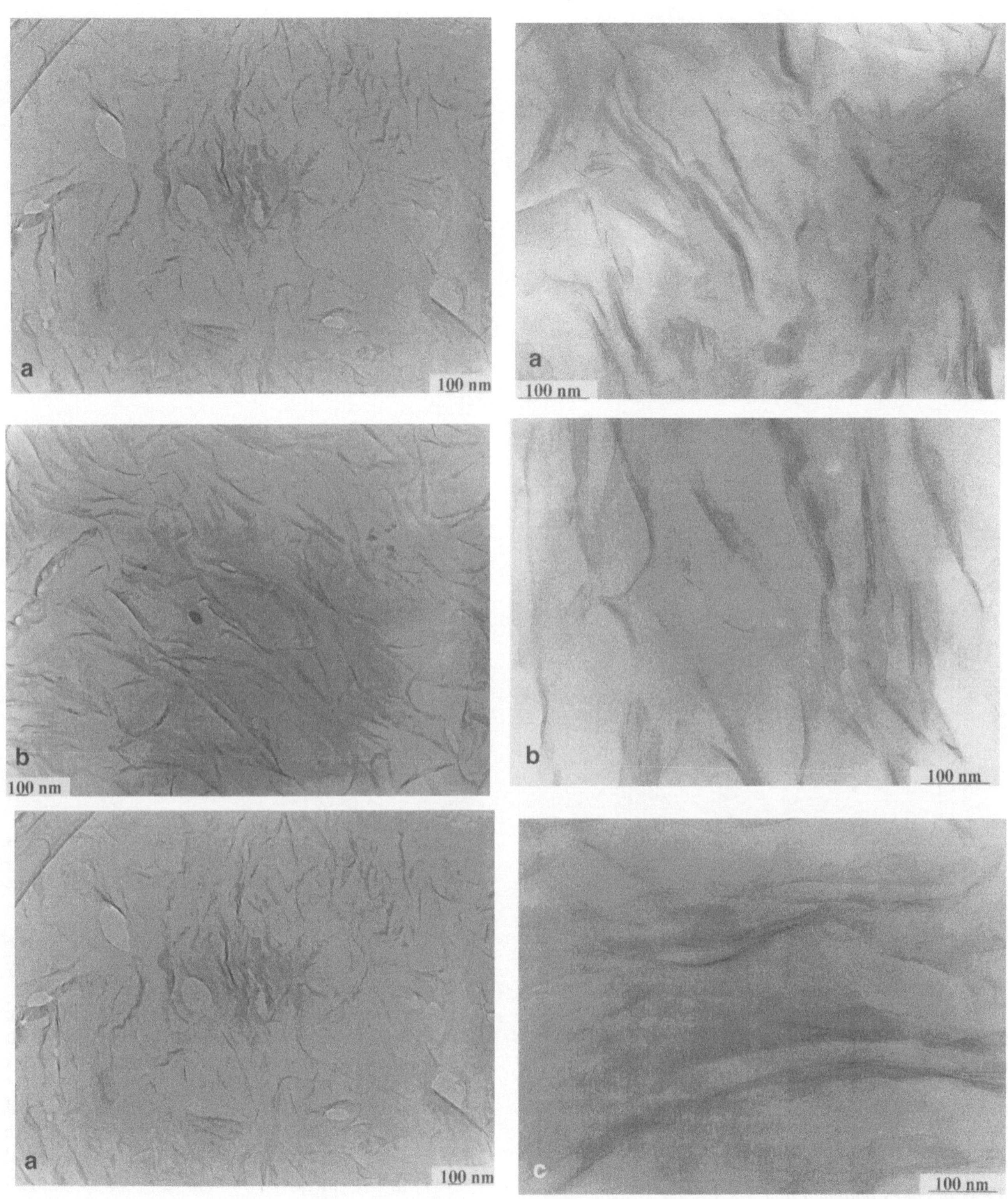

Fig. 6a–c Transmission electron microscopy (*TEM*) photographs of the cakes obtained at 1.5×10^5 Pa at low magnification. **a**: pH 4.23, **b** pH 11.75 and **c** pH 7.50

Fig. 7a–c TEM photographs of the cakes obtained at 1.5×10^5 Pa at high magnification. **a** pH 4.23, **b** pH 11.75 and **c** pH 7.50

particles are very disoriented, with many edge-to-face contacts. This is explained by the fact that particles with a small lateral extension can be easily disoriented, while the long particles, because of their size, cannot be easily rearranged. The edge-to-face attractions, which are well built-up at this relatively low pressure, are clearly shown on the TEM photograph. The distribution of the number of layers per particle is between 2 and 9 with many isolated layers.

- The right side of the photograph in Fig. 7b illustrates the tendency of the orientation and the anisotropy of the porosity in the basic cake. On the left side, we see the disorientation of particles; thus, the resistance of the "repulsive network" to the low applied pressure. The distribution of the number of layers per particle is mainly from 2 to 9 as in the acidic cake, but in this case the isolated layers are very scarce.
- The preferential orientation of particles of the natural sample (Fig. 7c) is clearly shown and there is, as expected, only one type of interparticular interaction, the face-to-face aggregation. The open porosity is also confirmed at this scale, and the density of the particles is the lowest. Isolated layers are scarce and the stacking of three layers per particle is most frequent and representative, but stackings of up to ten layers per particle are also observed.

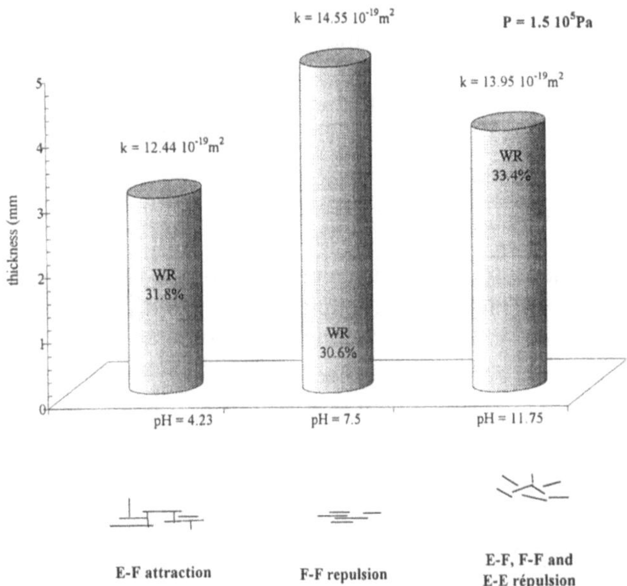

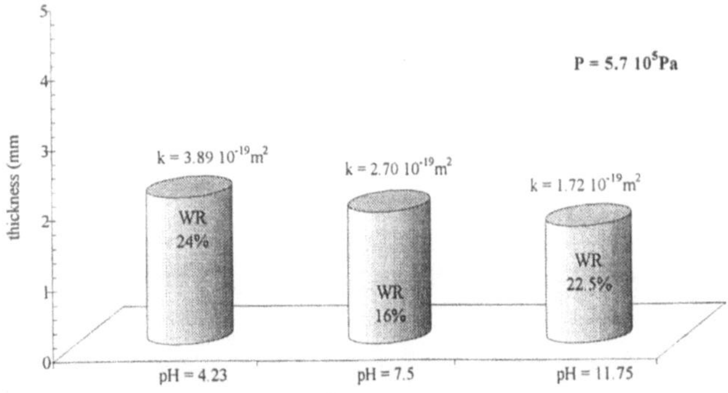

Fig. 8 Summary of the results of cake properties (thickness, *e*, water retention, *WR*, and permeability, *k*) as a function of pH and applied pressure

Discussion

According to our previous work [4] on the rheological and electrokinetic properties of bentonite as a function of pH, the three values of the pH chosen correspond to the maximum of face-to-edge attraction (pH 4.23), the isoelectric point of the edges (pH 7.50), which corresponds to zero charge on the edges and there is only the face-to-face double-layer repulsion, and all the contribution of edge-to-face, face-to-face and edge-to-face repulsion (pH 11.75) when all edges are negatively charged.

The variation of the filtration properties with pH is summarised in Fig. 8.

- At the relatively low applied pressure (1.5×10^5 Pa) the acidic cake is less permeable and retains more water but it is thinner than the natural one. The edge-to-face attractions (as hinges) attract the layers. At the natural pH, the diffuse double-layer repulsion prevents the aggregation of the clay mineral layers at this relatively low pressure. The basic cake is more permeable than the acidic one but it contains more water. In basic media, all faces and edges are negatively charged and water is probably adsorbed on faces and edges. In acidic media, the existence of edge-to-face connections prevents adsorption of water on these localised sites and consequently the basic cake contains more water than the acidic one.

At 5.7×10^5 Pa the applied pressure is probably sufficiently high to overcome the double-layer repulsion at natural pH and the natural cake retains less water than the other two. Also this pressure is sufficient to break down the three-dimensional "repulsive network" in basic media. As the particles in the initial basic gel are in all directions, the basic cake traps more water than the natural one, in which the particles are parallel; therefore, the basic cake is less permeable and thinner than the natural one.

At the acidic pH, the cake is the most permeable and its water retention is the highest, which proves that it is the most porous. This highest porosity is confirmed by the fact that the acidic cake is the thickest one. This displays once again the existence of edge-to-face attractions at acidic pH, which allow more water to be trapped between particles.

To summarise, the effect of pH on the filtration properties is explained in terms of interparticular interactions. If the edge-to-face attraction did not exist in acidic media and if the edge-to-face, face-to-face and edge-to-edge repulsions did not exist in basic media, otherwise if there was only face-to-face repulsion whatever the pH, there would be no variation of the filtration properties with pH. The experiments show clearly that permeability, thickness and water retention of the cakes vary with pH. Also, the filtration parameters change with applied pressure, which means that the interparticular interactions change when the pH is varied.

Conclusion

The microtexture of the cakes depends on the interparticular interactions in the initial suspension. The results obtained by different techniques (filtration, TEM and X-ray diffraction) are in good agreement with the literature data on the interactions between particles in the montmorillonite gels. Comparison of the filtration and textural properties of the bentonite studied allows the interparticular interactions between clay particles at different pH to be revealed. The existence of edge-to-face attraction was confirmed in acidic media. In highly basic media, a network based on edge-to-edge, edge-to-face and face-to-face repulsion instead of attraction was also confirmed.

Acknowledgements The authors wish to thank M. Crespin for the X-ray diffraction experiments and T. Cacciaguerra for the TEM sample preparation.

References

1. Benna M, Kbir-Ariguib N, Clinard C, Bergaya F (2001) Appl Clay Sci 19:103–120
2. Loeber L (1992) Thesis. University of Orléans
3. Li Y (1996) Thesis. University of Orléans
4. Benna M, Kbir-Ariguib N, Magnin A, Bergaya F (1999) J Colloid Interface Sci 218:442–455
5. Norrish K (1954) Discuss Faraday Soc 18:120–134
6. Pons CH, Rousseau F, Tchoubar D (1981) Clay Miner 16:23–42
7. Tessier D (1984) Doctoral thesis. University of Paris VII
8. Tessier D (1991) In: De Boodt M, Hayes M, Herbillon A (eds) Soils and their association in aggregates. Plenum, New York, pp 387–415
9. Gaboriau H (1991) Thesis. University of Orléans
10. Van Olphen H (1977) Introduction to clay colloid chemistry, 2nd edn. Wiley, New York
11. Plée D, Lebedenko F, Obrecht F, Letellier M, Van Damme H (1990) Cem Concr Res 20:45–61
12. Annabi-Bergaya F (1978) Doctoral thesis. University of Orléans

Progr Colloid Polym Sci (2001) 117: 211–216

Ottó Horváth
J. Hegyi

Light-induced reduction of heavy-metal ions on titanium dioxide dispersions

O. Horváth (✉) · J. Hegyi
University of Veszprém,
Department of General and Inorganic Chemistry, 8201 Veszprém,
P.O. Box 158, Hungary
e-mail: otto@vegic.sol.vein.hu
Tel.: +36-88-427915
Fax: +36-88-427915

Abstract Photoactive semiconductors such as TiO_2 can be applied for wastewater treatment; not only for degradation of organic pollutants, but also for removal of heavy-metal ions. Photoassisted reduction of different, mostly toxic metal ions, such as Hg(II), Pb(II), Bi(III), and Cu(II), has been realized in aqueous systems containing ethanol as a sacrificial electron donor. The efficiency of the photoreduction carried out under the same circumstances is primarily determined by the standard reduction potential of the species containing the metal ion to be removed as demonstrated by the examples of Hg(II), Bi(III), and Pb(II). The relatively lower efficiency for the deposition of Cu may be attributed to a short-circuiting effect. Surfactants of different type also function as potential electron donors in these systems. Negatively charged dodecyl sulfate proved to be much more efficient than ethanol, while cationic cetyltrimethylammonium is rather weak in this respect.

Key words Photocatalysis · Semiconductor · Heavy metal · Decontamination · Titanium dioxide

Introduction

Semiconductor catalysis has received considerable attention in the past two decades as an alternative for treating water polluted with hazardous organic chemicals [1–10]. To a lesser extent, it has also been studied for application to water containing metals (for decontamination and metal recovery) [11–15]. The utilization of semiconductor photocatalysis as a means of decontaminating heavy-metal (Hg, Pb) polluted water has several advantages compared to other methods. Toxic metals in their oxidation states can neither be biodegraded nor can they be chemically removed. However, under appropriate conditions quantitative removal of heavy-metal ions becomes possible with irradiated TiO_2 semiconductor particles [16–18]. Mostly, the metal ions are deposited on the surface of the semiconductor particles in a reduced, elemental form. The efficiency of this process depends on the standard reduction potentials of the metal ions, the Gibbs free-energy change of the overall reaction, the presence or absence of oxygen, the concentration of the sacrificial electron donor, the irradiation time, and the pH of the system. In this article, we present work on the photoassisted reduction of Hg(II), Pb(II), Bi(III), and Cu(II) on TiO_2. We show how the application of an added sacrificial electron donor (ethanol) can make the overall reaction catalytic, how the concentration of this reductant influences the efficiency of the process. The effects of commercial surfactants as potential electron donors were also studied for the purpose of simultaneous removal of heavy-metal ions and detergent contaminants from wastewater.

Experimental

Reagents

The following metal salts (reagent grade) were used: $HgCl_2$, $CuSO_4 \cdot 5\ H_2O$, $Pb(NO_3)_2$, $BiCl_3$ (Aldrich). TiO_2 (Fluka, predominantly in rutile form) was utilized as a photocatalyst. Triply distilled water was used for the preparation of solutions containing 1×10^{-3} M metal ion to be reduced and 0.1 wt % TiO_2 suspended. Analytical grade ethanol, sodium dodecyl sulfate(SDS), and cetyltrimethylammonium bromide (CTAB) (Aldrich) were applied as electron donors in various concentration. No pH adjustment of the solutions to be irradiated was done.

Instruments

A 150-W high-pressure mercury lamp was utilized as a light source. A 10-cm cylindrical glass cuvette filled with circulating cooling water was used to absorb the IR emission toward the photoreactor, which was a 1-cm cuvette with 3-cm^3 volume. It was open on top, and 1.2 cm^2 of its front surface was illuminated. Using a spectroradiometer, the integral lamp output between 310 and 390 nm was found to be 62 mW/cm^2 at the front surface of the reactor cuvette. A GBC UV–vis 911A spectrophotometer was applied for the analysis.

Procedure

Photolyses of the reaction mixtures were carried out for various time intervals up to 40 min. Oxygen-free experiments were carried out by bubbling nitrogen through the reaction solution for 15 min prior to and throughout the experiment. The contents of the reactor were magnetically stirred and thermostated (at 20 °C) by using an aluminum block with circulating water.

Analysis

After the irradiation the suspension was centrifuged and the supernatant solution was analyzed for metal ions spectrophotometrically by using halide ions as complexing agents. A 0.5 cm^3 aliquot of the clear part of the sample was taken and introduced into 2.5 cm^3 4 M NaCl solution in a 1-cm quartz cuvette. The absorbance of the chlorometallate complexes formed was measured at a wavelength appropriate for the metal ion studied. Only rather dilute halide solution could be used for the systems containing surfactants because of their precipitation at higher ionic strenghts. In these cases, 2 cm^3 clear sample was mixed with 1 cm^3 0.01 M NaI solution to form iodometallates. Absorbances of the halometallates obeyed the BLB law for each metal ion in the concentration range studied ($0 - 7 \times 10^{-4}$ M).

Results and discussion

Theory

The energetics and charge-transfer processes involved in a typical photocatalytic process are illustrated in Scheme 1. Bandgap illumination (*hv*) of a semiconductor particle suspended in water causes electronic transitions from the valence band to the conduction band, leaving holes in the former. These electrons and holes then either migrate to the particle surface and become involved in redox reactions or they recombine and simply liberate heat.

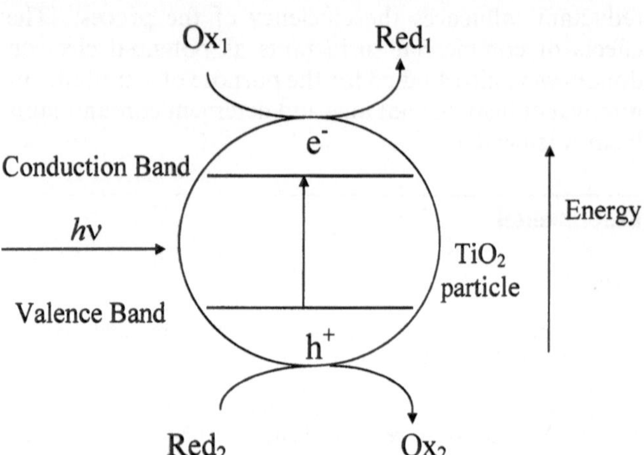

Conduction band electrons are consumed in reactions that reduce oxidants ($Ox_1 \rightarrow Red_1$), while holes are filled via oxidation reactions ($Red_2 \rightarrow Ox_2$).

The photogenerated hole (h^+) can oxidize H_2O as an initial step.

$$2H_2O + 4h^+ \xrightarrow[h\nu]{TiO_2} O_2 + 4H^+ \quad (1)$$

Oxygen molecule, however, can function as an electron acceptor to chemically reverse this process.

$$O_2 + 4H^+ + 4e^- \xrightarrow[h\nu]{TiO_2} 2H_2O \quad (2)$$

If the carrier fluxes owing to the reactions in Eqs. (1) and (2) are exactly balanced, there is no net chemistry, and O_2 has functioned as a chemical surface state for mediating the e^-–h^+ recombination process. Net chemistry can only occur at the TiO_2 surface if either reaction is intercepted at an intermediate stage. Alternatively, the $e^-$and h^+ at the TiO_2 surface must react with different redox couples in the contacting medium as illustrated in Scheme 1.

In pollutant destruction, either Eq. (1) or Eq. (2) constitutes half the conjugate reaction pair; the other half of the pair or partner comprises the pollutant molecule, ion, or microorganism.

$$4OH^- + 4h^+ \xrightarrow[h\nu]{TiO_2} 4OH^{\bullet} \quad , \quad (3)$$

$$O_2 + 2H_2O + 4e^- \xrightarrow[h\nu]{TiO_2} 4OH^- \quad (4)$$

Hydroxyl radicals are generated by the oxidation of water at the valence band of TiO_2 [1, 4, 6, 7]. This occurs at a standard potential of 2.8 V [19], which decreases with increasing pH. The very reactive OH radicals interact with the organic substrates, promoting their oxidation and, finally, mineralization.

During typical photocatalytic oxidation of organic components, oxygen is reduced to become superoxide (–0.56 V) and/or perhydroxyl (–0.13 V) radicals [19], depending on the pH. These radicals eventually form hydroxyl radicals, which enter into the oxidation cycle [1, 4, 6, 7].

Besides oxygen, any dissolved species with a reduction potential more positive than the conduction band of the photocatalyst can, in principle, consume electrons and complete the redox cycle. For example, with a toxic metal ion, M^{2+}:

$$2M^{2+} + 4e^- \xrightarrow[h\nu]{TiO_2} 2M \quad (5)$$

In this instance, Eqs. (1) and (5) form a conjugate pair, giving the overall reaction

$$2M^{2+} + 2H_2O \rightarrow 2M + O_2 + 4H^+ \qquad (6)$$

Considering the energetics of this reaction (Eq. 6), it is photocatalytic if its Gibbs free-energy change is negative ($\Delta G^0 < 0$) [20]. In this case, the reaction is driven in the spontaneous direction by the light, and the radiant energy simply serves to overcome the activation barrier for the process. If Eq. (6) involves a positive ΔG^0, the reaction is photosynthetic, the light drives the system in the thermodynamically uphill direction. Photoassisted reduction of metal ions is photosynthetic in most cases, except for those with very positive potentials, i.e., when they have standard reduction potentials that are more positive relative to the O_2/H_2O redox couple. Also these reactions can be made photocatalytic with the addition of appropriate hole scavengers, resulting in a negative free-energy change for the overall process.

Photocatalytic reduction of Hg(II)

Hg(II) is the only one of the metal ions studied for which the standard reduction potential is more positive than that of the O_2/H_2O pair at pH 7 (0.85 V versus 1.229–0.059pH = 0.816 V [21]). The difference between the redox potentials, however, is not significant. Besides, this reduction potential (0.85 V) is valid only for Hg^{2+}. In aqueous solutions and natural pH, $HgCl_2$ dissolves undissociated; about 1% $HgCl_2$ exists as Hg^{2+} [19]. The redox potential for the $HgCl_2/Hg$ couple is 0.41 V [22], which is too low for a catalytic reaction. Owing to this fact, photolysis of an air-saturated solution did not lead to an appreciable deposition of mercury on TiO_2, i.e., less than 10% of the initial concentration (1×10^{-3} M) of Hg(II) was reduced after 20-min irradiation. Photolysis carried out under similar circumstances but in a deoxygenated system with continuous purging with nitrogen resulted in 40% efficiency of reduction after 20 min (Fig. 1). Addition of 5×10^{-4} M ethanol (as a hole scavenger) dramatically accelerated the photoinduced deposition of mercury. More than 85% of the initial amount of Hg(II) was reduced within 10 min (Fig. 1). It clearly indicates that the overall reaction

$$HgCl_2 + CH_3CH_2OH \xrightarrow[h\nu]{TiO_2} Hg + 2HCl + CH_3CHO \qquad (7)$$

is photocatalytic. A sharp decrease in the reduction rate can be noticed in the reduction efficiency versus irradiation time plot after 10-min irradiation. This phenomenon can be attributed to the limiting effect of the low initial concentration of ethanol and the decrease in the actual reduction potential of Hg(II). Theoretically, according to Eq. (7), 5×10^{-4} M ethanol stoichiometrically correponds to 1×10^{-3} M $HgCl_2$. At about 90% deposition, however, the reduction potential of Hg(II) in this system is about 0.6 V lower than its value at the beginning of the photolysis. Besides, most of the electron donor ethanol is used up by that time. These simultaneous effects decrease the rate of the photoinduced reduction by more than 1 order of magnitude. Of course, at higher ethanol concentration (above 0.03 M) no similar effect can be experienced, and photoreduction and elimination of mercury from the solution is more than 99% complete in under 20 min of illumination, even in the air-saturated system.

Photocatalytic reduction of Bi(III)

The standard reduction potential of the Bi(III)/Bi couple is 0.317 V [23], which is more negative than that of the O_2/H_2O pair in neutral and, especially, in acidic solution.

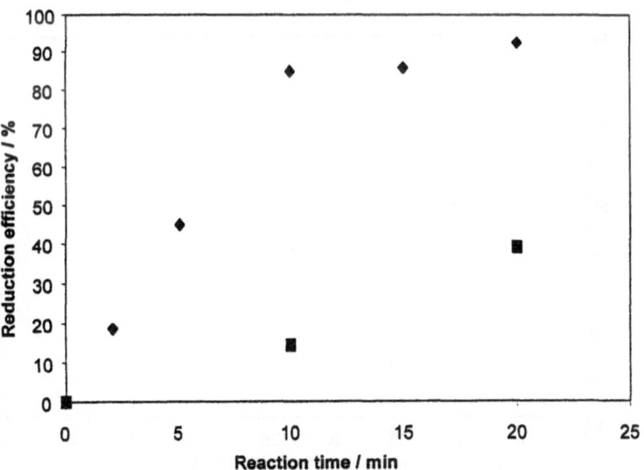

Fig. 1 Efficiency of Hg(II) reduction versus irradiation time in the absence (■) and in the presence (◆) of (5×10^{-4} M) ethanol

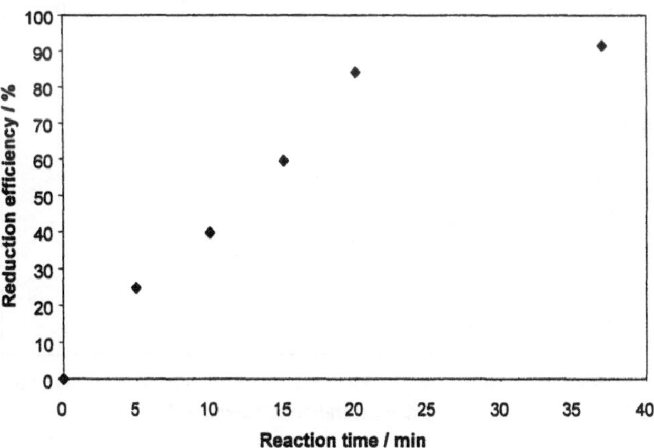

Fig. 2 Efficiency of Bi(III) reduction versus irradiation time. The concentration of added ethanol is 5×10^{-4} M

Thus, the reduction of Bi(III) can only be photocatalytic in the presence of ethanol as a sacrificial electron donor. With the use of this hole scavenger, Bi(III) is also reduced and deposited on the surface of TiO_2 particles. The reduction efficiency is over 90% after 30-min irradiation in a deaerated system initially containing 5×10^{-4} M ethanol (Fig. 2); hence, Bi(III) is quantitatively removable if the concentration of the ethanol added is increased.

Effect of ethanol concentration

As mentioned for the photoreduction of Hg(II), an increase in the concentration of the sacrificial electron donor added enhances the rate of the photoinduced reduction and, thus, the efficiency of the deposition after a certain period of irradiation time. The efficiency of the photoreduction of Hg(II), Bi(III), and Cu(II) was measured after 10-min illumination at various ethanol concentrations. As can be seen in Fig. 3, the efficiency versus concentration of ethanol functions are in accordance with the relation of the standard reduction potentials in the case of Hg(II) (0.41 V for $HgCl_2$/Hg) and Bi(III) (0.317 V). At $c_{ethanol} > 5 \times 10^{-4}$ M the plots for these metal ions are close to each other, corresponding to similar redox potentials. Although the standard reduction potential of the Cu(II)/Cu pair (0.34 V) is between those of Hg(II) and Bi(III), a higher concentration of hole scavengers than in the case of Bi(III) was needed to reach the same efficiency. Moreover, only an efficiency of about 60% was reached even at 0.1 M ethanol concentration. (For a quantitative deposition of Cu, 1 M ethanol had to be applied.) Such deviating behavior of Cu may be interpreted by its short-circuiting effect on the surface of the TiO_2 particles [20], which promotes the electron–hole recombination.

Photocatalytic reduction of Pb(II)

Besides Hg, one of the most toxic heavy metals is Pb, so its removal from wastewater is of basic importance. The standard reduction potential of Pb(II) is −0.125 V [23]; however, it is partly removable with the use of ethanol. In this case, however, an extremely high ethanol concentration is necessary (40 v/v%≈7 M). Even so, a maximum efficiency of 84% could be reached (Fig. 4). It is worth mentioning that in earlier experiments Pb(II) was deposited in the form of PbO_2 as a result of photoinduced oxidation on Pt-loaded TiO_2 particles [24]. Metal-free TiO_2, however, was not effective in this respect.

Effects of anionic and cationic surfactants

SDS is one of the most frequently used detergents in domestic and industrial laundry operations. As a primary pollutant in municipal wastewater, SDS has attracted environmental concerns [25]. Since it is toxic to aquatic and animal life, and is rather resistant to biodegradation, its photocatalytic destruction, which proved to be feasible and reasonable, has become of practical importance. The degradation of SDS on TiO_2 is an oxidation reaction [10]; hence, this process must promote the photoreduction of Hg(II). This conclusion is also supported by the photolysis of the $HgCl_2/TiO_2$ system containing 0.03 M SDS. As it can be seen in Fig. 5, practically a quantitative deposition of Hg could be achieved after 10-min illumination. If ethanol is added to the initial solution, the efficiency of the reduction decreases with the increase in the ethanol concentration (Fig. 6); it is higher, however, than the efficiency with ethanol alone. This phenomenon clearly indicates that SDS as a hole scavenger is significantly better than ethanol. This synergic effect suggests the simultaneous removal of toxic metals and anionic surfactants from wastewater by photocatalytic processes.

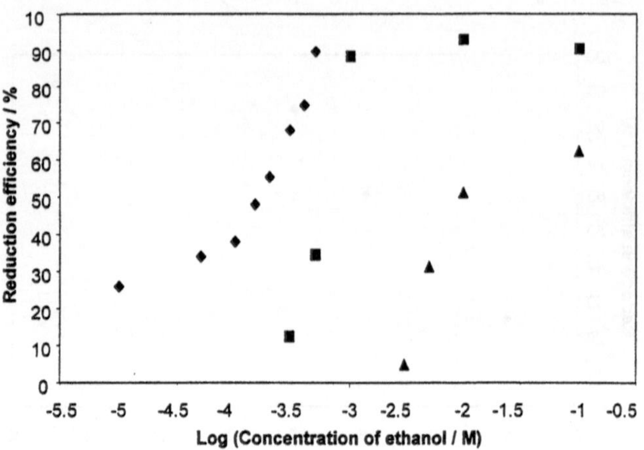

Fig. 3 Efficiency of Hg(II) (◆), Bi(III) (■), and Cu(II) (▲) reduction versus the logarithm of the concentration of ethanol. The irradiation time is 10 min

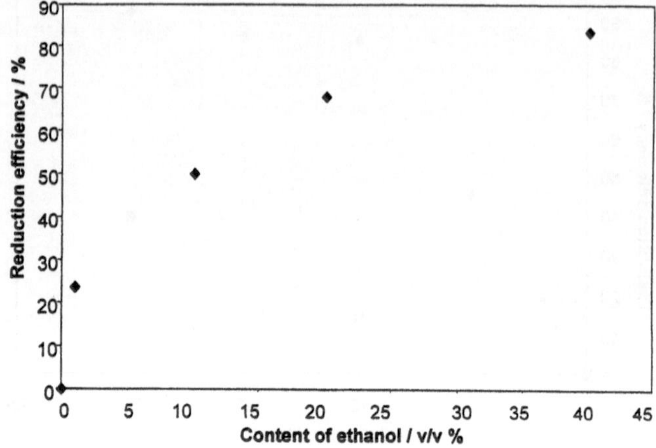

Fig. 4 Efficiency of Pb(II) reduction versus ethanol concentration. The irradiation time is 10 min

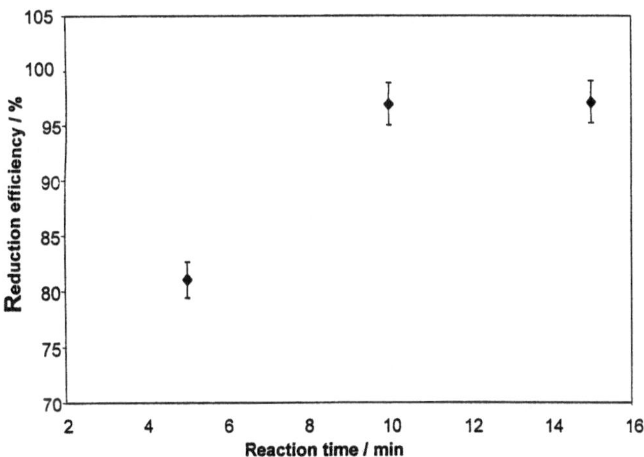

Fig. 5 Efficiency of Hg(II) reduction versus reaction time in the presence of 0.03 M sodium dodecyl sulfate (*SDS*)

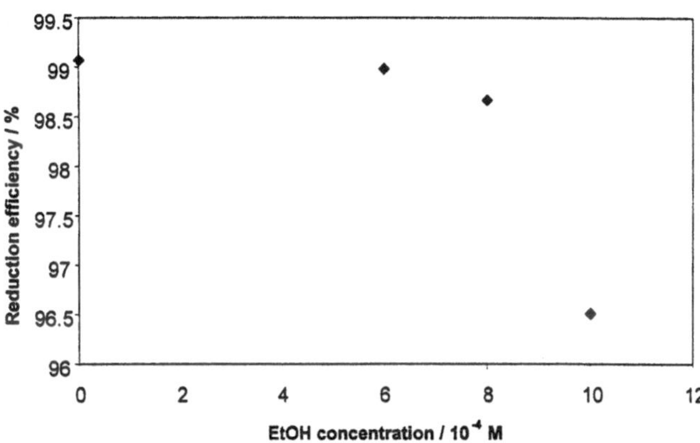

Fig. 6 Efficiency of Hg(II) reduction versus ethanol concentration in the presence of 0.03 M SDS

CTAB as a cationic surfactant was also tested as potential electron donor for the photoreduction of Hg(II). Deviating from SDS, CTAB (0.01 M) only slightly promotes the photoreduction of Hg(II) (Fig. 7). If 0.01 M CTAB and ethanol are used together, the reduction efficiency is higher than in the presence of CTAB alone, but lower than in the presence of ethanol alone; about 40% after 20-min irradiation even with 1 M ethanol (and 0.01 M CTAB). This phenomenon indicates that cetyltrimethylammonium ions compete with ethanol for the holes on the surface of the TiO_2 particles, but the positive charge of the surfactant hinders its efficient reaction. As mentioned, the pH was not adjusted in the solutions studied in this work; however, the pH can considerably influence both the redox reactions of the sacrificial electron donors and the surface charge of the TiO_2 particles and, thus, the adsorption of charged reactants, such as detergents. In order to explore the pH effects in the presence of surfactants, further experiments are in progress.

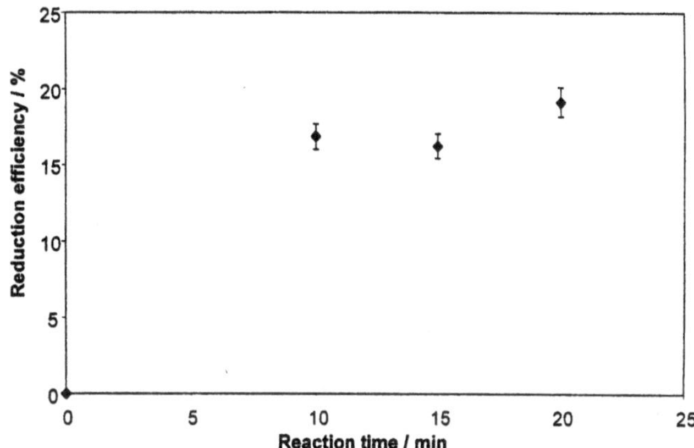

Fig. 7 Efficiency of Hg(II) reduction versus irradiation time in the presence of 0.01 M cetyltrimethylammonium bromide

Conclusion

Light-induced reduction of most of the toxic heavy-metal ions on TiO_2 particles can be made photocatalytic by the application of a sacrificial electron donor at an appropriate concentration. Ethanol proved to be suitable for this purpose. In the case of originally photocatalytic systems, addition of electron donors enhance the rate of the reduction. The presence of oxygen diminishes the photoinduced deposition of metals, because of the competitive scavenging of conductance band electrons and reoxidation of the metals deposited. The efficiency of the photoreduction carried out under the same circumstances is primarily determined by the standard reduction potential of the species containing the metal ion to be removed. It has been demonstrated by the examples of Hg(II), Bi(III), and Pb(II). Cu(II) showed deviating behavior; the relatively lower efficiency of its deposition may be attributed to a short-circuiting effect. Commercially used surfactants also proved to be potential hole scavengers. Negatively charged dodecyl sulfate is much more efficient than ethanol, while cationic cetyltrimethylammonium is rather weak in this respect.

Acknowledgement Support of this work by the Hungarian Ministry of Education (FKFP 0082/1999) is gratefully acknowledged.

References

1. Fox MA (1983) Acc Chem Res 16:314
2. Hsiao C, Lee C, Ollis DF (1983) J Catal 82:418
3. Ahmed S, Ollis DF (1984) Sol Energy 32:597
4. Okamoto K, Yamamoto Y, Tanaka H, Tanaka M (1985) Bull Chem Soc Jpn 58:2015

5. Ollis DF (1985) Environ Sci Technol 19:480
6. Matthews RW (1986) Water Res 20:569
7. Matthews RW (1988) J Catal 111:264
8. Terzian R, Serpone N, Minero C, Pelizetti E (1991) J Catal 128:352
9. Heller A, Brock JR (1994) In: Helz GR, Zepp RG, Crosby DG (eds) Aquatic surface photochemistry. Lewia, Boca Raton, pp 427–436
10. Lea J, Adesina AA (1998) J Photochem Photobiol A 118:111
11. Curran JS, Domenech J, Jaffrezic-Renault N, Philippe R (1985) J Phys Chem 89:957
12. Borgarello E, Serpone N, Emo G, Harris R, Pelizetti E (1986) Inorg Chem 25:4499
13. Domenech J, Munoz J (1987) Electrochim Acta 32:1383
14. Herrmann J-M, Disdier J, Pichat P (1988) J Catal 113:72
15. Herrmann J-M, Guillard C, Pichat P (1993) Catal Today 17:7
16. Serpone N, Ah-You YK, Tran TP, Harris R, Pelizetti E, Hidaka H (1987) Sol Energy 139:491
17. Prairie MR, Evans LR, Stange BM, Martinez SL (1993) Environ Sci Technol 27:1176
18. Tennakone K, Wijayantha KGU (1997) J Photochem Photobiol A 113:89
19. Latimer WL (1952) Oxidation potentials, 2nd edn. Prentice-Hall, Englewood Cliffs
20. Rajeshwar K, Ibanez JG (1995) J Chem Educ 72:1044
21. Greenwood NN, Earnshaw A (1984) Chemistry of the elements. Pergamon, Oxford, pp 736, 1409
22. Bard AJ (ed) (1982) Encyclopedia of electrochemistry of the elements, vol 9, 4th edn. Dekker, New York
23. Shriver DF, Atkins PW, Langford CH (1992) Anorganische Chemie. VCH, Weinheim, pp 707–722
24. Tanaka K, Harada K, Murata S (1986) Sol Energy 36:159
25. Kroschwitz JI, Howe-Grant M (eds) (1994) Kirk-Othmer Encyclopedia of chemical technology, vols 7–8, 4th edn. Wiley, New York

Progr Colloid Polym Sci (2001) 117:217–222

György Fetter
Tamás Horányi
Attila Bóta

In situ structural investigations of the Synperonic(A7)–water system under shear

G. Fetter
Department of Chemical Information Technology,
Budapest University of Technology and Economics,
1521 Budapest, Hungary

T. Horányi · A. Bóta (✉)
Institute of Physical Chemistry,
Budapest University of Technology and Economics, 1521 Budapest,
Hungary
e-mail: bota.fkt@chem.bme.hu

Abstract A sample holder was constructed for small-angle X-ray scattering(SAXS) measurements that can be used as a shear cell. The lamellar structure of a nonionic surfactant (Synperonic A7)–water system was investigated in this cell. The SAXS curves of the sample were detected in the steady state and under shear for a certain time, and finally in the steady state once again. The Bragg profiles of the SAXS curves shift to smaller characteristic periodic distances under the shear stress. After shear the Bragg profiles do not return to the starting position but drastic changes can be detected in their shapes which indicate significant changes in the lamellar arrangements. These structural changes also explain the tixotropic behaviour of the system which was observed in rheological tests.

Key words Small-angle X-ray scattering · Lamellar systems · Shear · Tixotropy · Surfactant

Introduction

The association of colloid systems containing amphiphilic molecules (among them the surfactants, also called tensides, which are of great practical importance) form nanometre-sized structures in different arrangements (micelles, lamellar, hexagonal, or inverse hexagonal) depending on the concentration. They are the so-called lyotropic systems, having a wide range of application in industry and the household. They can be produced, for example, from mixtures of fatty alcohols and water [1].

The concentration of the tenside and the temperature have a major influence on the rheological characteristics, which are very important in practical applications [2]. A nonionic surfactant (Synperonic A7)–water system was studied intensively by Tadros and coworkers [3–5]. They reported that this system exhibits lamellar structure in the concentration range between 55 and 80% (w/w). Németh et al. [6–8] characterized the structural behaviour by rheological measurements and the experimental data were interpreted by using a slip-plane model. The rheological test revealed that the system was tixotropic. This rheological behaviour shows that the shear stress may affect the lamellar structure. The change in the lamellar structure can be attributed to two reasons: the destruction of the lamellar arrangement and the formation of new structures.

Simultaneous structural and rheological investigations permit in situ examinations to reveal and feature of the structure which are connected with the influence of the shearing stress [9]. Small-angle X-ray scattering (SAXS) and small-angle neutron scattering are powerful methods to study lamellar arrangement. Synchrotron rays are especially useful because of the high-power radiation, which allows the identification of structural changes with short lifetimes (seconds or milliseconds) [10]. We have constructed a shear cell which allows in situ SAXS measurements. Scanning the structural changes of the Synperonic A7–water lamellar system under shear stress in our cell allows essential information for the interpretation of the rheological characteristics and properties of the system to be obtained.

Experimental

Material

The sample was prepared from a surfactant Synperonic (ICI product, Brussels, B) and water. The surfactant was used as supplied. Synperonic was made by ethoxylation of Synperonic alcohol (consisting of 66% C_{13} and 34 % C_{15} alkyl chains). The ethoxylation process yields a wide distribution of poly(ethylene oxide) chains; therefore, only the average ethoxylation number can be given, which in the case of A7 is 7. The sample which was investigated contained 80% (w/w) surfactant and 20% (w/w) water. Before homogenization the sample must be heated to about 60 °C because its viscosity decreases, and the higher-temperature sample was easily homogenized with water. Of course, the sample must be bubble-free; therefore, it was stirred intensively for 30 min. Then, the mixture was left to cool to room temperature and was stored for a 1-week period before any measurement.

Equipment

The shear cell developed for the SAXS equipment is presented in Fig. 1. It consists of three main parts: the sample holder bordered partly by moveable X-ray windows made of Plexiglass (without characteristic SAXS pattern), the electromechanical mover fastened to the sample holder, and the digital control unit connected by wires to the mover.

In the sample holder one of the Plexiglass windows can be moved perpendicularly to the X-ray source to induce stress in the sample. The thickness of sample layer is adjustable between 0 and 1 mm. The width of the window is 16 mm and its height is 6 mm. The sample holder can be thermostated.

The most important parts of the electromechanical mover are two solenoids in different positions. One of the solenoids is fixed, the other is displaceable in the direction of the vibration of the Plexiglass window. The possible amplitude of the vibration is 0–5 mm. The resolution of the amplitude adjustment is 0.05 mm. There is a soft steel sheet between the solenoids which is caused to vibrate by the alternating magnetic field and which transmits the motion towards the Plexiglass window.

The digital control unit switches a voltage to the solenoids for 0–100 s. The resolution of the time adjustment is 0.001 s. The working times of the solenoids need not be the same; they can be adjusted separately. The control unit is directed by a crystal oscillator.

SAXS apparatus

The SAXS measurements were performed by a Kratky camera and a proportional counter (Anton Paar, Graz, Austria). The intensity of Ni-filtered Cu Kα radiation (1.542 Å) was recorded in the 10^{-3}–10^{-1} Å^{-1} range of the scattering variable, defined as $s=(2\sin\theta)/\lambda$, where 2θ is the scattering angle. The primary beam was slit-collimated. The intensity curves were corrected by considering the geometry of the beam profile in order to obtain point-focused curves. The sample was filled into thin-walled quartz capillaries (Hilgenberg, Germany) with a diameter of 1 mm for the preliminary SAXS investigations or directly into the shear cell for the "in situ" SAXS measurements.

Rheology

Rheological measurements were made with a Haake RS 100 oscillatory rheometer. A stainless steel cone–plate sensor was used (diameter of 20 mm and cone angle of 4°) and the sample thickness was 0.134 mm in the top of the sensor. The sample was inserted into the centre of the plate of the sensor, then the plate of the sensor was slowly elevated with constant velocity to the measurement position. The sample were kept under saturated water vapour for the whole time of the measurements, because the structure and the rheological properties of the lamellar system are highly dependent on the water concentration. Before the measurements, the sample was thermostated for 10 min; this was also a relaxation–saturation period. The maximum allowed deviation in the temperature during the measurements was ±0.2 °C.

Results

The lamellar structure of the system was investigated at different temperatures. The sample has a centrosymmetric two-dimensional SAXS pattern because of its macroscopical nonoriented arrangements. The shapes of the scattering curves differ strongly, depending on the temperature as can be seen in Fig. 2. In the SAXS curves, a single Bragg reflection peak can be observed which is characteristic of the lamellar structure. At 20 °C the Bragg diffraction peak is very sharp, reflecting a highly ordered lamellar arrangement. This structural arrangement changes drastically when the sample is heated to 60 °C. The broadening of the Bragg profile reveals a significantly less ordered structure than that formed at 20 °C. The change of the lamellar structure is in agreement with the thermotropic behaviour of the system [6]. The existence of the lamellar structure at

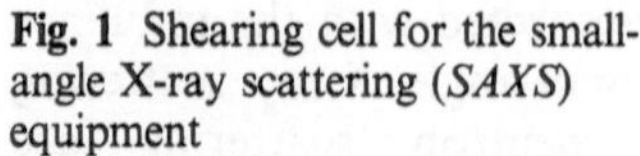

Fig. 1 Shearing cell for the small-angle X-ray scattering (*SAXS*) equipment

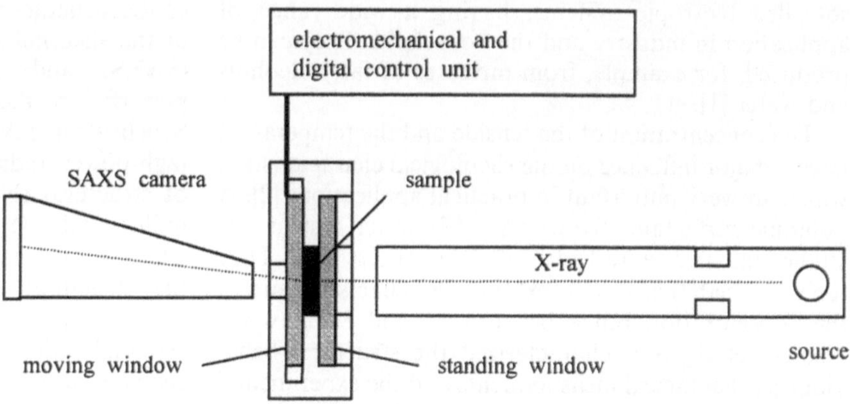

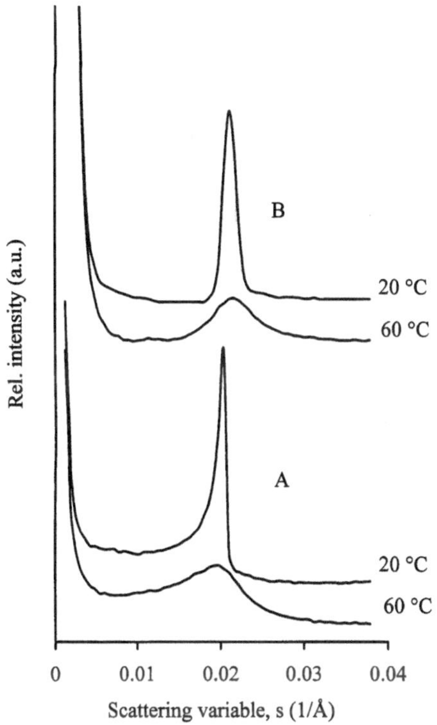

Fig. 2 SAXS patterns of the 80% (w/w) Synperonic A7–water lamellar system at 20 and 60 °C (measured *A*, desmeared *B*)

20 °C can be proved by the freeze–fracture method, which is an excellent tool for direct visualization of colloid structures. The freeze–fracture electron micrographs of the sample give unambiguous information on the layer structures. Micrographs of two different parts of the sample showing typical large multilamellar bodies consisting of closely packed parallel layers with smooth surfaces are shown in Figs. 3 and 4. The parallel layers form large sheets (Fig. 3); however, the planes of some stacks are crumpled and some stacks show different irregular shapes (Fig. 4). The latter morphology was already present and embedded in the large sheets.

The sample was filled into the shearing cell and incubated at 20 °C. The characteristic Bragg profiles of the lamellar arrangement were detected as a function of time at 20 °C. These diffractograms are shown in Fig. 5. In the steady state, the Bragg profile has the same position and the same full width at half maximum as detected with the sample filled in a quartz capillary, indicating that the regularity of the lamellar arrangement does not depend on the surface character of the sample holder. After starting the shear process, the Bragg profile was measured four times. The time of data collection in each measurement was extended to 5 min. The time of the measurement of the maximum of each Bragg peak is defined as the shear time and is marked on each curve in Fig. 5. The lamellar arrangement is affected by the shear process as the Bragg profile is shifted. During the process the periodicity is changed from 48.3 to 46.8 Å. During the shear process the widths of the profiles do not change, which means that the average layer number of the stacks remains constant. After the shear ended, broadening of the Bragg profile was observed. After 1 h the shape of the Bragg profile changed entirely and the peak was split into two diffuse peaks. The positions of the peaks correspond to layer distances of about 50 and 45 Å, the first longer

Fig. 3 Freeze–fracture electron micrograph of the 80% (w/w) Synperonic A7–water lamellar system

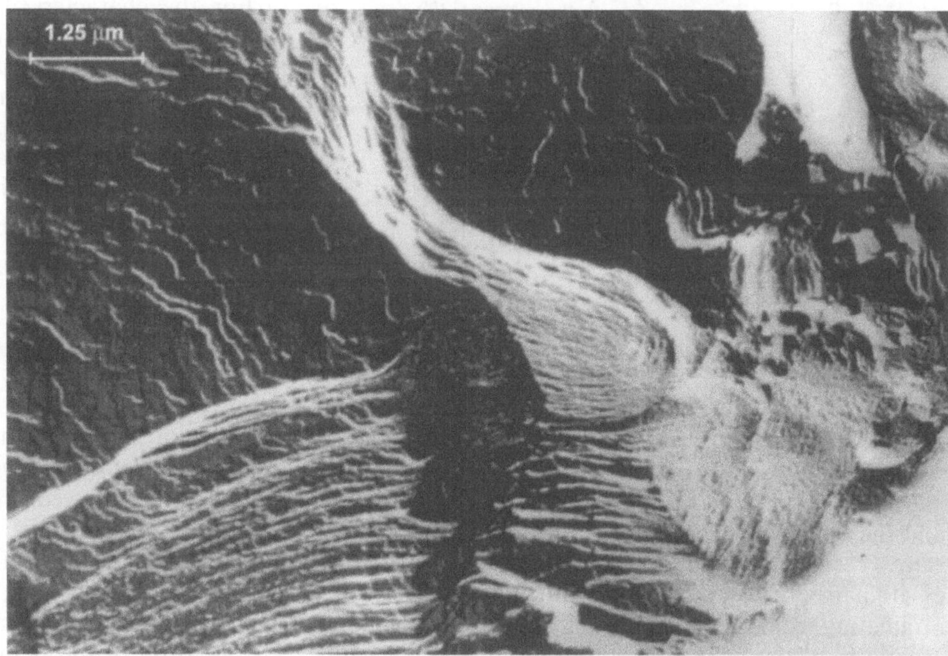

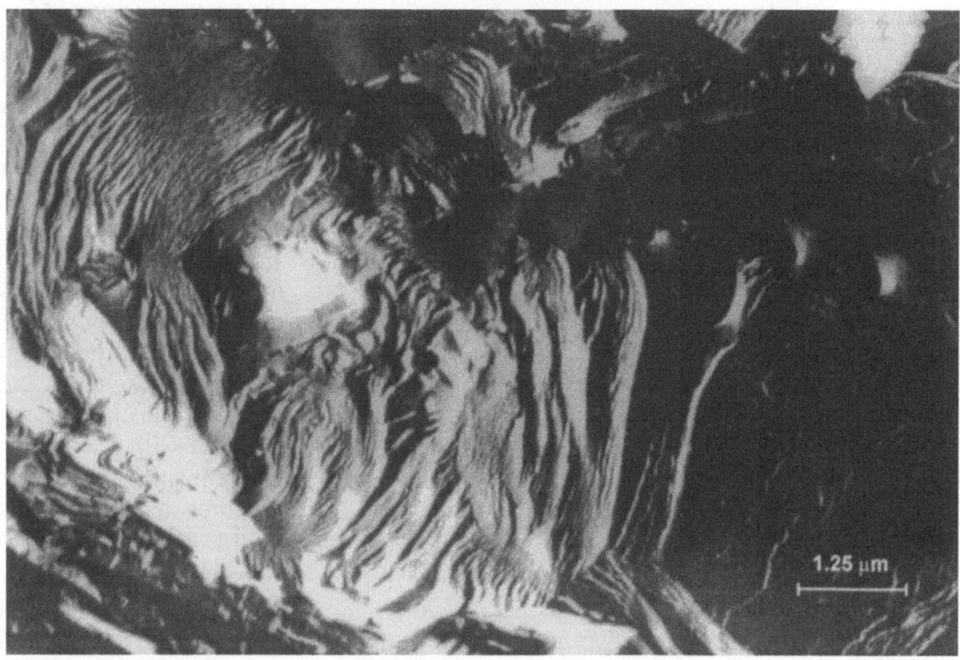

Fig. 4 Freeze–fracture electron micrograph of the 80% (w/w) Synperonic A7–water lamellar system

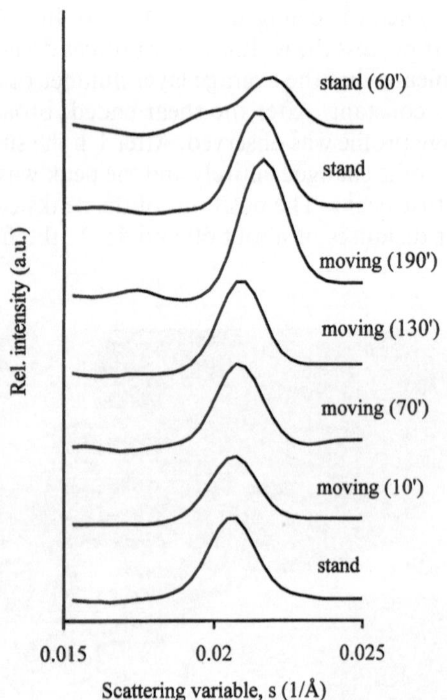

Fig. 5 Change in the profiles in the first Bragg reflexion of the SAXS curve at 20 °C with shearing time

and the second shorter than the distance in the starting state. Presumably, phase separation occurs and two types of domain form, which have different compositions originating from the inhomogeneity of the sample, more exactly from the wide distribution of ethoxylate chains and from the presence of two types of alkyl chains (C_{13} and C_{15}).

Another series of measurements was made at 60 °C. The Bragg profiles obtained are presented in Fig. 6. The diffuse peak was observed to broaden after a short shear process. After a long time of shear the Bragg peak disappeared entirely, indicating that the lamellar structure ceased to exist. When the shear ended the Bragg peak did not appear again, which means the layer structure was not restored.

For the characterization of the structural changes measured during the shear process, the layer distance and a control parameter are plotted as function of the shear time in Fig. 7. The reciprocal value of the full width at half maximum of the Bragg peak was defined as a control parameter because this value is related to the number of layers of each domain. The changes in the layer distance and the control parameter are affected by the temperature very significantly. In the temperature domain of the layer structure around 20 °C, the control parameter remains nearly constant and the layer distance diminishes only in a narrow range. At 60 °C, the destroyed lamellar structure may be metastable because it disappears after a short shear. At this high temperature the existence of the layer structure can be explained by the thermal prehistory. Namely, after preparation at room temperature followed by heating, domains with destroyed layer structures can remain in the metastable form which built up after short shear.

The rheological test carried out at 20 °C reveals significant tixotropic behaviour as can be seen in Fig. 8. The viscosity of the samples decreases and at the same

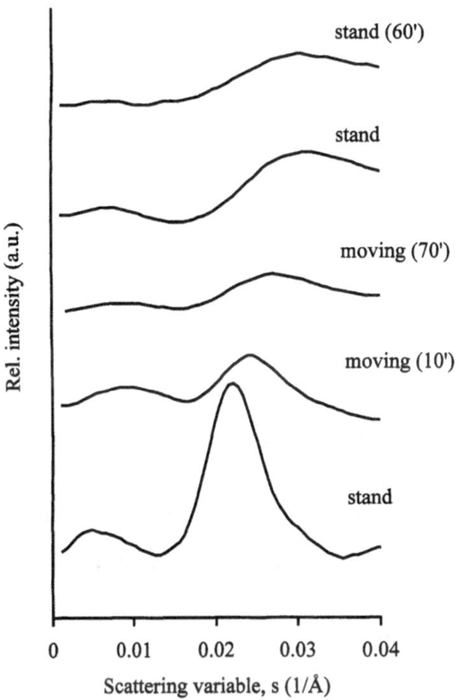

Fig. 6 Change in the profiles in the first Bragg reflexion of the SAXS curve at 60 °C with shearing time

time the stress increases with an increase in the shear rate. Finally, neither the viscosity nor the stress returns to the starting values after the end of the measurement. The same test measurement cannot be done at temperatures higher than 50 °C because the sample is in the fluid state and it cannot be filled into the cell of the rheometer; however, a drastically destroyed lamellar arrangement was detected by SAXS even at 60 °C.

Conclusion

The lamellar arrangement is inhomogeneous as can be seen in the freeze–fracture pictures. The thickness of the layer stacks can be estimated to be more than 10^2 nm and in the lateral direction the extension of the homogeneous domain is 1 or 2 orders of magnitude larger. The freeze–fracture pictures throw light on the fact that a great number of structural defects are present, which can influence the rheological behaviour of the system. Without doubt the conditions of the two shear processes (in the shear cell and the rheometer) are different, but the application of the shear cell constructed reveals that the tixotropic behaviour of the Synperonic A7–water system is connected to only small changes in the lamellar structure. The changes that are connected to the

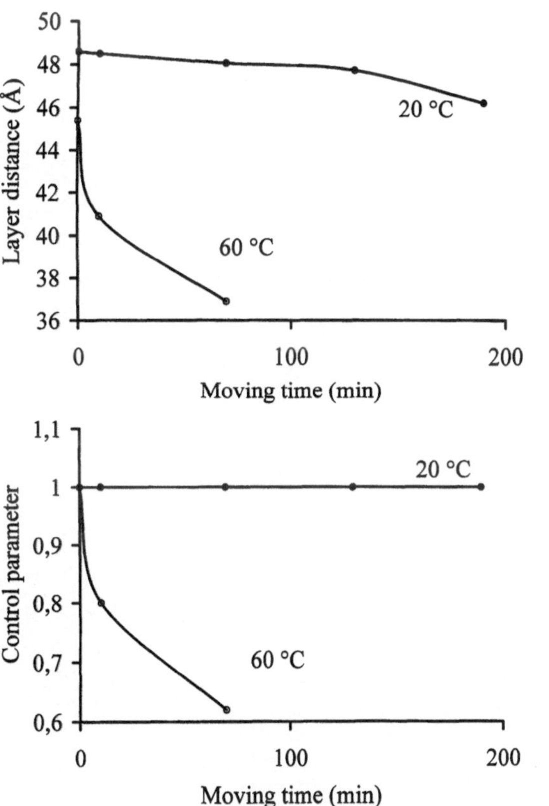

Fig. 7 Changes in the layer distance and the control parameter (reciprocal value of the full width at half maximum of the Bragg profile) during the time of the stress

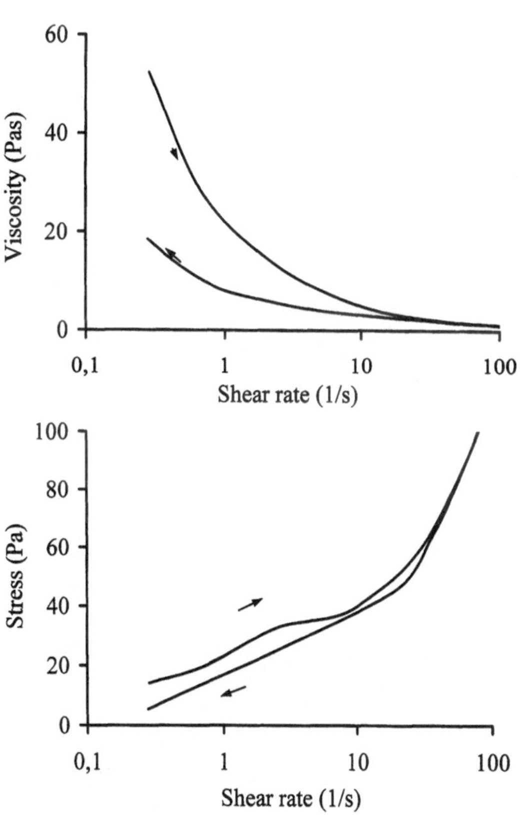

Fig. 8 Rheological test of the 80% (w/w) Synperonic A7–water lamellar system at 20 °C

rheological behaviour may occur in larger dimensions and cannot be observed by the SAXS method in the range of the scattering variable applied (10^{-3}–10^{-1} Å^{-1}). The shear results in changes with long relaxation times in both the layer arrangement and the domain formation. One hour after the ending of the shear process the Bragg profiles do not return to the starting position detected in the starting steady state but drastic changes appear in their shapes, indicating simultaneous processes: significant changes in the lamellar arrangements; formation of new domain structures. We can assume that these structural changes may be in progress in the shear test at some level and result in the tixotropic behaviour of the the Synperonic A7–water system.

Acknowledgements This work was supported by the Hungarian Scientific Funds OTKA (T 014396, T 21781) and a bilateral German-Hungarian Program TÉT (D-42/1998).

References

1. Mchick MJ (ed) (1987) Nonionic surfactants, physical chemistry. Surfactant science series. Dekker, New York
2. Sherman P (1970) Industrial rheology, Academic, London
3. Tadros TF, Dimitrova GT, Luckman P, Ftaelman MC, Loll P (1995) Euro Cosmet 3:17
4. Dimitrova GT, Tadros TF, Luckham PF, Kipps MR(1996) Langmuir 5:315–318
5. Dimitrova GT, Tadros TF, Luckham P (1995) Langmuir 11:1101–1111
6. Németh Z, Halász L, Pálinkás J, Bóta A, Horányi T (1998) Colloid Surf A 145:107–119
7. Németh Z, Pálinkás J, Halász L, Bóta A, Horányi T (1999) Tenside Surfactants Deterg 36:88–95
8. Németh Z (1998) PhD thesis. Institute of Physical Chemistry,Budapest University of Technology and Economics
9. Penfold J, Staples E, Khan Lodhi A, Tucker I, Tiddy GJ (1997) J Phys Chem B 101:66–72
10. Schwarzenbacher R, Kriechbaum M, Amenitsch H, Laggner P (1998) J Phys Chem 102:9161–9167

AUTHOR TITLE INDEX

KEY WORD INDEX